Dynamic Vulnerability

Dynamic Vulnerability Assessment and Intelligent Control for Sustainable Power Systems

Edited by

Professor José Luis Rueda-Torres
Delft University of Technology
The Netherlands

Professor Francisco González-Longatt
Loughborough University
Leicestershire, United Kingdom

The right of José Luis Rueda-Torres and Francisco González-Longatt to be identified as the authors of the editorial material in this work has been asserted in accordance with law.

Registered Offices
John Wiley & Sons, Inc., 111 River Street, Hoboken, NJ 07030, USA
John Wiley & Sons Ltd, The Atrium, Southern Gate, Chichester, West Sussex, PO19 8SQ, UK

Editorial Office
The Atrium, Southern Gate, Chichester, West Sussex, PO19 8SQ, UK

For details of our global editorial offices, customer services, and more information about Wiley products visit us at www.wiley.com.

Wiley also publishes its books in a variety of electronic formats and by print-on-demand. Some content that appears in standard print versions of this book may not be available in other formats.

Library of Congress Cataloging-in-Publication Data:

Names: Rueda-Torres, José Luis, 1980- author. | González-Longatt, Francisco, 1972- author.
Title: Dynamic vulnerability assessment and intelligent control for sustainable power systems / edited by Professor José Luis Rueda-Torres, Professor Francisco González-Longatt.
Description: First edition. | Hoboken, NJ : John Wiley & Sons, 2018. | Includes bibliographical references and index. |
Identifiers: LCCN 2017042787 (print) | LCCN 2017050856 (ebook) | ISBN 9781119214977 (pdf) | ISBN 9781119214960 (epub) | ISBN 9781119214953 (cloth)
Subjects: LCSH: Electric power distribution–Testing. | Smart power grids.
Classification: LCC TK3081 (ebook) | LCC TK3081 .D96 2018 (print) | DDC 621.31/7–dc23
LC record available at https://lccn.loc.gov/2017042787

Cover design by Wiley
Cover image: © agsandrew/Gettyimages

Set in 10/12pt WarnockPro by SPi Global, Chennai, India
Printed and bound in Malaysia by Vivar Printing Sdn Bhd

10 9 8 7 6 5 4 3 2 1

Contents

List of Contributors

Jaime C. Cepeda
Operador Nacional de Electricidad
(CENACE), and Escuela Politécnica
Nacional (EPN)
Quito
Ecuador

Dirk Van Hertem
ESAT – Electa
University of Leuven
Belgium

Steven De Boeck
ESAT – Electa
University of Leuven
Belgium

Hakan Ergun
ESAT – Electa
University of Leuven
Belgium

Evelyn Heylen
ESAT – Electa
University of Leuven
Belgium

Tom Van Acker
ESAT – Electa
University of Leuven
Belgium

Marten Ovaere
Department of Economics
University of Leuven
Belgium

Bart W. Tuinema
Delft University of Technology
The Netherlands

Nikoleta Kandalepa
TenneT TSO B.V
Arnhem
The Netherlands

Qing Liu
Kyushu Institute of Technology
Kitakyushu
Japan

Hassan Bevrani
University of Kurdistan
Sanandaj
Iran

Yasunori Mitani
Kyushu Institute of Technology
Kitakyushu
Japan

Delia G. Colomé
Universidad Nacional de San Juan
Argentina

István Erlich
University Duisburg-Essen
Duisburg
Germany

Florin Capitanescu
Luxembourg Institute of Science and
Technology, Belvaux
Luxembourg

Da Wang
Delft University of Technology
The Netherlands

Worawat Nakawiro
King Mongkut's Institute of Technology
Ladkrabang
Bangkok
Thailand

Adedotun J. Agbemuko
Institut de Recerca en Energia de
Catalunya (IREC)
Barcelona
Spain

Mario Ndreko
TenneT TSO GmbH
Bayreuth
Germany

Marjan Popov
Delft University of Technology
The Netherlands

Mart A.M.M. van der Meijden
TenneT TSO B.V
Arnhem
The Netherlands and Delft University of
Technology
The Netherlands

Hoan Van Pham
Power Generation Corporation 2
Vietnam Electricity and School of
Engineering and Technology
Tra Vinh University
Vietnam

Sultan Nasiruddin Ahmed
FGH GmbH
Aachen
Germany

Gustavo Valverde
University of Costa Rica
San Jose
Costa Rica

Hamid Soleimani Bidgoli
Université de Liège
Belgium

Petros Aristidou
University of Leeds
United Kingdom

Mevludin Glavic
Université de Liège
Belgium

Thierry Van Cutsem
Université de Liège
Belgium

Nelson Granda
Escuela Politécnica Nacional
Quito
Ecuador

Rommel P. Aguilar
Universidad Nacional de San Juan
Argentina

Fabián E. Pérez-Yauli
Escuela Politécnica Nacional
Quito
Ecuador

Pablo X. Verdugo
Operador Nacional de Electricidad
(CENACE)
Quito
Ecuador

Aharon B. De La Torre
Operador Nacional de Electricidad
(CENACE)
Quito
Ecuador

Diego E. Echeverría
Operador Nacional de Electricidad
(CENACE)
Quito
Ecuador

Foreword

Over the last decades, the electrical power system has gone through a fundamental transformation never seen before. The liberalisation of the power industry that set the whole process in motion has opened up the possibility of electricity trading across utility and even national boundaries. The distance between where power is generated and where the final consumption takes place and with it the power transit through the high voltage transmission lines has increased immensely. A further development compounding the competitive electricity market and power transmission over long distances has been the large-scale installation of renewables-based power generation units. In addition to the volatility and stochasticity of the power outputs of these units, utilities now also have to contend with possible bi-directional power flows in the distribution networks.

Due to the different dynamic characteristics of renewable generation units compared with conventional power plants, the increasing share of renewables-based generation capacity in the system can give rise to new dynamic phenomena that can reduce the existing security of the whole system. Additionally, restrictions regarding expansion or reinforcement of the existing network mean that lines have to be loaded up to or near their maximum current carrying capabilities. It can thus be safely concluded that the increasing uncertainty regarding load flows and the use of power plants in a heavily loaded network, together with the new power generation technologies such as wind and solar as well as transmission technologies such as VSC-HVDC, would necessarily lead to the reduction of existing security levels unless appropriate countermeasures are implemented.

This book takes up this most up-to-date topic and provides valuable contributions in the areas of both vulnerability assessment and intelligent control. The use of many of the methods under discussion has been made possible by the powerful computers and communication technologies that are now available. Also, in the last decade, significant advances in the area of computational intelligence have been made. These results are now mature enough for use in the planning and operation of power systems. During a contingency, for example, the operator is often overwhelmed by the rapidly changing situation and the associated flood of information, on the basis of which appropriate steps have to be taken. Clearly, the dispatcher cannot be expected to form an objective judgment on the unfolding situation based on his/her observation and experiences alone. The uncertainties must be assessed by suitable analytical tools in order to make the best possible decision within the shortest time possible, and computer-based decision support systems come in handy here. Other promising techniques in this context are the model based predictive control approaches. If a contingency or unfavourable operating condition is predicted some time ahead of its occurrence, a suitable countermeasure can be devised over the intervening period taking prior experience into account. Also, since the available time for decision and control actions is typically very short, real-time applications are required.

The current challenges, and particularly those ahead in the upcoming years, urgently require the introduction of new methods and approaches to ensure the preservation of the existing level of system security, which is taken for granted and assumed so far to be self-evident. The approaches described in this book grew out of the work of talented and committed young scientists working in the area. On the one hand, the contributions serve as a thought-provoking impulse for practising engineers who are looking for new ways to cope with the challenges of today and the future. However, many of these forward-looking ideas are already ready for implementation. On the other hand, this book also allows graduate students to get an overview of modern mathematical and computational methods. Certainly, the book presupposes a thorough knowledge of power system analysis, dynamics and control. Building on this, however, it introduces the reader to an exciting world of new approaches. The combination of practice-orientation and introduction of modern methods for vulnerability assessment and control applications make this book particularly valuable, and recommended reading for a wide audience in the area of power engineering.

January 2017

Prof. István Erlich
Chair Professor of Department of Electrical
Engineering and Information Technologies
Head of the Institute of Electrical Power Systems
University Duisburg-Essen

Preface

Traditionally, electrical power systems worldwide have been planned and operated in a relatively conservative manner, in which power system security, in terms of stability (i.e. dynamic performance under disturbances), has not been considered a major issue. Most of the tools developed and applied for these tasks were conceived to deal with reduced levels of uncertainty and have proven to be helpful to identify optimal developmental and operational strategies that ensure maximum net techno-economic benefits, in which only the fulfilment of steady-state performance constraints has been tackled.

The societal ambition of a cleaner, sustainable and affordable electrical energy supply is motivating a dramatic change in the infrastructure of transmission and distribution systems in order to catch up with the rapid and massive addition of evolving technologies for power generation based on renewable energy sources, particularly wind and solar photovoltaics. In addition to this, the emergence of the prosumer figure and new interactive business schemes entail operations within a heterogeneous and rapidly evolving market environment.

In view of this, power system security, and especially the analysis of vulnerability and possible mitigation measures against disturbances, deserves special attention, since planning and operating the electric power system of the future will involve dealing with a large volume of uncertainties that are reflected in highly variable operating conditions and will eventually lead to unprecedented events.

This book covers the fundamentals and application of recently developed methodologies for assessment and enhancement of power system security in short-term operational planning (e.g. intra-day, day-ahead, a week ahead, and monthly time horizons) and real-time operation. The methodologies are based on advanced data mining, probabilistic theory and computational intelligence algorithms, in order to provide knowledge-based support for monitoring, control and protection tasks. Each chapter of the book provides a thorough introduction to the intriguing mathematics behind each methodology as well as a sound discussion on its application to a specific case study, which addresses different aspects of power system steady-state and dynamic security.

In order to properly follow the content of the book, the reader is expected to have a basic background in power system analysis (e.g. power flow and fault calculation), power system stability (e.g. stability phenomena and modelling needs), and basics of control theory (e.g. Fourier transforms, linear systems). This background is usually acquired in graduate programs in electrical engineering and dedicated training courses and seminars. Therefore, the book is recommended for formal instruction, via advanced courses, of postgraduate students as well as for specialists working in power system operation and planning in industry. The content of the book is organised into two parts as follows:

Part I: Dynamic Vulnerability Assessment

Chapter 1 provides general definitions and rationale behind power system vulnerability assessment and phasor measurement technology, with special emphasis on the fundamental relationship between these concepts as seen in modern control centres.

Chapter 2 addresses power system reliability management and provides a broad discussion on the challenges for reliability management due to uncertainties in different time frames, ranging from long-term system development to short-term system operation.

Chapter 3 concerns the fundamentals of probabilistic reliability analysis, with emphasis on the study of large transmission networks. Two common approaches are presented: enumeration and Monte Carlo simulation. The chapter also provides a comprehensive study of the impact of underground cables on the Dutch extra-high-voltage (EHV) transmission network.

Chapter 4 introduces an enhanced data processing method based on the Hilbert–Huang Transform technology for studying low-frequency power system oscillations. Application to a real case study in Japan is overviewed and discussed.

Chapter 5 concerns the application of Monte Carlo simulation to recreate a statistical database of power system dynamic behaviour, followed by empirical orthogonal functions to approximate the dynamic vulnerability regions and a support vector classifier for online post-contingency dynamic vulnerability status prediction. The tuning of the classifier via a mean–variance mapping optimisation algorithm is also outlined.

Chapter 6 addresses the challenge of real-time vulnerability assessment. It introduces the notion of real-time coherency identification and vulnerability symptoms, for both fast and slow dynamic phenomena, and their identification from PMU data based on key performance indicators and clustering techniques.

Chapter 7 focuses on the security constrained optimal power flow problem, discussing the challenges and proposed solutions to leverage the computational effort in light of the more frequent use of risk-based security assessment and criteria for massive integration of renewable generation and the associated volumes of uncertainty.

Chapter 8 presents the various reliability management actions (preventive and corrective) as well as their modelling and integration into a security constrained optimal power flow problem. The different actions are represented by using a suitable linearized formulation, which allows keeping the computational costs low while retaining a sufficiently accurate approximation of the behaviour of the system.

PART II: Intelligent Control

Chapter 9 is devoted to damping control to mitigate oscillatory stability threats by using model-based predictive control. This is an emerging method that is receiving increasing interest in the control and power engineering community for the design of adaptive and coordinated control schemes. In this chapter, a hierarchical model-based predictive control scheme is proposed to calculate supplementary signals that are superimposed on the inputs of the damping controllers that are usually attached to different devices such as synchronous generators and FACTS devices.

Chapter 10 introduces a combined approach of an artificial neural network and ant colony optimisation to provide a fast estimation of voltage stability margin and to define the necessary adjustments of set-points of controllable reactive power sources based on voltage stability constrained optimal power flow.

Chapter 11 presents a control scheme for voltage and power control in high-voltage multi-terminal DC grids used for the grid connection of large offshore wind power plants. The proposed control scheme employs a computational intelligence technique in the form of a fuzzy controller for primary voltage control and a genetic algorithm for the secondary control level.

Chapter 12 concerns the application of model-based predictive control for reactive power control to adjust power system voltages during normal (i.e. quasi-steady state) conditions. This kind of control scheme has a slow response from, say, 10 to 60 seconds, to small operational changes and does not provide any fast reaction during large disturbances to prevent undesirable adverse implications.

Chapter 13 proposes an optimisation approach in which the objective function is augmented to incorporate the global optimisation of a linearized large scale multi-agent power system using the Lagrangian decomposition algorithm. The aim is to maintain centralised coordination among agents via a master agent leaving loss minimization as the only distributed optimisation, which is analysed while protecting the local sensitive data.

Chapter 14 presents a basic formulation of model-based predictive control for voltage corrective control, as well as the management of congestion and thermal overloads in distribution networks in the presence of high penetration of distributed generation units.

Chapter 15 addresses the interplay between transmission and distribution networks from the point of view of long-term voltage stability. It introduces the notion of Volt-Var Control (VVC) and the application of model-based predictive control for coordination of reactive power support between distribution and transmission.

Chapter 16 overviews an approach for power system controlled islanding. The approach is based on the development and integration of novel algorithms and procedures for graph partitioning and frequency behaviour estimation. It helps in avoiding a system collapse by splitting the system into electrical islands with adequate generation-load balance.

Chapter 17 provides insight into the application and value of empirical orthogonal functions as a promising alternative for signal processing applied to fault diagnosis. A comprehensive case study evidences that fault signals decomposed in terms of these orthogonal basis functions exhibit well-defined patterns, which can be used for recognising the main features of fault events such as inception angle, fault type and fault location.

Chapter 18 presents the main developmental aspects and lessons learnt so far concerning the implementation of a real phasor based vulnerability assessment and control scheme in the Ecuadorian National Interconnected System.

The book has intentionally been designed to allow some overlap between the chapters; it is desired to illustrate how some of the presented approaches could share some common elements, implementations or even developments and applications, despite being conceived for different purposes and uses.

We hope that the book proves to be a useful source of information on the understating of dynamic vulnerability assessment and intelligent control, but at the same time provides the basis for discussion among readers with diverse expertise and backgrounds. Given the great variety of topics covered in the book, which could not be completely covered in a single edition, it is expected that a second edition of the book will be made available soon.

José Luis Rueda-Torres, Delft University of Technology, The Netherlands.

Francisco González-Longatt, Loughborough University, UK.

1

Introduction: The Role of Wide Area Monitoring Systems in Dynamic Vulnerability Assessment

Jaime C. Cepeda[1] and José Luis Rueda-Torres[2]

[1] Head of Research and Development and University Professor, Technical Development Department and Electrical Energy Department, Operador Nacional de Electricidad CENACE, and Escuela Politécnica Nacional EPN, Quito Ecuador
[2] Assistant professor of Intelligent Electrical Power Systems, Department of Electrical Sustainable Energy, Delft University of Technology, The Netherlands

1.1 Introduction

Currently, most social, political, and economic activities depend on the reliability of several energy infrastructures. This fact has established the necessity of improving the security and robustness of Electric Power Systems [1]. In addition, the lack of investment, the use of congested transmission lines, and other technical reasons, such as environmental constraints, have been pushing Bulk Power Systems dangerously close to their physical limits [2]. Under these conditions, certain sudden perturbations can cause cascading events that may lead to system blackouts [1, 3]. It is crucial to ensure that these perturbations do not affect security, so the development of protection systems that guarantee service continuity is required. In this regard, Special Protection Schemes (SPS) are designed in order to detect abnormal conditions and carry out corrective actions that mitigate possible consequences and allow an acceptable system performance [4].

However, the conditions that lead the system to a blackout are not easy to identify because the process of system collapse depends on multiple interactions [5, 6]. Vulnerability assessment (VA) is carried out by checking the system performance under the severest contingencies with the purpose of detecting the conditions that might initiate cascading failures and may provoke system collapse [7]. A vulnerable system is a system that operates with a "reduced level of security that renders it vulnerable to the cumulative effects of a series of moderate disturbances" [7]. The concept of vulnerability involves a system's security level (static and dynamic security) and the tendency of its conditions changing to a critical state [8] that is called the "Verge of Collapse State" [5]. In this context, vulnerability assessment assumes the function of detecting the necessity of performing global control actions (e.g., triggering of SPSs).

In recent years, emerging technologies such as Phasor Measurement Units (PMUs), which provide voltage and current phasor measurements with updating periods of a few milliseconds, have allowed the development of modern approaches that come closer to the target of real time vulnerability assessment [6, 7]. Most of these real time applications have been focused on identifying signals that suggest a possibly insecure steady state.

Dynamic Vulnerability Assessment and Intelligent Control for Sustainable Power Systems, First Edition.
Edited by José Luis Rueda-Torres and Francisco González-Longatt.
© 2018 John Wiley & Sons Ltd. Published 2018 by John Wiley & Sons Ltd.

This kind of VA is capable of alerting the operator to take appropriate countermeasures, with the goal of bringing the system to a more secure operating condition (i.e., preventive control) [9]. Nevertheless, the use of PMUs has great potential to allow the performance of post-contingency Dynamic Vulnerability Assessment (DVA) that could be used to trigger SPSs in order to implement corrective control actions. In this connection, a Wide Area Monitoring System (WAMS), based on synchrophasor technology, constitutes the basic infrastructure for implementing a comprehensive scheme for carrying out real time DVA and afterwards executing real time protection and control actions. This comprehensive scheme is called a Wide Area Monitoring, Protection, and Control system (WAMPAC) [10]. This chapter presents a general overview of concepts related to power system vulnerability assessment and phasor measurement technology and subsequently highlights the fundamental relationship between these in modern control centers.

1.2 Power System Vulnerability

A vulnerable system is a system that operates with a "reduced level of security that renders it vulnerable to the cumulative effects of a series of moderate disturbances." Vulnerability is a measure of system weakness regarding the occurrence of cascading events [7].

The concept of vulnerability involves a system's security level (i.e., static and dynamic security) and its tendency to change its conditions to a critical state [8] that is called the "Verge of Collapse State" [5].

A vulnerable area is a specific section of the system where vulnerability begins to develop. The occurrence of an abnormal contingency in vulnerable areas and a highly stressed operating condition define a system in the verge of collapse state [5].

In this chapter, vulnerability is defined as *"**the risk level presented by a power system during a specific static or dynamic operating condition regarding the occurrence of cascading events.**"* This concept makes vulnerability an essential indicator of system collapse proximity.

Although there are a lot of vulnerability causes, which vary from natural disasters to human failures, system vulnerability is characterized by four different symptoms of system stress: angle instability, voltage instability, frequency instability, and overloads [5]. So, vulnerability assessment should be performed through analyzing the system status as regards these symptoms of system stress.

1.2.1 Vulnerability Assessment

Vulnerability assessment has the objective of preventing the occurrence of collapses due to catastrophic perturbations [11]. Performing VA requires specific mathematical models capable of analyzing the multiple interactions taking place between the different power system components [11]. These models have to consider the varied phenomena involved in the vulnerability condition and also the diverse timeframes in which the corresponding phenomena occur.

Many methods have been proposed for vulnerability assessment, which have been classified based on various criteria [6, 7]. However, in terms of their potential implementation in control centers, the techniques to assess vulnerability can be classified into

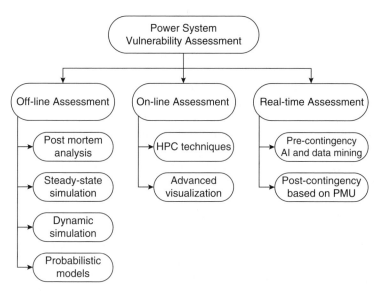

Figure 1.1 Power system vulnerability assessment methods.

off-line, on-line, and real time methods. Figure 1.1 depicts the proposed classification of vulnerability assessment methods.

- *Off-line assessment:* Off-line assessment is done using conventional methods that are based on different complex model simulations; these usually involve time-consuming tasks, which restricts on-line applications. The high complexity is provoked by the huge number and diversity of the components that constitute an electric power system and their particular performance during dynamic phenomena. Among this class of methods are those that assess vulnerability via power flow computations. These approaches are based on the hypothesis that a typical contingency provokes minor changes in the Bulk Power Systems, so they migrate from one quasi-steady state to another. The simulations do not include any dynamic response analysis [7]. This type of evaluation is commonly called Static Security Assessment (SSA). As a complement of SSA, dynamic simulation is commonly used in order to consider all the static and dynamic power system components, simulating any type of contingency and any type of instability [9, 12]. In this case, the complexity of the modeling and mathematical computations demands high processing time [7, 9]. This evaluation is usually called Dynamic Security Assessment (DSA). Usually, both in SSA and DSA, several "what-if" contingencies are simulated in order to determine the most critical disturbances (i.e., N-x contingency analysis) [9].
- *On-line assessment:* In an on-line assessment, the information must be available in the SCADA/EMS (Supervisory Control and Data Acquisition/Energy Management System). The input data are updated using adequate tools and equipment (e.g., IEDs, PMUs, WAMS), but the output is not necessarily obtained as quickly as the real time events occur [12].
- *Real time assessment:* The input data reflect the most recent picture of the system conditions in a real time analysis, and the entire process is performed within very short time, typically not exceeding a couple of seconds [12]. Emerging technologies, such

as Phasor Measurement Units (PMUs), have allowed the development of modern approaches to vulnerability assessment [7, 12] capable of being updated practically in real time. Additionally, some novel mathematical methods permit processing the data obtained in real time through identification of indicators or patterns that show system vulnerability, using artificial intelligence (AI) or modern data mining techniques [7]. Tools based on AI allow the analysis of power system dynamic performance patterns in real time, using the knowledge obtained from off-line learning [7]. On the other hand, data mining techniques permit uncovering valuable hidden information immersed in the electric signals [7], which can exhibit certain regularities (patterns) signaling a possibly vulnerable condition [10, 13]. VA methods based on AI and data mining are oriented to evaluate both pre-contingency quasi-steady state data (i.e., real time DSA for coordinate preventive control actions) as well as post-contingency dynamic data (i.e., post-contingency Dynamic Vulnerability Assessment—DVA—for triggering corrective control actions). In the particular case of post-contingency DVA, it is noteworthy that PMUs are able to provide time-synchronized phasor data, which contain valuable dynamic information that could indicate system vulnerability status and potential collapses [7]. These post-contingency data offer a new framework for system vulnerability assessment that is known as Dynamic Vulnerability Assessment (DVA), which is oriented to coordinating corrective control actions. In this connection, post-contingency DVA requires even a quicker response than pre-contingency DSA, so that AI and data mining techniques are, possibly, the most prominent mathematical tools to be applied for accomplishing this type of real time assessment.

1.2.2 Timescale of Power System Actions and Operations

As mentioned, an important aspect to be considered in vulnerability assessment is the duration of the events involved in the VA timeframe of interest. Due to the complex tasks related to power system operation, which comprise modeling, analysis, simulation, and control actions, the timescale varies from microseconds to several hours [1]. Table 1.1 presents some power system actions and operations and their corresponding timeframes. Most of this book tackles the Dynamic Vulnerability

Table 1.1 Actions and operations within the power system [1].

Action or operation	Timeframe	
Electromagnetic transients	μs – ms	
Switching overvoltage	Ms	
Fault protection	100 ms	
Electromagnetic effects in machine windings	ms – s	
Electromechanical transients– stability	ms – s	
Electromechanical oscillations	ms – min	*DVA*
Frequency control	1 s – 10 s	*timeframe*
Overloads	5 s –h	
Economic load dispatch	10 s – 1 h	
Thermodynamic effects	s –h	
Energy Management System applications	Steady state; ongoing	

Table 1.2 Grid blackouts registered around the world [14–16].

Place	Date	Cascade duration	Disconnected customers	Disconnected power
Northwestern America	10/08/1996	6 min	7.5 millions	30 GW
Northeastern America	14/08/2003	1 h	50 millions	62 GW
Southern Sweden and eastern Denmark	23/09/2003	5 min	4 millions	6.6 GW
Italy	28/09/2003	24 min	56 millions	24 GW
Ecuador	01/03/2003	20 s	3 millions	1.2 GW

Assessment (DVA) problem, focusing on post-contingency short-term phenomena, which develop in a timeframe of 15 to 20 seconds after a contingency occurs. This time window includes so-called short-term stability phenomena, which comprise transient stability, short-term voltage stability, oscillatory stability, and short-term frequency stability; and also, the possible overloads provoked by a change of system topology (these are the power system symptoms of stress).

Cascading events are the mechanism whereby failures propagate within a power system, thus leading the system to a possible collapse [14]. These events are usually caused by the conventional system protection schemes triggering in response to one or more symptoms of stress. These phenomena typically occur within timeframes between some milliseconds and several seconds, which means that a complete cascading event usually takes place in timescales from seconds up to hours. Therefore, the DVA timeframe definition specified in most of this book (i.e., 15 s to 20 s) matches the period in which an N-1 critical contingency might drive the system to further undesirable events (i.e., N-2 contingencies), which could be considered as the beginning of a cascading event.

A summary of some of the major blackouts registered in the last years is presented in Table 1.2. Several of the vulnerability assessment methodologies that will be presented in this book take a supervisory role whose response should be in the order of a few milliseconds to a few seconds. The use of monitoring methods such as the ones presented in this book, being capable of predicting the risk of initiating cascading events and whose response is in the order of seconds, might prevent the future occurrence of collapses such as those depicted in Table 1.2, whose total evolution is in the order of seconds to minutes.

1.3 Power System Vulnerability Symptoms

System vulnerability is characterized by four different symptoms of system stress, namely angle instability, voltage instability, frequency instability, and overloads [5].

Thus, most of the phenomena tackled throughout this book, based on the previously defined VA timeframe, comprise four short-term stability phenomena and post-contingency overloads. These phenomena include: transient stability, oscillatory stability, short-term voltage stability, short-term frequency stability, and post-contingency overloads initiated by topological modifications and power injection changes that follow an N-1 contingency.

1.3.1 Rotor Angle Stability

Rotor angle stability refers to the ability of synchronous machines of an interconnected power system to remain in synchronism after being subjected to a disturbance [17].

This type of stability depends on the ability to maintain or restore equilibrium between electromagnetic torque and mechanical torque of each synchronous machine in the system [17]. The electromagnetic torque of a synchronous machine, after a perturbation, can be resolved into two components: the "synchronizing torque" (in phase with rotor angle deviation—$\Delta\delta$), and the "damping torque" (in phase with the speed deviation—$\Delta\omega$), as shown by (1.1) [18].

$$\Delta T_e = T_S\Delta\delta + T_D\Delta\omega \tag{1.1}$$

where $T_S\Delta\delta$ is the synchronizing torque, T_S being the synchronizing torque coefficient, and $T_D\Delta\omega$ is the damping torque, T_D being the damping torque coefficient.

Instability in power systems is the result of a lack of one or both torque components: lack of sufficient synchronizing torque results in aperiodic or non-oscillatory instability (which characterizes transient stability phenomenon), whereas lack of damping torque results in oscillatory instability [17].

1.3.1.1 Transient Stability

Transient stability (TS) is a type of rotor angle stability that occurs when the system is subjected to a severe disturbance (e.g., a short circuit on a transmission line). The time-frame of interest in this phenomenon is usually 3–5 seconds following the disturbance [17]. This type of instability is usually characterized by a lack of synchronizing torque.

1.3.1.2 Oscillatory Stability

The rotor angle stability problem involves the study of the electromechanical oscillations inherent in power systems. Oscillatory instability occurs when there is lack of damping torque [17].

Oscillation problems may be either local or global in nature [18]. Local problems (local mode oscillations) are associated with oscillations among the rotors of a few generators close to each other. These oscillations have frequencies in the range of 0.7 to 2.0 Hz. Global problems (inter-area mode oscillations) are caused by interactions among large groups of generators. These oscillations have frequencies in the range of 0.1 to 0.7 Hz [18].

There are other two types of oscillatory problems caused by controllers of different system components (control modes), or by the turbine-generator shaft system rotational components (torsional modes) [18]. These types of oscillations present a large associated range of frequencies.

Modal analysis [18] is the most commonly used tool to analyze oscillations in a power system. It consists on the determination of the oscillatory modes and the analysis of their corresponding complex modal frequencies (i.e., the mode eigenvalues: $\sigma_i \pm j\omega_i$). Oscillatory stability is satisfied when all the modes present a positive damping. The frequency and damping of the oscillation can be determined from the eigenvalue as follows:

$$f_i = \frac{\omega_i}{2\pi} \tag{1.2}$$

$$\zeta_i = \frac{-\sigma_i}{\sqrt{\sigma_i^2 + \omega_i^2}} \tag{1.3}$$

where f_i and ζ_i are the oscillation frequency and the damping ratio of i-th mode, respectively.

1.3.2 Short-Term Voltage Stability

Voltage stability (VS) refers to the ability of a power system to maintain steady voltages at all buses in the system after being subjected to a disturbance. Short-term voltage stability involves the dynamics of fast acting load components such as induction motors, controlled loads, and HVDC converters. The timeframe of interest is in the order of several seconds. Load dynamic modeling is often essential. Short circuits near loads could cause relevant effects in this type of stability [17].

1.3.3 Short-Term Frequency Stability

Frequency stability (FS) refers to the ability of a power system to maintain a steady frequency following a severe system upset resulting in a significant imbalance between generation and load.

Short-term frequency instability is characterized by the formation of an under-generated area with insufficient under-frequency load shedding such that frequency decays rapidly causing the area blackout within a few seconds [17]. On the other hand, when a large amount of load is disconnected, an over-generated area might be formed. In these cases, over-frequency issues might provoke the uncontrolled outage of generation units.

1.3.4 Post-Contingency Overloads

Power flows through system branches can vary after a contingency because a disturbance modifies the grid topology. High electric currents can provoke overloads, and this could trigger protection systems that may separate elements from the system and increase the system vulnerability problem [9]. So, a system may be stable following a contingency, yet insecure due to post-fault system conditions resulting in equipment overloads [17].

In this book, the possible overloads that should be controlled within few seconds are tackled. This type of overload commonly corresponds to special series equipment which presents fast overload control requirements (e.g., electronic devices). On the other hand, equipment that presents slow overload control requirements (i.e., minutes or hours) would permit an on-line assessment via contingency analysis as part of SCADA/EMS applications. In these cases, a quick vulnerability assessment might not be required.

Power system series equipment presents different overload timeframes due to the variety of particular features such as physical limits, thermal capability, and stability constraints, among others. For instance, AC transmission line overload is usually specified by three types of limits: thermal, angle stability, and voltage, from which the temperature commonly does not rise so fast to reach the maximum designed limit. However, angle stability and voltage limits might be quickly reached after a contingency [19].

The physical limits present particular interest in some series electronic equipment because of the special care required for the semiconductor devices of the power converters of HVDCs, FACTS devices, and photovoltaic or wind facilities. In these types of equipment, a quick assessment of possible overloads has relevant importance [19].

1.4 Synchronized Phasor Measurement Technology

1.4.1 Phasor Representation of Sinusoids

A phasor or phase vector constitutes the most commonly applied analytic representation of steady state sinusoidal waveforms of fundamental power frequency. Phasors are used as the basic mathematical tool of AC circuit analysis [20].

A pure sinusoidal equation given by:

$$x(t) = X_m \cos(\omega t + \phi) \tag{1.4}$$

where ω is the signal angular frequency in radians per second, ϕ is the phase angle in radians, and X_m is the peak amplitude of the signal; this can also be written as (1.5) based on Euler's formula [21].

$$x(t) = \text{Re}\left\{ X_m e^{j(\omega t + \phi)} \right\} = \text{Re}\left\{ X_m e^{j\phi} [e^{j\omega t}] \right\} \tag{1.5}$$

The factor $e^{j\omega t}$ of (1.5) can be suppressed from this equation when considering that the frequency ω is a constant parameter in steady state. Under this assumption, (1.5) can be simply represented by a complex number $\overline{X}$ rotating at the angular speed ω, known as its phasor representation or simply phasor, as shown by (1.6) [21].

$$x(t) \Leftrightarrow \overline{X} = \left(\frac{X_m}{\sqrt{2}} \right) e^{j\phi} = \left(\frac{X_m}{\sqrt{2}} \right) (\cos\phi + j\sin\phi) \tag{1.6}$$

where $X_m / \sqrt{2}$ is the RMS (root mean square) signal value.

Figure 1.2 illustrates the phasor representation of a pure sinusoidal equation given by (1.4).

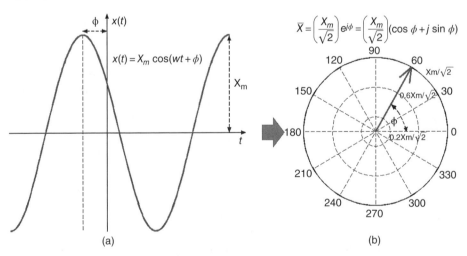

Figure 1.2 Phasor representation of sinusoids: (a) sinusoidal function, (b) phasor representation.

1.4.2 Synchronized Phasors

A *synchronized phasor* or *synchrophasor* is defined by [22] as "a phasor calculated from data samples using a standard time signal as the reference for the measurement."

The synchronized phasors are angularly referenced to a cosine wave at nominal system frequency synchronized to Coordinated Universal Time (UTC, according to the French language). The phase angle ϕ is defined to be $0°$ when the maximum of the function $x(t)$ (1.4) occurs at the UTC second rollover (1 pulse per second (PPS) time signal), and $-90°$ when the positive zero crossing occurs at the UTC second rollover [22, 23]. This fact means that synchronized phasors from remote sites have a defined common phase relationship [22].

The synchronizing signal must have uninterrupted and continuous availability and it has to offer easy access to every place where the electrical variables are measured. Additionally, enough accuracy has to be guaranteed in order to accomplish the maximum error requirements (Total Vector Error (TVE)). Thus, the most used synchronizing source is the Global Positioning System (GPS) based on the 1-PPS synchronizing signal, with a maximum error of 1 µs [22].

The concept of TVE and related compliance tests are expanded in [23]. In addition, the definition of frequency and the rate of change of frequency error limits, for both steady state and dynamic conditions, is also considered in [23].

1.4.3 Phasor Measurement Units (PMUs)

Phasor Measurement Units are devices that allow estimation of the synchrophasors of AC voltage and current sinusoidal waves. In order to calculate a synchrophasor, a PMU needs to read the electrical wave $x(t)$ as well as a reference cosine wave synchronized to UTC.

For calculating a synchrophasor, a PMU uses a phasor estimation algorithm. These algorithms employ N samples in a specific time period for carrying out the phasor estimation. The most commonly applied algorithm is the Discrete Fourier Transform (DFT) [21].

Their high accuracy and response speed and their time synchronization ability make PMUs appropriate for wide area monitoring in steady and dynamic states, as well as for control and protection applications [2]. Figure 1.3 schematizes the basic structure of a PMU.

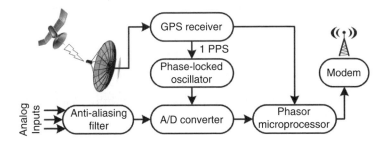

Figure 1.3 PMU basic structure [21].

1.4.4 Discrete Fourier Transform and Phasor Calculation

The DFT is a method for computing the Fourier transform $X(f)$ of a few samples belonging to a specific signal $x(t)$. In this case, the Fourier transform is calculated in discrete intervals in the frequency domain when the signal is sampled in discrete points in the time domain.

Consider a periodic function $x(t)$ with period T_0, and N samples: $x(k\Delta T)$, $k = 0, 1, 2, \ldots, N-1$, where ΔT is the sampling interval; then $T_0 = N\Delta T$. Under these considerations, the Fourier transform $X(f)$ has only N different values corresponding to the frequencies $f = n/T_0$, where n is between 0 and N/2 (if N is an even number). The following equation represents the DFT definition considering N samples and ΔT sampling intervals [21].

$$X\left(\frac{n}{T_0}\right) = \sum_{k=0}^{N-1} x(k\Delta T)e^{-\frac{j2\pi kn}{N}}, \; n = 0, \pm 1, \pm 2, \ldots, \pm\frac{N}{2} \tag{1.7}$$

The DFT is a useful tool for Fourier series spectral analysis. The Fourier series coefficients of a periodic function can be computed from the DFT of the sampled data by dividing the DFT corresponding to the frequency of interest by N. Since each frequency always appears in two positions ($\pm n$), the coefficients of the Fourier series are:

$$c_n = \frac{2}{N}X\left(\frac{n}{T_0}\right) = \frac{2}{N}\sum_{k=0}^{N-1} x(k\Delta T)e^{-\frac{j2\pi kn}{N}}, \; n = 0, 1, 2, \ldots, \frac{N}{2} \tag{1.8}$$

Considering that a phasor represents a pure sinusoidal function, it is necessary to extract the *fundamental frequency component* f_0 (coefficient c_1 of the Fourier series) of the sampled signal [21]. Thus, the phasor representation is:

$$\overline{X} = \frac{c_1}{\sqrt{2}} = \frac{2}{N\sqrt{2}}X\left(\frac{1}{T_0}\right) = \frac{\sqrt{2}}{N}X(f_0) \tag{1.9}$$

$$\overline{X} = \frac{\sqrt{2}}{N}\sum_{k=0}^{N-1} x(k\Delta T)e^{-\frac{j2\pi k}{N}} \tag{1.10}$$

1.4.5 Wide Area Monitoring Systems

Wide area monitoring systems (WAMS) permit access to distributed phasor measurements throughout the network by means of adequately located PMUs. For this purpose, in addition to PMUs, advanced signal processing algorithms, specialized communication systems, and a real time dynamic infrastructure are required. This technology includes applications related to the real time supervision and control of electric power systems. WAMS are mainly composed of PMUs, Phasor Data Concentrators (PDCs), and communication systems [10, 24].

WAMS provide direct voltage and current phasor measurements with updating periods of few milliseconds, ideally avoiding the necessity for state estimators. The fast updating intervals also allow the monitoring of power system dynamic behavior and offer the possibility of detecting potential insecure conditions that might lead the systems to blackouts [12].

It is possible to distinguish three different applications for wide area measurement systems:

- Wide area monitoring systems (WAMS)
- Wide area control systems (WACS)
- Wide area protection systems (WAPS)

Several monitoring, control, and protection applications are depicted in Figure 1.4. These applications are ranked depending on different operational levels. It is possible to note that synchronized phasor measurement technology corresponds to the fourth operational level, which is established by the so-called *Wide area monitoring, protection, and control systems (WAMPAC)* [2].

The overall WAMPAC scheme comprises several interrelated stages, each one having embedded applications to perform different relevant tasks in order to achieve the objective of allowing a "self-healing grid," as illustrated in Figure 1.5.

Measurements of the actual system condition from PMU/SCADA-EMS are used to perform the following applications:

i) Update of system behavior knowledge database (i.e., operational statistics) and execution of pattern recognition routines, which supports the Dynamic Security Assessment (DSA) and real time Dynamic Vulnerability Assessment (DVA) tasks.
ii) DSA in normal operating states to determine whether or not the system security level is to be degraded in future time when considering a selected set of credible contingencies.
iii) Real time DVA to determine the actual system security level as well as its tendency to change its conditions to a critical state in post-disturbance immediate time.

The information obtained from DSA and DVA is passed to the control center (for real time monitoring and decision making) and potentially directly to the control systems as well.

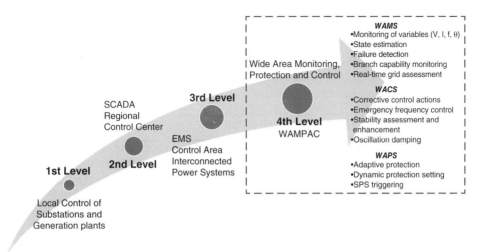

Figure 1.4 Wide area monitoring, protection and control systems [2].

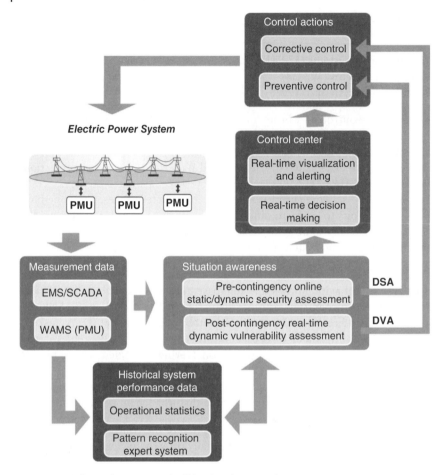

Figure 1.5 Scheme for integrated self-healing functionalities to support secure system operation in real time.

Several chapters of this book focus on presenting methodologies capable of performing post-contingency real time DVA based on the development and application of pattern recognition expert systems in combination with the evaluation of various performance indices.

1.4.6 WAMPAC Communication Time Delay

In order to perform global control actions in a timely manner, the WAMPAC scheme must act prior to the tripping of local protection systems. This fact highlights the importance of getting a quick response from all the process involved in the WAMPAC structure, which comprise: the time delay for phasor measurement, fiber-optic communications, PDC throughput including wait time for slowly arriving packets, transfer trip, and circuit breaker tripping or closing [24].

In this connection, it is highly important to consider the total WAMPAC communication time delay in order to be sure the vulnerability assessment will orient timely control actions.

Table 1.3 Time delay of a WAMPAC scheme per process [24].

Process	Time delay in 60-Hz cycles	Time delay (ms)
PMU measurement	3	50
Fiber-optic communications	2	33
PDC throughput	2	33
Transfer trip	1	17
Circuit breaker	2–5	33–83
Total WAMPAC time delay	10–13	167–217

This time delay depends on the complexity of each power system, and it has to be determined prior the development of the WAMPAC scheme. For instance, Table 1.3 presents some reference values that orient a typical WAMPAC communication time delay [24].

1.5 The Fundamental Role of WAMS in Dynamic Vulnerability Assessment

As described in the previous section, PMUs are able to provide time-synchronized phasor data with delays of less than 200 ms, which contain valuable dynamic information that could indicate, in real time, the system vulnerability status and potential collapses [7]. These post-contingency data are fundamental to developing methodologies capable of assessing dynamic vulnerability in real time. The huge potential of the dynamic information immersed in PMU data actually allows the assessment of both tasks involved in the concept of vulnerability: (i) the system security level (static and dynamic security), and (ii) the tendency to change the system conditions to a critical state that is called the "Verge of Collapse State."

For instance, synchronized measurements obtained from Wide Area Monitoring Systems (WAMS) can be used to detect a possible system separation, to monitor the voltage and frequency of critical nodes, or to check power flows in important branches in order to analyze congestion. PMU data could also be used to analyze poorly-damped power system oscillations [25], to carry out transient stability assessment (TSA) [26, 27], or to develop closed loop control schemes in order to improve system stability [12].

A characteristic ellipsoid (CELL) method to monitor the power system dynamic behavior and provide wide area situational awareness using PMU data is presented in [13]. A logic-based method to identify a contingency severity and to assess voltage stability from PMU data is proposed in [28]. These approaches permit the assessment of post-contingency system security level, but they do not consider the tendency of the system to change its conditions to a critical state.

Some methods to perform real time TSA are summarized in [26]. From these methods, Emergency Single Machine Equivalent (E-SIME) seems attractive for real time applications. This method uses the multi-machine rotor angles (δ) and the system accelerating power (Pa) to establish a One Machine Infinite Bus equivalent (OMIB) in order to predict transient instability, using data obtained from PMUs. In this approach, the calculus of the mechanical variables and the prediction of the Pa–δ curves provoke time delays and accuracy problems. In this same context, [26] presents a method for TSA based on post-contingency area-based center-of-inertia (COI) referred rotor angles estimation from PMU measurements. To this aim, several off-line tasks were addressed to deal with issues regarding the selection of PMU monitoring locations, reduction of data volume, and definition, training, and tuning of a suitable support vector regressor (SVR).

A scheme based on phase-space visualization, which allows finding patterns that could indicate a possible collapse, is presented in [29]. This method permits monitoring of phase-space curves of critical variables and providing early warning, but it needs to draw images in order to find the similarities among them, which makes an automatic response difficult.

A technique for prediction of the transient stability status of a power system following a large disturbance such as a fault, based on synchronously measured samples of the magnitudes of voltage phasors at major generation centers, is presented in [30]. The voltage samples are taken immediately after a fault is cleared and used as inputs to a binary classifier based on a support vector machine (SVM) to identify the transient stability status. This approach considers that there exist PMUs installed in all generators' terminal buses, and does not contemplate any feature extraction technique in order to reduce the amount of classifier input data; this increases the dimensionality of the problem and might provoke mistakes and delays in the training stage.

A data mining-based framework for performing power system vulnerability assessment and predicting system instabilities that may lead to possible system cascading failure is presented in [31]. Based on Monte Carlo methods, power system contingency data of dynamic and steady state simulations are generated. These data are then analyzed via advanced data mining methods, based on a combination between Graph Theory, Local Correlation Network Pattern (LCNP) and Kernel Based Classification, to find the most relevant system information or patterns with respect to system instability; this is then used for achieving instability prediction based on PMU measurements as well as generator variable measurements. This method suggests using the first 10 or 20 system variables (including generator and system variables) with the greatest chi-square (obtained from LCNP) as the features of the prediction model. This method presents the drawbacks of using too many variables, including those belonging to generators that are not easily to be available (such as rotor angle, speed, mechanical power, etc.) Additionally, results of classification accuracy show a recall (i.e., dependability) of around 93%, which might be improved by using a different data analysis technique.

A vulnerability assessment method for quickly assessing the multi-area power system post-contingency vulnerability status, mainly as regards rotor angle stability, is developed in [32]. This method uses PMU data to compute Wide Area Severity Indices (WASI), as an extension of the severity indices firstly presented in [33]. These WASI are calculated in the frequency domain through applying the Short Time Fourier Transform (STFT) to data obtained from PMUs located in specific system buses, determined by a coherency criterion defined in [34] and improved in [35]. A systematic scheme for rapidly classifying the stability status resulted from WASI using fuzzy rule-based classifiers is presented in [36]. This classifier uses large-size decision trees (DTs) to generate initial accurate classification boundaries for decision making as early as 1 or 2 s after fault clearing. A WASI-based model-predictive framework for determining fast catastrophe precursors in a bulk power system has been presented in [37]. These fast catastrophe predictors use random-forest (RF) learning in order to increase the classification accuracy, showing excellent performance as regards typical DTs. In [38], a comparison between several data mining classifiers, such as artificial neural network (ANN), SVM, DT, Fuzzy DT, Fuzzy ID3 (Iterative Dichotomiser 3), and RF is performed; from this, the RF-based predictor is the only one to achieve more than 99.0% accuracy based on WASI. However, the RF has poor transparency characteristics and might exhibit over-fitting problems. In this connection, a more transparent predictor is suggested to be used in [38] at the expense of loss of accuracy. WASI offer a novel concept in power system real time VA that reasonably permits estimating the tendency of the system to change its conditions to a critical state. Nevertheless, these indices present the problem of showing a considerable large overlapping zone. For instance, test results in [37] conclude that "75% of stable cases verify the relationship FastWASI300ms $<$ -2.5, while for 75% unstable cases, FastWASI300ms $>$ -2.5." This fact denotes that the overlapping zone is formed by 25% of stable cases belonging to the region of the unstable cases and vice versa. This large overlapping range is the main reason for classification difficulties.

A data-mining-based approach for predicting power system post-contingency vulnerability status in real time is proposed in [10]. To this aim, system dynamic vulnerability regions (DVRs) are first determined by applying singular value decomposition, by means of empirical orthogonal functions (EOFs), to a post-contingency database obtained from Phasor Measurement Units (PMUs) adequately located throughout the system. In this way, the obtained pattern vectors (EOF scores) allow mapping the DVRs within the coordinate system formed by the set of EOFs, which permits revealing the main patterns buried in the collected PMU signals. The database along with the DVRs enable the definition, training, and identification of a support vector classifier (SVC), which is employed to predict the post-contingency vulnerability status as regards three short-term stability phenomena, that is: transient stability, short-term voltage stability, and short-term frequency stability (TVFS). Enhanced procedures for feature extraction and selection as well as heuristic optimization-based parameter identification are proposed to ensure a robust performance of the SVC. This method allows the assessment of system vulnerability for three different phenomena, estimating the system tendency of changing its conditions to the Verge of Collapse State, and overcomes the drawback of large overlapping caused by applying other types of signal processing tools.

The potential applications for DVA make PMUs fundamental in the basic architecture of a self-healing grid. However, despite the valuable dynamic data obtained from PMUs, they are not able to give any information about system health by themselves. So, it is necessary to develop mathematical tools capable of quickly analyzing the data in order to accurately assess dynamic vulnerability in real time. This is a very novel research area in which further investigation is needed to improve tools and methods, so that in the future, Dynamic Vulnerability Assessment would be performed with confidence as an important element of a self-healing grid structure.

An additional aspect to be considered is PMU data quality. In fact, synchronized measurements will be used as input to control and protection systems after the DVA has determined the necessity of performing corrective control actions; therefore, it is fundamental to have enough confidence about data obtained from the PMUs. In this connection, it is necessary to carry out an evaluation of the performance of PMUs under dynamic conditions for previous application in DVA and further control schemes. This is because system perturbations provoke different transients that cause current and voltage wave distortions, such as peaks, sags, frequency variations, oscillations, harmonics, DC components, and so on, which affect the accuracy of synchronized measurements [21]. In this connection, prior to the development and enforcement of dynamic applications, it is necessary to evaluate the performance of PMUs under possible dynamic conditions, which might be used in order to specify the minimum requirements of the devices. This need is essential in order to guarantee the PMUs satisfactorily fulfill the guidelines suggested by the IEEE C37.118.1-2011 standard [23] to assure dynamic compliances, depending on the required application. For instance, the PMUs have to accomplish one of the two performance class filters: P class or M class. P class is for applications requiring fast response whereas M class better adjusts for applications of greater precision. So, depending on the specific application, a different combination of errors from the magnitude and phase angle measurements might be specified in order to get the desired TVE [23].

In summary, based on the presented literature review, WAMS has been determined as the fundamental source of data useful for being the input to adequate mathematical tools capable of assessing vulnerability in real time. To accomplish this function, some basic requirements are necessary, such as the implementation of PMUs with confident enough phasor estimation algorithms, robust communication and information technology systems and, mainly, the development and application of proper mathematical methodologies capable of analyzing huge volumes of data in real time, in order to give confident early warning indicators as soon as the different vulnerability phenomena evolve in real time.

1.6   Concluding Remarks

The lack of investment, the use of congested transmission lines, and other technical reasons, such as environmental constraints, have been dangerously pushing Bulk Power Systems close to their physical limits. Under these conditions, certain sudden perturbations can cause cascading events that may lead to system blackouts. In this connection, it is crucial to ensure system security, so the development of protection systems that allow guaranteeing of service continuity is required. Nevertheless, these remedial action schemes are usually set to operate when specific pre-established operational conditions are reached, so they are unable to work under unconsidered contingencies that might entail cascading events. Thus, the control of the system and the protection triggering should be adjusted depending on the real time progress of a given event, emphasizing the need to develop a "self-healing grid." Real time Dynamic Vulnerability Assessment (DVA) is a fundamental task within this self-healing grid structure, and PMU-based methods show a huge potential for performing this smart grid application, since they have the capability to detect the necessity of performing the remedial action. DVA has to be carried out in order to trigger adequate corrective control actions that mitigate the consequences of the disturbances and reduce the possibility of system blackouts. This task has opened a very novel research area, where further investigation is needed.

References

1 M. Amin, "Toward self-healing infrastructure systems", Electric Power Research Institute (EPRI), *IEEE Computer*, vol. **33**, no. 8, pp. 44–53, 2000.

2 C. Martinez, M. Parashar, J. Dyer, and J. Coroas, "Phasor data requirements for real time wide-area monitoring, control and protection applications", *CERTS/EPG, White Paper – Final Draft, for: EIPP – Real Time Task Team*, vol. **26**, pp. 8, January, 2005.

3 U. Kerin, G. Bizjak, E. Lerch, O. Ruhle, and R. Krebs, "Faster than real time: Dynamic security assessment for foresighted control actions", 2009 IEEE Bucharest Power Tech Conference, June 28th–July 2nd, Bucharest, Romania.

4 J. D. McCalley, and Fu Weihui, "Reliability of special protection systems", *IEEE Transactions on Power Systems, IEEE Power & Energy Society*, vol. **14**, no 4, pp. 1400–1406, November 1999.

5 D. McGillis, K. El-Arroudi, R. Brearley, and G. Joos, "The process of system collapse based on areas of vulnerability", Large Engineering Systems Conference on Power Engineering, pp. 35–40, Halifax, NS, July 2006.

6 I. Dobson, P. Zhang, et al, "Initial review of methods for cascading failure analysis in electric power transmission systems", IEEE PES CAMS Task Force on Understanding, Prediction, Mitigation and Restoration of Cascading Failures, IEEE Power Engineering Society General Meeting, Pittsburgh, PA USA July 2008.

7 Z. Huang, P. Zhang, et al, "Vulnerability assessment for cascading failures in electric power systems", Task Force on Understanding, Prediction, Mitigation and Restoration of Cascading Failures, IEEE PES Computer and Analytical Methods Subcommittee, IEEE Power and Energy Society Power Systems Conference and Exposition 2009, Seattle, WA.

8 A. Fouad, Qin Zhou, and V. Vittal, "System vulnerability as a concept to assess power system dynamic security", *IEEE Transactions on Power Systems*, vol. **9**, no 2, p. 1009–1015, May 1994.

9 J. Gimenez, and P. Mercado, "Online inference of the dynamic security level of power systems using fuzzy techniques", *IEEE Transactions on Power Systems*, vol. **22**, no. 2, pp. 717–726, May 2007.

10 J. Cepeda, J. Rueda, G. Colomé, and I. Erlich, "Data-mining-based approach for predicting the power system post-contingency dynamic vulnerability status", *International Transactions on Electrical Energy Systems*, vol. **25**, issue 10, pp. 2515–2546, October 2015.

11 Chen-Ching Liu, Juhwan Jung, Gerald T. Heydt, Vijay Vittal, and Arun G. Phadke, "The strategic power infrastructure defense (spid) system, a conceptual design", *IEEE Control System Magazine*, vol. **20**, no 4, pp. 40–52, August 2000.

12 S. C. Savulescu, et al, *Real-Time Stability Assessment in Modern Power System Control Centers*, IEEE Press Series on Power Engineering, Mohamed E. El-Hawary, Series Editor, a John Wiley & Sons, Inc., Publication, 2009.

13 Yuri Makarov, Carl Miller, Tony Nguen, and Jian Ma, "Characteristic ellipsoid method for monitoring power system dynamic behavior using phasor measurements", VII Symposium on Bulk Power System Dynamics and Control, Charleston, USA, August, 2007.

14 I. Dobson, B. Carreras, V. Lynch, and D. Newman, "Complex systems analysis of series of blackouts: cascading failure, critical points, and self-organization", *Chaos: An Interdisciplinary Journal of Nonlinear Science*, 2007, vol. **17**, no 2, p. 026103, June 2007.

15 G. Andersson, P. Donalek, R. Farmer, and I. Kamwa, "Causes of the 2003 major grid blackouts in North America and Europe, and recommended means to improve system dynamic performance", *IEEE Transactions on Power Systems*, vol. **20**, no. 4, pp. 1922–1928, November 2005.

16 V. Venkatasubramanian, and Yuan Li, "Analysis of 1996 western American electric blackouts", Bulk Power System Dynamics and Control – VI, Italy, August 2004, pp. 685–721.

17 P. Kundur, J. Paserba, V. Ajjarapu, et al, "Definition and classification of power system stability", IEEE/CIGRE Joint Task Force on Stability: Terms and Definitions. *IEEE Transactions on Power Systems*, vol. **19**, no. 3, pp. 1387–1401, August 2004.

18 P. Kundur, *Power System Stability and Control*, McGraw-Hill, Inc., Copyright 1994.

19 J. Cepeda, D. Ramirez, and G. Colome, "Probabilistic-based overload estimation for real-time smart grid vulnerability assessment", Proc. Sixth IEEE/PES Transmission and Distribution: Latin America Conference and Exposition (T&D-LA), Montevideo, Uruguay, September 2012.

20 A. Phadke, "Synchronized phasor measurements in power systems", *IEEE Computer Applications in Power*, vol. **6**, no. 2, pp.10–15, April 1993.

21 A. Phadke, and J. Thorp, *Synchronized Phasor Measurements and Their Applications*, Virginia Polytechnic Institute and State University, Springer Science + Business Media, 2008, ISBN 978-0-387-76535-8.

22 IEEE Power Engineering Society, "IEEE standard for synchrophasors for power systems", IEEE Std. C37.118-2005, March 2006.

23 IEEE Power Engineering Society, "IEEE standard for synchrophasors for power systems", IEEE Std. C37.118.1-2011, December 2011.

24 C. W. Taylor, D. C. Erickson, K. E Martin, R. E. Wilson, and V. Venkatasubramanian, "WACS – wide-area stability and voltage control system: R&D and online demonstration", *Proceedings of the IEEE*, vol. **93**, no. 5, May 2005.

25 J. Cepeda, G. Argüello, P. Verdugo, A. De La Torre, "Real-time monitoring of steady-state and oscillatory stability phenomena in the ecuadorian power system", IEEE Transmission and Distribution Latin America (T&D-LA) 2014, Medellín, Colombia, Septiembre 2014.

26 J. Cepeda, J. Rueda, G. Colomé, D. Echeverría, "Real-time transient stability assessment based on centre-of-inertia estimation from PMU measurements", *IET Generation, Transmission & Distribution*, vol. **8**, issue 8, pp. 1363-1376, August, 2014.

27 D. Echeverría, J. Cepeda, G. Colome, "Critical Machine Identification for Power Systems Transient Stability Problems using Data Mining", IEEE Transmission & Distribution Latin America (T&D-LA) 2014, Medellín, Colombia, Septiembre 2014.

28 A. Tiwari, and V. Ajjarapu, "Event Identification and Contingency Assessment for Voltage Stability via PMU", 39th North American Power Symposium, pp. 413–420, Las Cruces, October 2007.

29 Kai Sun, and S.T. Lee, "Power system security pattern recognition based on phase space visualization", Third International Conf. on Electric Utility Deregulation and Restructuring and Power Technologies, Nanjuing pp. 964–969, April 2008.

30 F. Gomez, *Prediction and Control of Transient Instability Using Wide Area Phasor Measurements*, Doctoral Thesis, University of Manitoba, September 2011.

31 P. Zhang, Y. D. Zhao, et al, *Program on Technology Innovation: Application of Data Mining Method to Vulnerability Assessment*, Electric Power Research Institute (EPRI), Final Report, July 2007.

32 I. Kamwa, J. Beland, and D. Mcnabb, "PMU-based vulnerability assessment using wide-area severity indices and tracking modal analysis", IEEE Power Systems Conference and Exposition, pp. 139–149, Atlanta, November, 2006.

33 I. Kamwa, R. Grondin, and L. Loud, "Time-varying contingency screening for dynamic security assessment using intelligent-systems techniques," *IEEE Transactions on Power Systems*, vol. **16**, no. 3, pp. 526–536, Aug. 2001.

34 I. Kamwa, A. K. Pradham, and G. Joos, "Automatic segmentation of large power systems into fuzzy coherent areas for dynamic vulnerability assessment", *IEEE Transactions on Power Systems*, vol. **22**, no. 4, pp. 1974–1985, Nov. 2007.

35 I. Kamwa, A. K. Pradham, G. Joos, and S. R. Samantaray, "Fuzzy partitioning of a real power system for dynamic vulnerability assessment", *IEEE Transactions on Power Systems*, vol. **24**, no. 3, pp. 1356–1365, August 2009.

36 I. Kamwa, S. R. Samantaray, and G. Joos, "Development of rule-based classifiers for rapid stability assessment of wide-area post-disturbance records", *IEEE Transactions on Power Systems*, vol.**24**, no. 1, pp.258–270, Feb. 2009.

37 I. Kamwa, S. R. Samantaray, and G. Joos, "Catastrophe predictors from ensemble decision-tree learning of wide-area severity indices", *IEEE Transactions on Smart Grid*, vol.**1**, no.2, pp. 144–158, Sept. 2010.

38 I. Kamwa, S. R. Samantaray, and G. Joos, "On the accuracy versus transparency trade-off of data-mining models for fast-response PMU-based catastrophe predictors", *IEEE Transactions on Smart Grid*, vol. **3**, no. 1, pp. 152–161, March 2012.

2

Steady-State Security

Evelyn Heylen[1], Steven De Boeck[1], Marten Ovaere[2], Hakan Ergun[1], and Dirk Van Hertem[1]

[1] *Research group ELECTA, Department of Electrical Engineering, University of Leuven, Belgium*
[2] *Department of Economics, University of Leuven, Belgium*

The power system can be seen today as one of the most critical infrastructures. The reliability of the power system influences the economics and social well-being of a modern society and has a direct impact on the quality of life. A one day blackout could lead to costs that are about 0.5% of the GDP of a country, which have to be added to possible social consequences of longer interruptions, such as diseases, deaths and injuries [1]. Also short term interruptions cause loss of production, frozen foods gone bad, traffic accidents, and so on. Moreover, the reliability of local energy provisions might be key to the selection of an industrial site, which is important for particular industries such as foundries and large IT providers.

Power system reliability is managed in order to minimize the impact to society. Ideally, the achieved power system reliability represents the optimal balance between the value of reliability and its cost. In practice, an "adequate level of reliability" is aimed for [2]. In order to obtain an adequate reliability level, appropriate actions can be taken throughout the lifetime of the power system, from years ahead (i.e., system development) up to real time (i.e., operation). These actions are the result of power system reliability management based on a well defined reliability criterion.

Power system reliability is determined by the existence of sufficient facilities in the system to satisfy the electrical power requirements of consumers at all times and by the system's ability to handle disturbances arising in the system. The latter is also defined as power system security and will be the main topic of this chapter. The focus is on steady-state security.

Section 2.1 introduces power system reliability management, while Section 2.2 explains the challenges due to uncertainties in reliability management in different timeframes ranging from long-term system development up to short-term system operation. Section 2.3 goes more into the shortcomings of currently used reliability criteria, while Section 2.4 elaborates on the socio-economic evaluation of power system reliability management considering the trade-off between the cost of reliability management and interruption costs. Section 2.5 concludes the chapter.

Dynamic Vulnerability Assessment and Intelligent Control for Sustainable Power Systems, First Edition.
Edited by José Luis Rueda-Torres and Francisco González-Longatt.
© 2018 John Wiley & Sons Ltd. Published 2018 by John Wiley & Sons Ltd.

2.1 Power System Reliability Management: A Combination of Reliability Assessment and Reliability Control

Power system reliability is defined as the probability that an electrical power system can perform a required function under given conditions for a given time interval [3]. It quantifies the ability of a power system to provide an adequate supply of electrical energy in order to satisfy the customer requirements with few interruptions over an extended period of time. The degree of reliability is measured by the frequency, duration and magnitude of the adverse effects on consumer service [4].

Power system reliability consists of *power system security* and *power system adequacy* [4]. An adequate power system has sufficient generating, transmission and distribution facilities in the system to satisfy the aggregate electric power and energy requirements of customers at all times, taking into account scheduled and unscheduled outages of the system components [4]. System security on the other hand describes the ability of the system to handle disturbances, such as the loss of major generation units or transmission facilities [4]. Sudden disturbances in power systems may result from factors external to the system itself, such as weather and environment, or internal factors such as insulation failure or failure of a plant item [5]. Power system security and adequacy are strongly interdependent, because adequacy is subject to transitions between different states, which are strictly not part of adequacy analysis but of security analysis [6].

Power system reliability expresses whether the system is able to fulfil its required function. It depends on the system's vulnerability to external threats that can lead to unwanted events and the expected function of the system specified by the reliability criterion. The *reliability criterion* imposes a basis to determine whether or not the power system is managed in an adequate and secure way. The *vulnerability* of a power system is an expression of the problem the system faces to maintain its function if a threat leads to an unwanted event and the problems the system faces to resume its activities after the event occurred. It is an internal characteristic of the system and is composed of its susceptibility to external threats and its coping capacity. The power system is *susceptible* to a threat if a realization of the threat leads to an unwanted event in the power system [7]. The *coping capacity* describes the ability of the operator and the power system itself to cope with an unwanted event, limit negative effects and restore the power system's function to a normal state [8]. It is determined by the adequacy and the security of the system. *Criticality* of unwanted events, such as power system failure, represents consequences for end-users, such as generators, consumers, traders, suppliers, and so on. [7]. The interaction between vulnerability and reliability with their different aspects and the reliability criterion is graphically shown in Figure 2.1.

Power system reliability management aims to serve load with a very high probability and this at the required quality and with a very low frequency of experiencing spectacular system failures such as blackouts [4]. Nowadays, reliability is mainly managed according to the N−1 criterion.[1] The power system reliability management process of transmission system operators is illustrated graphically in Figure 2.2. It consists of two

---

1 According to the ENTSO-E system operation guideline, the N−1 criterion means "the rule according to which elements remaining in operation within a TSOs responsibility area after a contingency from the contingency list must be capable of accommodating the new operational situation without violating operational security limits." [9]

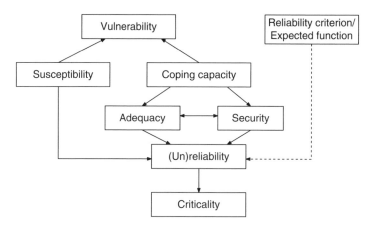

Figure 2.1 Interactions between the aspects determining reliability of power systems.

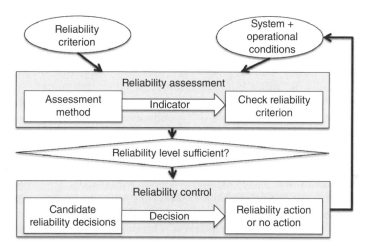

Figure 2.2 Overview of reliability management.

main tasks: (i) reliability assessment and (ii) reliability control. *Reliability assessment* aims at identifying and quantifying the actual reliability level, while *reliability control* consists of taking a sequence of decisions under uncertainty that satisfy the reliability criterion, while minimising the socio-economic costs of doing so.

2.1.1 Reliability Assessment

Reliability assessment methods allow verification of whether reliability criteria are satisfied and to quantify the reliability of the system in terms of reliability indicators based on frequency, duration or probability of malfunctioning. The two main reliability assessment methods are analytical contingency enumeration and simulation techniques, such as Monte Carlo simulation [10, 11]. Advantages and disadvantages of both techniques are summarized in Table 2.1. Due to the deterministic nature of currently used reliability criteria, contingency enumeration is mostly applied nowadays. Operational limit violations of system variables are verified for a predefined set of contingencies. Probabilistic

Table 2.1 Advantages and disadvantages of methods for steady-state security assessment.

Analytical	Simulation (Monte Carlo)
System represented as mathematical model	Simulate actual process and random behavior of the system
Mathematically complex	Higher computation time
Focus on expected values	Information about underlying distributions
State selection: Probability-based or order of contingency level	State selection: Probability-based random numbers

security assessment on the other hand should in theory assess all possible system states with their respective probability. This is not possible in practice, especially not in large systems. Alternatively, simulation techniques, such as Monte Carlo simulation, can be used for probabilistic reliability assessment [6]. Hybrid techniques combine aspects of both approaches [12].

Reliability assessment is a combination of security assessment and adequacy assessment. Adequacy assessment verifies whether the system is capable of supplying the load under specified contingencies without operating constraint violations. Power system security assessment on the other hand determines whether the immediate response of the system to a disturbance generates potential reliability problems [13]. A distinction can be made between dynamic (time-dependent) and static or steady-state (time-independent) security assessment, depending on whether transients after the disturbance are neglected or not. Steady-state security assessment should evaluate whether a new equilibrium state exist for the post-contingency system, while dynamic security assessment investigates the existence and security level of the transient trajectory in the state space from the original pre-contingency equilibrium point to the post-contingency equilibrium point. The power system model in dynamic security assessment is given by non-linear differential equations whose boundary conditions are given by the non-linear power flow equation. Static security can be considered as a first-order approximation of the dynamic power system state [14]. Alternatively, pseudo-dynamic evaluation techniques using sequential steady-state evaluation to assess the impact at several post-contingency stages exist [15].

2.1.2 Reliability Control

To be able to provide the desirable level of reliability, the power system needs to be able to cope with a certain number of contingencies. In current practice, power systems are designed and operated to withstand a number of "credible" contingencies. If the N−1 criterion is applied, the list of credible contingencies consists of all contingencies in which one component or no components are out of service. During the reliability assessment, the state of the power system is determined. Based on that state, the reliability control mechanism selects a reliability decision from the list of candidate decisions. Decisions imply either a reliability action or no action. Executed reliability actions aim to change the state of the system such that the reliability criterion is satisfied and system security is ensured. Ideally, reliability control performs these actions while minimizing total system cost. Available reliability actions depend on the time horizon considered.

2.1.2.1 Credible and Non-Credible Contingencies

A credible contingency is a disturbance which has been specifically foreseen in the planning and operation of the power system, and against which specific measures have been taken to ensure that limited consequences would follow its occurrence [5]. In general, a contingency is considered as the trip of one single or several network elements that cannot be predicted in advance [16]. A scheduled outage, such as outage of a line for maintenance, is not classified as a contingency. In continental Europe, a further subdivision of credible contingencies is made to distinguish between "normal contingencies" and "exceptional contingencies," as is depicted in Figure 2.3 [16]. Based on the currently used N−1 framework, the normal contingency is the loss of a single network element. Exceptional contingencies consist of single events that effect multiple network components, for example the falling of a tower resulting in the loss of a double circuit line. Each transmission system operator determines, based on its own risk assessment, which exceptional contingencies are included in the contingency list in order to prevent cascading events. A system that is able to maintain all system parameters within acceptable limits for each of the contingencies on this contingency list is considered to be N−1 secure. On the other hand we have non-credible contingencies, which are often also described as out-of-range or high impact low probability contingencies. These are rare contingencies that often result from exceptional technical malfunctions, force majeure conditions, common mode failures or human errors [17]. As well as being rare, out-of-range contingencies vary significantly with respect to their causes and consequences and thus are hardly ever predictable. Often they are accompanied by the removal of multiple components, cascading of outages and loss of stability leading to a wide area impact. As such these out-of-range contingencies, as well as a sequence of normal contingencies, pose a real threat to power system security [18, 19].

2.1.2.2 Operating State of the Power System

The condition the power system is operating in can be subdivided into a number of different operating states, describing the health of the system. The operating state is a

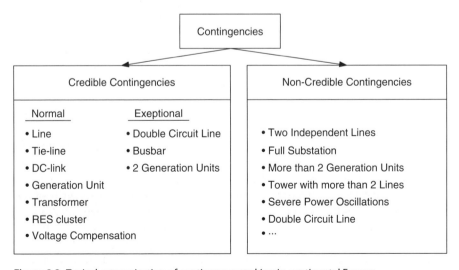

Figure 2.3 Typical categorization of contingency ranking in continental Europe.

concise statement on the viability of the system in its current operating mode [5]. A given operating state can be seen as the collection of system states (combinations of generation, load and grid settings) for which that operating mode is valid. Depending on the state in which the system is operating, the operator manages the system in a particular manner. In order to operate the system securely, it is important to correctly identify the operating state. Due to different events in the system, for example a contingency, the state of the power system can deteriorate. Based on the current state, the expected state of the system and possible future contingency states, the operator will undertake control actions. In an interconnected power system, it is of utmost importance that each system operator communicates the operating state of their area, certainly when not operating in the normal state. Depending on the interconnected area, different classifications have been agreed to. The classification which has been adopted here is the classification of continental Europe. In this classification, a distinction is made between four different states, namely: normal, alert, emergency and blackout state [20, 21]. The security criterion used is the well-known N–1 criterion.

Normal: The system is considered to be in the normal state when all the system parameters (frequency, voltages, thermal loadings) are within secure operational limits and following any event of the contingency list, taking into account effects of predefined remedial actions, all operational criteria are still fulfilled. The system is operating within its security criterion. The system is said to be operating N–1 secure.

Alert: All system parameters are still within secure operational limits, but at least for one contingency state, the security criterion (e.g., N–1 security) cannot be maintained. The system is still in a secure state. In case of occurrence of a contingency, it is uncertain whether the system can come back to the normal or alert state. Neighboring systems potentially endure a significant risk. Corrective actions are required without delay in order to comply with the reliability criterion. In case such actions are not available, the system may enter into a more dangerous state once the system operating conditions change. This change can arise from a new contingency or the gradual change of system variables such as load or generation.

Emergency: The system is strongly disturbed and system parameters are no longer within secure operational limits. As such, the power system state is not viable and without timely intervention will result in full or partial system collapse. Available and prepared actions are undertaken immediately without guarantee of total efficiency to limit propagation to neighboring systems. These actions are defined in the defense plan and strive to avoid full system collapse and to limit the risk of spreading disturbances to other parts of the system or the neighboring systems. These actions can be manual or automatic. The automatic actions are often described in special system protection schemes and try to maintain the integrity of the backbone of the power system. As a result of these emergency actions, it is likely that part of the system is islanded or even disconnected. In this case, system integrity is not maintained and restoration is needed.

Blackout: This state is characterized by almost total absence of voltage in a certain area of the transmission system as a consequence of the tripping or islanded operation of generating units due to abnormal variations of voltage and/or frequency which occurred during an emergency state. The blackout can be partial (only part of the system is affected) or total (the whole system is affected).

The different states of the power system are shown in Figure 2.4. The transition between the different states can lead either to system deterioration or to system restoration. System deterioration is the sequence of contingencies that lead to the transition from normal to a less secure state (alert, emergency) or even blackout. This process can be halted by either manual or automatic control actions.

During normal operation, the system is operated to meet accepted security and reliability standards at a cost as low as possible, while making provisions to guarantee future operation [5]. The speed associated with these control actions is of relatively low importance. For these remedial actions, distinction can be made between preventive and corrective actions. A *preventive action* is launched prior to the occurrence of the contingency. This to anticipate the need that may occur due to the uncertainty to cope with the contingency effectively, within the given time span resulting from the system constraints. This action ensures additional margin to the security boundary. The preventive actions will impose a cost, even though it is unsure whether the contingency will take place and hence whether the margin was required. The preventive controls will avoid the system going beyond system limits as long as only considered contingencies occur. For the N–1 criterion, this means that the power system controls are set such that no security boundary is violated after the outage of any single credible contingency. Rather, a corrective action is activated after the occurrence of the contingency. Typically this only imposes a cost once the contingency takes place (activation), unless reservation costs are required (e.g., keeping generation capacity on hot standby). As such the corrective actions realizes the transition from the alert state to the normal state.

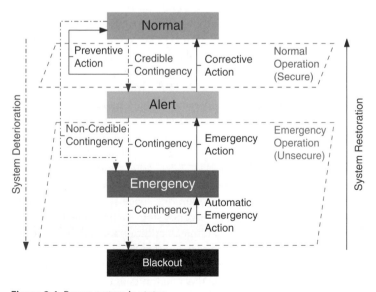

Figure 2.4 Power system in states.

Emergency operation consists of actions undertaken to prevent further system degeneration and to halt this process in the alert or emergency state. In general, the cost of these operations is of secondary consideration, as it is of utmost importance to return to normal operation as quickly as possible, even if only for a part of the system, and to avoid system collapse. This set of emergency actions, which are collected in the defense plan, can be both manual or automatic, depending on the type of phenomena and magnitude of the operational limit violation. In general, the defense plan includes a set of coordinated and mostly automatic measures to ensure fast reaction to large disturbances and to avoid their propagation through the system [22].

Once part of the system enters the blackout state, the restoration plan is initiated. The restoration plan aims to reduce the duration of power system interruptions by reenergizing the backbone transmission system as fast as possible, to allow gradual reconnection of generating units and, subsequently, supply to customers. Prompt and effective power system restoration is essential for the minimization of downtime and costs to the utility and its consumers.

A full blackout due to a foreseen power imbalance can be avoided using rolling blackouts. During these rolling blackouts, the electricity supply will be intentionally switched off in indicated areas for a fixed time period. An alternative for small power imbalances are brownouts, which are deliberate decreases of system voltage for a short period of time to avoid rolling blackouts.

2.1.2.3 System State Space Representation

An abstract visualization of the system states is provided by the system state representation in Figure 2.5 [23]. In this visualization, the current system operation point is depicted relative to the different security borders defined by each of the states. The current operation point is indicated by the point A and is the result of all the input and control variables and parameters present in the power system. The position of the operation point in this state space representation depends on the active power injections and off take P, the reactive power injections and off take Q, the set points of the phase shifting transformers (α) and the set points of the HVDC connections in the system ($P_{HV\ DC}$ and $Q_{HV\ DC}$). As most of these input variables are continuously subjected to small changes such as load and generated power variations, the operating point is continuously moving in an uncertainty cloud (Figure 2.5a). In the traditional power system, vertically integrated and mainly depending on fully controllable generation, this uncertainty cloud was relatively small and mainly characterized by load variations. However, due to power system unbundling and ongoing integration of more renewable generation sources, the uncertainty has increased. Now, both generation and load variations are bigger and interdependent with the power market and weather conditions. As can be seen in Figure 2.5b, even though the operating point obtained during steady state power system analysis was considered to be in the normal operating space, these variations lead to a more likely transition into the alert or even emergency state. Depending on how the system is managed, this is usually addressed by adding an additional reliability margin.

The outer black line in Figure 2.5 represents the security threshold of the system. This is the theoretical limit, determined by all the security limits (boundary conditions) of the system. Such security limits can consist of the thermal limits of the branches in the system, the voltage and transient stability limits. In theory, it is possible to operate the

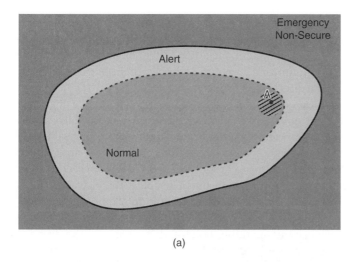

(a)

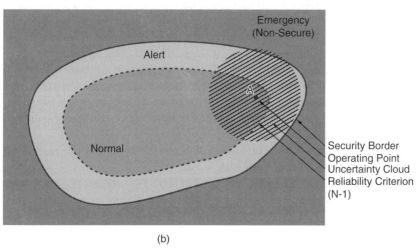

(b)

Figure 2.5 State space representation of system states. (a) Limited uncertainty (b) Increased uncertainty.

system up to this threshold without directly resulting into a loss of load or even a black-out. The space outside this region is unsecured, and sustained operation in this space will eventually lead to partial or full system collapse as some of the systems operating limits are violated. As such these operating points are not steady state secure. The space confined by the dashed line is the space defined by the $N-1$ criterion, or any other reliability criteria in use by the operator. It represents the boundary between the normal state and the alert state. During normal operation, the operating point can temporarily shift between normal and alert space. However, the operator tries to maintain the operating point in the normal space. The alert state results in an operating point between the dashed and full line. The state space should not be considered to be static, as both the boundary conditions and the operating point change continuously. However, each of the operating points in the secure space can be considered as steady-state-secure.

In general, the operator tries to maximize the distance between the operating point and the secure threshold, which can be considered as the security margin. But simultaneously all stakeholders of the power system, including system operators, try to make optimal use of the system and minimize generation and operation cost; this typically moves the operating point in the direction of the security margin.

A credible contingency can either result in a change of operating point or a change of a system limit. In Figure 2.6 the tripping of a transmission line is depicted, which causes a reduction of the secure space. As a consequence, the operating point may shift into the alert space as the normal state space is reduced. At this point, the system operator has a number of possible corrective actions to bring the system to a more secure state. A first option is the re-dispatch of generation or the use of flexible demand, which causes the operation point to move away from the secure threshold to a more secure location. A second solution consists of topology changes (line or busbar switching), which can enlarge the secure space near the operating point and can increase the distance between this point and the secure threshold. Thirdly, as more and more controllable devices (phase-shifting transformers, HVDCs) are being integrated into the power system, changing the set points of power flow controllable devices may shift the operation point to the normal state space. All possible actions come at a cost and have a certain effectiveness in influencing the operating point. Depending on the available resources and their cost, the system operator chooses the most appropriate action, either as a preventive action, or as a corrective action.

In Figure 2.7, the effect of a generation outage is shown in the state space representation. An outage of a generator results in a change in input variables and can shift the operating point into the alert state. After this outage, several corrective actions (re-dispatch, load disconnection, etc.) remain available to the operator to bring the system back to the normal state. After activation of the corrective action, the operating point is shifted back to the normal state as depicted in Figure 2.7. In practice, the outage of a generator is proactively managed by the operator through preventive actions, such as system topology changes.

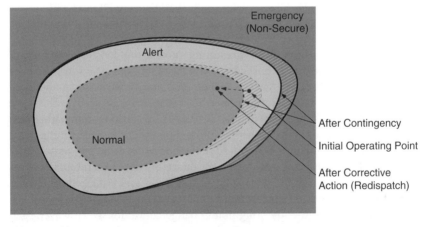

Figure 2.6 Line outage in state space representation.

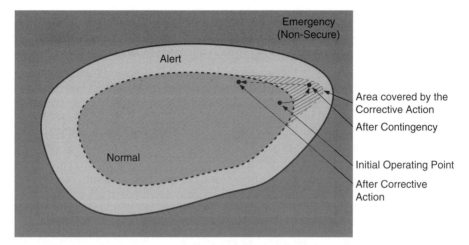

Figure 2.7 Generation outage in state space representation.

2.2 Reliability Under Various Timeframes

The level of uncertainty in system operation increases with increased distance to actual operation. As such, the system planner is faced with substantially larger uncertainties than the operator in the control room (Figure 2.8). Both the state of the grid and the generation and load injections in the system are more uncertain. At the same time, the system planner has more possibilities to accommodate the future power system (e.g., building new transmission lines), while the operator in the control room can only use that equipment which is available at that moment. Investment plans and operational decisions taken by system operators are driven by the uncertainty space the operator perceives at the time of planning. Because of these differences, reliability management

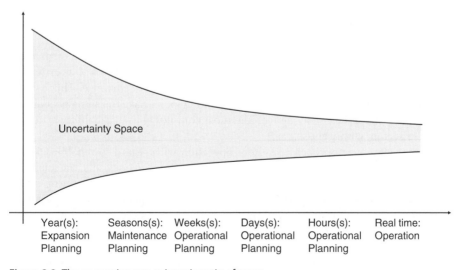

Figure 2.8 The uncertainty space in various timeframes.

is performed differently in the different timeframes. In order to deal with the uncertainty and avoid potentially operating in an operating state with insufficient system reliability, system operators usually define either a set of extreme cases or analyze a large number of operating states and contingencies.

The traditional approach to long-term system design is to define maximum and minimum system load cases and apply a fixed set of contingencies to these extreme cases. While there is a higher uncertainty due to more volatile power flows from renewable generation, computational speed has improved significantly. As a consequence, most system operators have moved toward analysis of a larger set of generation and demand samples in order to find more accurate boundaries for the system state space. New investments and topological changes to the system are based on planning cases using these generation and demand samples. This helps system operators to achieve cost savings in long-term system design as well as daily operation, while better covering the uncertainty space around nodal generation and demand values. During system planning, the analysis is mainly focused on active power flows.

The generation and demand samples used by the system operator typically differ between investment and operational planning, both in number of samples used, the granularity involved and the actual values of these samples. As the planning time horizon reduces, more information becomes available and the uncertainty around the system state reduces. This allows the operator to make better guesses of where the borders of the system state will be. This also means that the actions taken closer to real time are in general better fitted to addressing the reliability of the system. Close to real time, the operator usually has fewer control actions at their disposal and less time to evaluate and to decide which actions to take in case of contingencies. As a consequence, the operator chooses to operate the system in a state with sufficient security margin. This is done through the use of preventive actions, using corrective actions only if necessary, resulting in higher operational costs.

The number of possible contingencies in a power system scale exponentially with the number of considered elements. For instance, in a system with 10 elements, 2^{10} different combinations of outages, thus contingencies, exist. In long term and operational planning, the system is designed to be functional for a specific set of contingencies as the computation of all possible contingencies is not feasible. Hence, the set of contingencies needs to be defined very carefully, as it is a design parameter in the long term and a limiting factor during operation. System operators commonly use the $(N - x)$ criterion, assuming that all contingencies have the same probability of occurrence. This usually leads to an overestimation of the risk as not all contingencies have the same probability of occurrence or the same consequence. The most commonly used reliability criterion is the N–1 criterion, with x chosen as 1.

A more cost-effective approach to system operation with regard to reliability is the application of dynamic contingency sets. Using dynamic contingency sets, the analyzed contingencies are updated over time as more information about the operating states of the system becomes available. For example, in long-term planning, the contingencies can be selected based on the contingency probability, as the uncertainty around the operating state is high and the estimation of the consequence more difficult. Closer to real-time operation, the contingency set can be narrowed down and reconstructed using the expected impact of the contingencies, given that the uncertainty around the operational state will have reduced.

2.3 Reliability Criteria

The first reliability criteria used in practical applications, and still used nowadays, are deterministic in nature. The currently used criteria are typically derived from the deterministic N−1 criterion, which states that at any moment, the system should be able to withstand the loss of any one of its main elements (lines, transformers, generators, etc.) without significant degradation of service quality. The N−1 criterion and its derivates, however, have various shortcomings [1, 10, 24–27]:

- They can be interpreted in many ways. In practice, neither the number of elements to be considered ("N") nor the type of contingencies considered ("−1") is dealt with equally amongst system operators, or even within a single organization.
- Different contingencies are assumed to be equally severe and equally likely to occur.
- They do not give an incentive based on economic principles, as they do not take outage costs into account.
- All grid elements are assumed equally important and all generators and consumers have equal weight.
- They do not consider the stochastic nature of failures of grid components, generation and demand, the interdependencies between different events or the interaction between the frequency of the contingencies and the exposure time to high stress conditions.
- They are binary criteria: the system is either reliable or not reliable. Therefore, an accurate reliability level cannot be obtained. This results in over- or under-investments and operating costs that do not correspond with the required reliability requirements.
- They only take into account single contingencies. Single contingencies are much more probable than simultaneous contingencies if the outages are independent events. Nevertheless, simultaneous contingencies are not impossible. For example, hidden failures in the protection system can trigger additional outages, in addition to the original fault. Furthermore, due to the significant increase in the rate of outages during bad weather conditions, the probability of two quasi-simultaneous but independent outages may no longer be negligible.

Nevertheless, N−1 criteria, or adapted formulations of the N−1 criterion, are still commonplace, and this for various reasons. Operational experiences using these methods have been very good, due to the predictability and controllability of electrical power system operation and the experience of the system operator. Interconnections were initially aimed at reducing risks in terms of short-term adequacy, while keeping cross-border flows limited in normal operation. N−1 criteria could be easily satisfied due to the conservative design of the interconnections at the initial stage, although this could lead to non-optimal operation. Currently, cross-border interconnections are utilized differently, with significantly increased power flows due to the development of the European electricity market. The deterministic N−1 approach is also easy to understand, transparent and straightforward to implement in contrast to the complexity associated with implementing probabilistic approaches.

However, many probabilistic aspects are inherent in the power system due to both internal and external events. Firstly, outages are stochastic events, both in terms of their frequency and duration. Events occur randomly, for example, uncontrolled vegetation can lead to sudden short circuits with overhead lines, power system components can fail

Table 2.2 Advantages and challenges of probabilistic reliability criteria.

Advantages	Challenges
Uncertainties included	Number of states to consider
Probability and severity of contingencies considered in decision making process	Selection of appropriate indicators as different indicators imply different decisions [29]
Quantified reliability level	Practical meaning of indicator values still unknown

in an unpredictable manner, and so on. Secondly, demand and generation fluctuate over time, resulting in uncertainties in operating point, both real-time and during forecasting. The variability of (particularly renewable) generation is linked to weather behavior and influences the market behavior in the system.

A lot of research has already been done on development and application of new approaches incorporating those probabilistic and stochastic effects in reliability analysis [28]. Probabilistic approaches are already used in reliability calculations for power system planning and development, for instance to determine the generation reserve in the system development phase. Some countries that use probabilistic reliability criteria for planning are Australia [10], New Zealand [10] and the province of British Columbia in Canada [26]. Transmission system operators (TSOs) rarely apply probabilistic approaches in the operational timeframe [27]. The limited use of probabilistic approaches is inter alia due to the transparency, straightforward characteristics, lower computational burden and the acceptable level of reliability that result from deterministic criteria, but on the other hand it also results from data limitations and the lack of quantified benefits of using probabilistic approaches. Advantages and challenges of probabilistic reliability criteria are summarized in Table 2.2.

2.4 Reliability and Its Cost as a Function of Uncertainty

To ensure a reliable power system, network operators apply different actions, each at their own cost. Since aiming for a completely reliable power system would cause the cost of these actions to be infinite, network operators need to determine an acceptable reliability level for the system. That is, a reliability level that balances the costs of reliability management – referred to as reliability costs – and interruption costs.

2.4.1 Reliability Costs

Reliability costs can be defined as the sum of all electricity market costs to provide a certain reliability level. The network operator incurs reliability costs at different time horizons:

- System expansion: construction, upgrading, replacement, retrofitting or decommissioning of assets like AC or DC high-voltage transmission lines, substations, shunt reactors, phase-shifting transformers, etc.
- Asset management: monitoring the health status of network components, planning maintenance activities, repairing the components in case of failure, etc.

- Operational planning: congestion management, system protection, reserve provision, preventive actions, voltage control, decisions on outage executions, etc.
- Real-time operation: corrective actions, activation of reserves, reliability assessment, etc.

The reliability level can be increased by either reducing the number of incidents or by reducing the impact of incidents. The former can be done through investing in less error prone solutions (e.g., assets with lower failure rates, construction less subject to faults, etc.) The latter can, for instance, be done through investing in more redundancy, or through the use of higher reliability margins. Both approaches can become costly. In addition, increasing the reliability level also leads to costs that are not borne by the system owner or operator, for example increased generation costs in the case of network congestion and (untaxed) environmental costs of additional spinning reserves. In summary, reliability costs increase with the reliability level and reach infinity for a system with 100% availability on all nodes.

2.4.2 Interruption Costs

If the power system is unable to supply its consumers with electricity of the required quality (blackout, brownout, rolling black-out, local electricity interruption), the consumers will incur a cost. For illustrative purposes we focus on continuity of supply. The cost of an electricity interruption is the product of the interruption extent and the consequence of the interruption. The interruption extent is measured as Energy Not Supplied (ENS) [MWh]. This is the product of the interrupted load [MW] and the interruption duration [h]. The consequences of an interruption are represented by the Value Of Lost Load (VOLL) [€/MWh]. This is the average consumer cost – for example, broken appliances, spoiled food, failed manufacturing, lost utility of electrical heating, and so on – of a one-MWh interruption. The interruption cost (IC) is thus the product of ENS and VOLL:

$$\text{Interruption cost} = \text{(interruption extent).(consequence)} = ENS \cdot VOLL \quad (2.1)$$

As an example, the interruption cost of a six-hour interruption of 2 MW, valued at a VOLL of 5000 €/MWh [30], is:

$$IC = ENS \cdot VOLL = 12 \text{ MWh} \cdot 5000 \text{ €/MWh} = 60\,000 \text{ €} \quad (2.2)$$

Interruption costs can also be calculated ex-ante – that is, before any real interruption takes place – in simulation studies. The Expected Interruption Cost (EIC) is the probability-weighted interruption cost over the set of all possible system states S (contingencies, load levels, forecast errors, weather, etc.)

$$EIC = EENS \cdot VOLL = \sum_{s \in S} p_s \cdot ENS_s \cdot VOLL \quad (2.3)$$

Calculating the expected interruption cost requires a large amount of technical and economic data. On the technical side one needs failure probabilities of all system components, ideally as a function of temperature and weather;[2] wind and solar data; demand

---

2 Nine out of the 10 most risky days in 2010–2014 in the North American bulk power system were caused by adverse weather [31].

data; forecast errors; maintenance planning and repair times; and so on in order to calculate interruption probabilities and the extent of interruptions. Depending on the system, these data might be non-linear, correlated and time dependent [32]. This allows one to calculate the expected energy not supplied (EENS).

On the economic side, one needs accurate estimates of the cost of interrupted load. Because of difficulties in correctly estimating the VOLL [30],[3] VOLL is usually assumed to be a constant value per TSO zone. In reality the VOLL depends on the type of interrupted consumer, the duration and region of interruption, the time of occurrence, advance notification of interruption, weather at the time of interruption, and so on.

If sufficient data and a detailed understanding of the system is available, *EENS* and *VOLL* could be differentiated in space (n), by consumer group (c) and in time (t):

$$EIC_t = \sum_{n=1}^{N} \sum_{c=1}^{C} EENS_{n,c,t} \cdot VOLL_{n,c,t} \tag{2.4}$$

2.4.3 Minimizing the Sum of Reliability and Interruption Costs

The objective of reliability management is to determine and execute those actions that maximize total socio-economic welfare. However, under two simplifying assumptions, minimization of the sum of reliability and interruption costs is a good proxy for welfare maximization. The assumptions are that changes in the electricity market should (i) not change the behavior of electricity market actors, such as producers and consumers, and (ii) have little effect on other markets. These two assumptions are off course never fully met. For example, if electricity becomes more expensive, consumers will (i) buy slightly less electricity and (ii) have less budget left to buy other goods. The optimal reliability level is then the level at which the sum of reliability and interruption costs is minimized.

Figure 2.9 plots expected total costs (solid line) of the electricity market as a function of the reliability level ρ. The dotted line represents expected interruption costs, decreasing

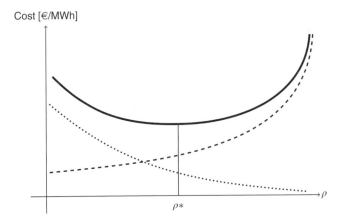

Figure 2.9 Total costs (solid line), interruption costs (dotted line) and reliability costs (dashed line) as a function of the reliability level ρ.

---

3 A rich literature exists on measuring the value of supply interruptions using stated-preference [33, 34], revealed preference, indirect analytical methods [35] or case studies.

Table 2.3 Illustrative example of reliability cost and interruption cost of different network operator actions.

Action	Reliability cost [€/hour]	ΔEENS [MWh/hour]	VOLL [€/MWh]	ΔEIC [€/hour]
Increased maintenance spending	350	−0.04	5000	−200
Transmission line investment	800	−0.1	5000	−500
Decreased generation reserve margin	−500	0.06	5000	300

with the reliability level, while the dashed line represents reliability costs, increasing with the reliability level. If the reliability level is too high, expected interruption costs are too low while reliability costs are too high. If the reliability level is too low, expected interruption costs are too high while reliability costs are too low.

The exact shape of the expected interruption costs and reliability costs function is difficult to assess. To determine if certain costly actions are increasing or decreasing total costs, one has to compare the marginal cost and benefits of all possible actions. That is, one has to trade off the cost of an action and the resultant decrease of expected interruption costs.

For example, assume that three possible actions are available:

1. increased maintenance spending,
2. an investment in a new transmission line,
3. decreasing the generation reserve margin.

Table 2.3 summarizes the reliability cost of these actions, expressed per hour, as well as their effect on expected energy not supplied (EENS). Assuming a VOLL of 5000 €/MWh, the change of expected interruption cost is also calculated. The example in this table shows that the reliability cost of both increased maintenance spending and a transmission line investment is higher than the benefit of decreased expected interruption cost. However, decreasing the generation reserve margin is cost-decreasing since the reliability cost decrease is higher than the EIC increase.

Clearly, the parameters used in this example might not be certain in advance. For example, if the VOLL were 10 000 €/MWh instead of 5000 €/MWh – that is, the societal consequences of the expected interruptions are higher – increasing maintenance spending would decrease total costs. Likewise, if the effect of the transmission line investment on EENS were -0.2 MWh/hour instead of -0.1 MWh/hour, this investment action would also decrease total costs. However, this does not imply that each action that decreases the sum of reliability and interruption costs is an efficient action. One should first determine and execute those actions that have the highest ratio of benefits versus costs, up to the point where the ratio of benefits and cost equals one for the most efficient actions. That is, up to the point where the marginal benefit equals the marginal cost.

2.5 Conclusion

An adequate reliability level in the power system is crucial, as it is one of the most critical infrastructures for the economy and social wellbeing of a modern society. Security is an important aspect determining a power system's reliability level, next to the adequacy of the system. Due to evolutions in the systems resulting in increased uncertainties, the currently used reliability management strategies based on a deterministic N−1 reliability criterion are being challenged. More advanced methodologies might be required in order to handle uncertainties in different timeframes in an appropriate way and to obtain a balance between reliability cost and interruption cost. Such criteria not only need to manage uncertainty − they also need to take into account the changing requirements over different time horizons, from planning to real time operations.

References

1 Kirschen, D. (2002) Power system security. *Power Engineering Journal*, **16** (5), 241–248.

2 NERC Operating Committee and NERC Planning Committee (2007), Definition of "adequate level of reliability", http://www.nerc.com/docs/pc/Definition-of-ALR-approved-at-Dec-07-OC-PC-mtgs.pdf.

3 International Electrotechnical Commission and others (2009). IEC/IEV 617-01-01. *Electropedia: The world's online electrotechnical vocabulary*, http://www.electropedia.org/.

4 Billinton, R. and Allan, R.N. (1984) Power system reliability in perspective. *Electronics and Power*, **30** (3), 231–236.

5 Knight, U.G. (2001) *Power systems in emergencies: From contingency planning to crisis management*, John Wiley & Sons, Ltd.

6 Allan, R. and Billinton, R. (2000) Probabilistic assessment of power systems. *Proceedings of the IEEE*, **88** (2), 140–162.

7 Hofmann, M., Kjølle, G., and Gjerde, O. (2012) Development of indicators to monitor vulnerabilities in power systems, in *PSAM11/ESREL2012, Helsinki*.

8 Kjølle, G., Gjerde, O., and Hofmann, M. (2013) Vulnerability and security in a changing power system. *SINTEF Energy Research, Trondheim*, p. 51.

9 ENTSO-E (2016) System operation guideline, *Tech. Rep.*, ENTSO-E. URL http://networkcodes.entsoe.eu/operational-codes/operational-security/.

10 Task Force C4.601 CIGRE (2010) Review of the current status of tools/techniques for risk based and probabilistic planning in power systems.

11 Allan, R. and Billinton, R. (2000) Probabilistic assessment of power systems. *Proceedings of the IEEE*, **88** (2), 140–162.

12 Billinton, R. and Wenyuan, L. (1991) Hybrid approach for reliability evaluation of composite generation and transmission systems using Monte-Carlo simulation and enumeration technique. *IEE Proceedings C-Generation, Transmission and Distribution*, **138** (3), 233–241.

13 Meliopoulos, S., Taylor, D., Singh, C., Yang, F., Kang, S. W., and Stefopoulos, G. (2005), Comprehensive power system reliability assessment. *PSERC Publication 05–13*.

14 Niebur, D. and Fischl, R. (1997) *Artificial intelligence techniques in power systems*, chapter Artificial neural networks for static security assessment, pages 143–191. Number 22, IET.

15 Monticelli, A., Pereira, M.V.F., and Granville, S. (1987), Security-constrained optimal power flow with post-contingency corrective rescheduling. *IEEE Transactions on Power Systems*, **2** (1), 175–180.

16 ENTSOE (2009) Policy 3: Operational security, *Tech. Rep.*, ENTSOE.

17 ENTSO-E subgroup: system protection and dynamics (2012) Special protection schemes, *Tech. Rep.*, ENTSO-E.

18 Ejebe, G.C. and Wollenberg, B.F. (1979) Automatic contingency selection. *IEEE Transactions on Power Apparatus and Systems*, **PAS-98** (1), 97–109.

19 Chen, Q. and McCalley, J.D. (2005) Identifying high risk N-k contingencies for online security assessment. *IEEE Transactions on Power Systems*, **20** (2), 823–834.

20 ENTSOE (2015) Policy 5: Emergency operations, *Tech. Rep.*, ENTSOE.

21 ENTSOE (2010) Appendix policy 5: Emergency operations, *Tech. Rep.*, ENTSOE.

22 ENTSOE (2010) Technical background and recommendations for defence plans in the continental europe synchronous area, *Tech. Rep.*, ENTSOE.

23 S. De Boeck and D. Van Hertem (2013) Coordination of multiple HVDC links in power systems during alert and emergency situations, in *PowerTech (POWERTECH), 2013 IEEE Grenoble*, pp. 1–6.

24 Reppen, N.D. (2004) Increasing utilization of the transmission grid requires new reliability criteria and comprehensive reliability assessment, in *Probabilistic Methods Applied to Power Systems, 2004 International Conference on*, IEEE, pp. 933–938.

25 Billinton, R. and Allan, R. (1984) Power system reliability in perspective. *Electronics and Power*, **30** (3), 231–236.

26 Li, W. and Choudhury, P. (2008) Probabilistic planning of transmission systems: Why, how and an actual example, in *Power and Energy Society General Meeting-Conversion and Delivery of Electrical Energy in the 21st Century, 2008 IEEE*, IEEE, pp. 1–8.

27 Kirschen, D. and Jayaweera, D. (2007) Comparison of risk-based and deterministic security assessments. *IET Generation, Transmission & Distribution*, **1** (4), 527–533.

28 Allan, R., Billinton, R., Breipohl, A., and Grigg, C. (1999) Bibliography on the application of probability methods in power system reliability evaluation. *IEEE Transactions on Power Systems*, **14** (1), 51–57.

29 Heylen, E. and Van Hertem, D. (2014) Importance and difficulties of comparing reliability criteria and the assessment of reliability, in *Young Researchers Symposium, Ghent*, EESA.

30 CEER (2010) Guidelines of Good Practice on Estimation of Costs due to Electricity Interruptions and Voltage Disturbances, *Tech. Rep. December*.

31 NERC (2013) State of Reliability 2015, *Tech. Rep. May*.

32 Heylen, E., Labeeuw, W., Deconinck, G., and Van Hertem, D. (2016) Framework for evaluating and comparing performance of power system reliability criteria. *IEEE Transactions on Power Systems*, **31** (3), 5153–5162.

33 Reichl, J., Schmidthaler, M., and Schneider, F. (2013) The value of supply security: The costs of power outages to Austrian households, firms and the public sector. *Energy Economics*, **36**, 256–261.

34 Pepermans, G. (2011) The value of continuous power supply for Flemish households. *Energy Policy*, **39** (12), 7853–7864.

35 de Nooij, M., Koopmans, C., and Bijvoet, C. (2007) The value of supply security. The costs of power interruptions: Economic input for damage reduction and investment in networks. *Energy Economics*, **29** (2), 277–295.

3

Probabilistic Indicators for the Assessment of Reliability and Security of Future Power Systems

Bart W. Tuinema[1], Nikoleta Kandalepa[2], and José Luis Rueda-Torres[3]

[1] *Researcher of Intelligent Electrical Power Grids, Department of Electrical Sustainable Energy, Delft University of Technology, The Netherlands*
[2] *Grid Strategist, Asset Management, TenneT TSO B.V., Arnhem, The Netherlands*
[3] *Assistant professor of Intelligent Electrical Power Grids, Department of Electrical Sustainable Energy, Delft University of Technology, The Netherlands*

3.1 Introduction

Current developments in the power system put increasing stresses on the transmission network. The transition towards a renewable energy supply, the integration of large-scale renewable energy sources (RES) and the liberalisation of the electricity market are some of the challenges for the transmission network. At the same time, the lowest cost of an ageing transmission network is strived for. To prevent large blackouts, reliability analysis and reliability management are of the utmost importance. While deterministic approaches and criteria have been used effectively in the past, probabilistic methods will be increasingly applied in the future as these provide more insight into the reliability of the transmission network.

This chapter describes how probabilistic reliability analysis is used to study the reliability of large transmission networks. We discuss how probabilistic reliability analysis is related to deterministic criteria and risk categories. Two common approaches for probabilistic reliability analysis will be presented, that is, state enumeration and Monte Carlo simulation. In a case study of the Dutch extra-high voltage (EHV) transmission network, these methods are applied to analyse the reliability impact of EHV underground cables.

The organisation of this chapter is as follows. In Section 3.2, the concept of time horizons in the planning and operation of a power system is introduced. Several reliability indicators that are commonly used in probabilistic reliability analysis are discussed in Section 3.3. In Section 3.4, two methods for reliability analysis are described: state enumeration and Monte Carlo simulation. Section 3.5 describes a case study of the Dutch transmission network. General conclusions are discussed in section 3.6.

Dynamic Vulnerability Assessment and Intelligent Control for Sustainable Power Systems, First Edition.
Edited by José Luis Rueda-Torres and Francisco González-Longatt.
© 2018 John Wiley & Sons Ltd. Published 2018 by John Wiley & Sons Ltd.

3.2 Time Horizons in the Planning and Operation of Power Systems

3.2.1 Time Horizons

In the planning and operation of power systems, Transmission System Operator (TSO) actions are performed in different processes and time horizons [1]. The main objective is to maintain a high level of reliability. Basically, three main processes can be distinguished: grid development, asset management and system operation. TSO actions are taken in time horizons ranging from long-term, mid-term and short-term to real-time. Table 3.1 shows some typical activities for each of the main TSO processes and for different time horizons. As can be seen, TSO actions range from long-term grid development (with a timescale of decades) to real-time system operation (with a timescale up to real-time).

3.2.2 Overlapping and Interaction

There is always some overlap between the three main TSO processes. For example, the installation of new assets can be a grid development activity or an asset management activity. In asset management, maintenance activities are scheduled, but these could be cancelled in system operation during critical situations. Traditionally, TSO activities were mostly performed sequentially, as shown in Figure 3.1. An example is the Dutch 380 kV-ring, which was designed to be n-2 redundant (n-1 redundant during maintenance). This gave enough room to plan maintenance activities in asset management while still providing n-1 redundancy during system operation. As can be seen in Figure 3.1, in the sequential approach there is only a small overlap between the TSO processes.

In the future, the developments as mentioned in the introduction of this chapter are expected to put more stress on the transmission network. The transmission network will become more heavily loaded, and the room to plan TSO activities will decrease. Consequently, the overlap between the processes will increase and the planning of TSO activities will be more interacted, as illustrated in Figure 3.1. A typical example is the development of offshore grids. Other studies showed that it is often not economical to apply full n-1 redundancy in offshore networks. However, if there is no redundancy, maintenance activities will have more consequences for the availability of the offshore grid and therefore maintenance planning can become challenging. Moreover, without offshore redundancy, failures of offshore networks can have a large impact on the onshore power system during system operation as it becomes more likely that a substantial amount of wind capacity is interrupted. Here, the interaction of grid development (n-1 redundancy), asset management (maintenance planning) and operational planning (remedial actions) can be clearly seen.

3.2.3 Remedial Actions

As there is always interaction and overlap between the three main TSO processes, it is important to model this in the reliability analysis as well. In the application example described in this chapter, the focus is on long-term grid development decisions and the consequences for (short-term and real-time) system operation. It will be discussed how

Table 3.1 Actions taken during different time horizons [1].

Main Processes	Time Horizons			
	Long-Term	Mid-Term	Short-Term	Real-Time
Grid Development	Onshore/offshore grid expansion plans (based on generation/load scenarios)	Investments in new grid components (connections, substations, etc.)	Small modifications of the grid (new phase shifters, protection systems, etc.)	- - -
Asset Management	Refurbishment, replacement and upgrading of existing assets	Maintenance scheduling, allocation or resources	Repair and condition monitoring of assets	Condition monitoring, outage management
System Operation	Operational policies	Day-ahead planning	Hour-ahead planning, preventive control actions (redispatch, transport restrictions, cancelling maintenance)	Corrective control actions (redispatch, switching actions, wind/load curtailment)

Time Scale: ←Decades ←Years ←Months ←Weeks ←Days ←Hours ←Minutes

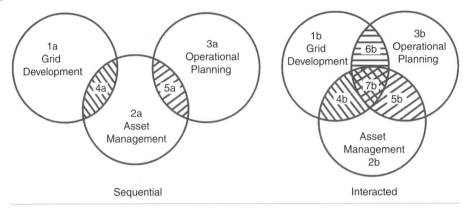

1. Investment in new infrastructure
2. Condition monitoring / maintenance
3. Operational planning / real-time operation
4. Small modifications of the network: refurbishment, replacement and upgrading
5. Maintenance planning / cancelling
6. Grid investments vs operational control actions
7. Integrated network design (grid development, asset management, and operational planning)

Figure 3.1 Interactions among three TSO processes [1].

the application of new underground cable connections in grid development is related to remedial actions like generation redispatch and load curtailment in system operation.

Remedial actions are operational interventions performed by the TSO to avoid/relieve overload of the network during critical situations. Typical remedial actions are: application of PSTs (Phase-Shifting Transformers), network reconfiguration, cancelling maintenance, local generation redispatch, cross-border-redispatch and load/wind curtailment [2]. With PSTs, the phase shift of the voltage can be varied such that the magnitude and direction of the power flow can be controlled. The network can be reconfigured by performing switching actions to reduce the loading of overloaded connections. This is mainly applied in the high voltage (HV) and lower voltage networks. If a connection is under maintenance in an overloaded part of the network, it can be decided to cancel this maintenance to relieve the loading of other connections. By performing local (or national) generator redispatch, the production of some generators is increased while it is decreased for other generators to reduce the loading of connections in overloaded parts of the network. If local generator redispatch is not sufficient, cross-border (or international) redispatch can be performed. Wind curtailment can also be used to relieve the overloading of the network. The last resort to relieve overloading is to disconnect load by load curtailment (or load shedding).

In system operation, a risk framework is often used in which the current risk status of the network is indicated. The reliability (measured as the level of redundancy) can be related to the risk categories of this risk framework [3]. Remedial actions can be related to these risk categories as well. Table 3.2 shows that remedial actions like PSTs and maintenance cancellation are often applied first, while load curtailment is the last resort. Although Table 3.2 shows how remedial actions are related to the risk categories and redundancy levels, the choice for a certain remedial action always depends on the situation.

Table 3.2 Risk categories, redundancy levels and remedial actions.

Risk category	Redundancy level		Remedial actions
Emergency	≥ n+1 (redundant)		Load curtailment Corrective redispatch
Alert	n-0 redundant		Preventive redispatch Cancellation of maintenance Phase-shifting transformers
Normal	n-1 redundant		Maintenance (re)scheduling
Robust normal	≤ n-2 redundant		Maintenance scheduling

3.3 Reliability Indicators

3.3.1 Security-of-Supply Related Indicators

The main function of a power system is to supply the load. If the power system is not able to supply its load, it is considered as unreliable. Therefore, the reliability indicator security-of-supply is often regarded as the most important Key Performance Indicator (KPI) of power system reliability. Several reliability indicators exist that are directly related to security-of-supply. Some reliability indicators are specially defined to measure the reliability of a part of the power system, while others are developed for combined generation/ transmission/distribution systems.

Reliability indicators developed to measure the reliability of the generation system [4, 5]:

- **LOLP** (*Loss of Load Probability*): probability that the demanded power cannot be supplied (partially or completely) by the generation system. The LOLP is often determined based on a per-hour study for a studied time period (usually a year)

$$LOLP = P[C < L] = \frac{\sum_{i=1}^{n} P_i[C_i < L_i]}{n} \quad [\text{-}] \tag{3.1}$$

where:
$P[C<L]$ = probability that the generation capacity is smaller than the load [-]
$P_i[C_i<L_i]$ = probability that the gen. capacity is smaller than the load for hour i [-]
C = total available generation capacity [MW]
C_i = total available generation capacity for hour i [MW]
L = total load level [MW]
L_i = total load level for hour i [MW]
n = total time of the studied period (8760 h for a whole year) [h]

- **LOLE** (*Loss of Load Expectation*): expected amount of time per period that the demanded power cannot be supplied (partially or completely) by the generation system

$$LOLE = \sum_{i=1}^{n} P_i[C_i < L_i] = n \cdot LOLP \quad [\text{h/y}] \tag{3.2}$$

where:
$P_i[C_i<L_i]$ = probability that the gen. capacity is smaller than the load for hour i [-]
C_i = total available generation capacity for hour i [MW]
L_i = total load level for hour i [MW]
n = total time of the studied period (8760 h/y for a whole year) [h/y]
$LOLP$ = loss of load probability [-]

- **LOEE** *(Loss of Energy Expectation)*: total amount of energy that is expected not to be supplied during a given time period (usually on a yearly basis) because of failures of the generation system

$$LOEE = \sum_{i=1}^{n} \sum_{j \in S_g} P_j L_{j,i} \quad [\text{MWh/y}] \tag{3.3}$$

where:
P_j = probability of generation capacity outage j [-]
L_j = not delivered load because of capacity outage j [MW]
n = total time of the studied period (8760 h/y for a whole year) [h/y]
S_g = set of possible generation capacity outages

Other reliability indicators are developed to measure the reliability of combined (generation)/transmission/distribution networks [2, 4, 5]. A few of these are:

- **PLC** *(Probability of Load Curtailment)*: probability that the demanded power cannot be supplied (partially or completely). The PLC is often determined based on a set of considered contingencies (single contingencies as well as higher-order contingencies) and a studied time period (usually on a yearly basis)

$$PLC = \sum_{i \in S_c} (P[load\,curt.|i] \cdot P_i) = \sum_{i \in S_c} \left(\frac{t_i}{T} \cdot P_i \right) \quad [\text{-}] \tag{3.4}$$

where:
S_c = set of considered contingencies
$P[load\,curt.|i]$ = probability that contingency i causes load curtailment [-]
P_i = probability of contingency i [-]
t_i = total time that there is load curtailment during the studied period, given that contingency i is present [h]
T = total time of the studied period [h]

- **EENS** *(Expected Energy Not Supplied)*: total amount of energy that is expected not to be supplied during a given time period (usually on a yearly basis) due to supply interruptions

$$EENS = \sum_{i \in S_c} E_i \cdot P_i \quad [\text{MWh}] \tag{3.5}$$

where:
S_c = set of considered contingencies

E_i = total curtailed energy in the study period if contingency i is present [MWh]
P_i = probability of contingency i [-]

- **SAIDI** *(System Average Interruption Duration Index)*: average outage duration of each customer during a given time period (usually on a yearly basis)

$$SAIDI = \frac{\sum_i^n (r_i \cdot N_i)}{N_t} \quad [\text{min}] \tag{3.6}$$

where:
n = number of interruptions [-]
r_i = duration of interruption i [min]
N_i = number of customers not served by interruption i [-]
N_t = total number of customers [-]

- **SAIFI** *(System Average Interruption Frequency Index)*: average number of interruptions per customer during a given time period (usually on yearly basis).

$$SAIFI = \frac{\sum_i^n N_i}{N_t} \quad [\text{-}] \tag{3.7}$$

where:
n = number of interruptions [-]
N_i = number of customers not served by interruption i [-]
N_t = total number of customers [-]

- **CAIDI** *(Customer Average Interruption Duration Index)*: average interruption duration per interrupted customer during a given time period (usually on yearly basis)

$$CAIDI = \frac{\sum_i^n (r_i \cdot N_i)}{\sum_i^n N_i} = \frac{SAIDI}{SAIFI} \quad [\text{min}] \tag{3.8}$$

where:
n = number of interruptions [-]
r_i = duration of interruption i [min]
N_i = number of customers not served by interruption i [-]

3.3.2 Additional Indicators

Traditionally, power system reliability studies concentrate on these security-of-supply related indicators. Often, it is assumed that there is one TSO that can take any remedial action to secure the load supply. The liberalisation of the electricity market has, however, led to several new actors such as producers, consumers, service providers and system operators. All of these actors have their own vision on power system reliability. For example, consumers wish to have a reliable power supply, while producers wish to be connected to the grid and sell their electricity. For a TSO, the transmission network is reliable if the customers (i.e. producers and consumers) are able to trade in electricity, while the network should also be maintainable.

In this sense, the reliability of a power system is not only reflected by the security-of-supply, but also related to indicators like the probability of wind curtailment, the probability of generation redispatch and the maintenance possibilities. These aspects can be included in reliability analysis by calculating several reliability indicators instead of one. As remedial actions are related to the risk states (see Table 3.2), calculating the probability of these states can provide more insight as well. Possible additional reliability indicators are then [2]:

- **Probability of Overload**: probability that one or more connections in the network are overloaded during a given time period (usually on a yearly basis)

$$P_{overload} = \sum_{i \in S_c} (P[overload|i] \cdot P_i) = \sum_{i \in S_c} \left(\frac{t_{ovl,i}}{T} \cdot P_i \right) \quad [\text{-}] \tag{3.9}$$

where:
S_c = set of considered contingencies
$P[overload|i]$ = probability that contingency i causes an overloaded connection [-]
P_i = probability of contingency i [-]
$t_{ovl,i}$ = total time that there is an overload during the studied period, given that contingency i is present [h]
T = total time of the studied period [h]

- **Probability of Generation Redispatch**: probability that generation redispatch is applied during a given time period (usually on a yearly basis)

$$P_{redispatch} = \sum_{i \in S_c} (P[redispatch|i] \cdot P_i) = \sum_{i \in S_c} \left(\frac{t_{rd,i}}{T} \cdot P_i \right) \quad [\text{-}] \tag{3.10}$$

where:
S_c = set of considered contingencies
$P[redispatch|i]$ = probability that contingency i causes generation redispatch [-]
P_i = probability of contingency i [-]
$t_{rd,i}$ = total time that redispatch is applied during the studied period, given that contingency i is present [h]
T = total time of the studied period [h]

- **Expected Redispatch Costs**: expected costs of the generation redispatch during a given time period (usually on a yearly basis)

$$C_{redispatch} = \sum_{i \in S_c} (C_{r,i} \cdot P_i) \quad [\text{-}] \tag{3.11}$$

where:
S_c = set of considered contingencies
$C_{r,i}$ = total redispatch costs during the study period, if contingency i is present [-]
P_i = probability of contingency i [-]

- **Probability of the Alert State**: probability that the power system is in the alert state during a given time period (usually on a yearly basis)

$$P_{alert} = \sum_{i \in S_c} (P[alert|i] \cdot P_i) = \sum_{i \in S_c} \left(\frac{t_{al,i}}{T} \cdot P_i \right) \quad [\text{-}] \quad (3.12)$$

where:

S_c = set of considered contingencies
$P[alert|i]$ = probability that contingency i leads to the alert state [-]
P_i = probability of contingency i [-]
$t_{al,i}$ = total time that the network is in the alert state during the studied period, given that contingency i is present [h]
T = total time of the studied period [h]

The application example in Section 3.5 of this chapter will illustrate how a combination of reliability indicators provides more insight into the reliability of the power system.

3.4 Reliability Analysis

Based on various input information (such as load/generation scenarios and network parameters), the reliability of the transmission network can be calculated. Figure 3.2 shows an overview of this process. The different parts of the calculation process are discussed in the following sections.

3.4.1 Input Information

The following input information is needed to perform the reliability analysis:

- **Load scenario**: In the load scenario, the load is specified for every hour of a study year. The load is aggregated per substation in the studied network. The load scenario

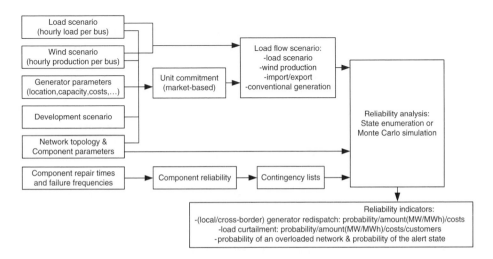

Figure 3.2 Overview of reliability analysis.

is based on the expected development of (domestic and industrial) load and assumptions about the development of RES and conventional generation.

- **Wind scenario**: In the wind scenario, the production of large-scale (offshore) wind parks is specified per hour of the study year. The wind scenario is based on a time series of the wind speed, taking parameters like the size and location of the wind farms into account.
- **Generator parameters**: For all generators in the studied power system, information such as the location, fuel type, costs and capacity is needed. This information also takes into account developments such as the installation or closure of units. National generators are included, as well as some generators in surrounding countries.
- **Development scenario**: In the development scenario, assumptions are made about the development of RES and conventional generation (e.g. [12]). Also, assumptions can be made about preferences for the generator units (e.g. preference for fuel type or location).
- **Network topology & component parameters**: This information includes the network topology, as well as all required information to perform a load flow calculation (e.g. component impedances and admittances, component capacities).
- **Component repair times and failure frequencies**: This information is needed to calculate the reliability of network components. Failure frequencies and repair times can be obtained from databases of historical failures. If the volume of available failure statistics is too limited (e.g. for new component types), estimations of failure frequency and repair time can be made.

3.4.2 Pre-calculations

The input information is now pre-processed in order to be used in the reliability analysis. The following calculations/analyses are performed:

- **Unit commitment**: In this analysis, the production of the generator units is scheduled such that all the load in the power system is supplied. The load scenario, wind scenario and generator parameters are the input information to this analysis. Network information and the assumptions of the development scenario can be used as restrictions for the unit commitment. The result of the unit commitment study is a generation scenario for the national generators and an import/export scenario, which is based on the generators in the surrounding countries.
- **Component reliability**: The reliability behaviour of network components is described based on the failure frequencies and repair times of the components. This information is then collected in contingency lists, which contain the contingencies to be considered in reliability analysis together with their parameters (such as failure frequency, repair time, unavailability [4, 5], component status). Independent as well as dependent failures, in which one cause leads to a simultaneous failure of multiple components, can be included in the contingency lists too.

3.4.3 Reliability Analysis

Based on the input information and the results of the pre-calculations (i.e. the load flow scenario, the network information and the contingency lists), the reliability of the network can now be analysed. There are two main approaches to probabilistic power system reliability analysis: state enumeration and Monte Carlo simulation [4, 5]. The first is a structured analytical study of possible contingencies, the second is a computer simulation. Both approaches are discussed in this section.

- **State enumeration**: In state enumeration, the reliability analysis is based on the possible contingency states. The general algorithm for state enumeration is shown in Figure 3.3. It can be seen that for every contingency state, the probability of this contingency state is calculated and a load flow is performed. If there is an overload in the network, remedial actions are applied. From the application of these remedial actions, information is gathered that will be used to calculate the reliability indicators. After applying remedial actions, the load flow is calculated again and if there still is an overload, more (or other) remedial actions are applied. The state enumeration stops when all contingency states to be considered have been studied.
- **Monte Carlo simulation**: In a Monte Carlo simulation, the reliability of the power system is studied by performing a computer simulation. The general algorithm of a Monte Carlo simulation is shown in Figure 3.4. The Monte Carlo simulation uses the

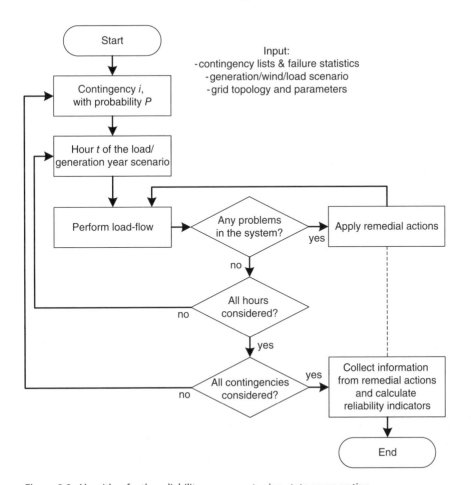

Figure 3.3 Algorithm for the reliability assessment using state enumeration.

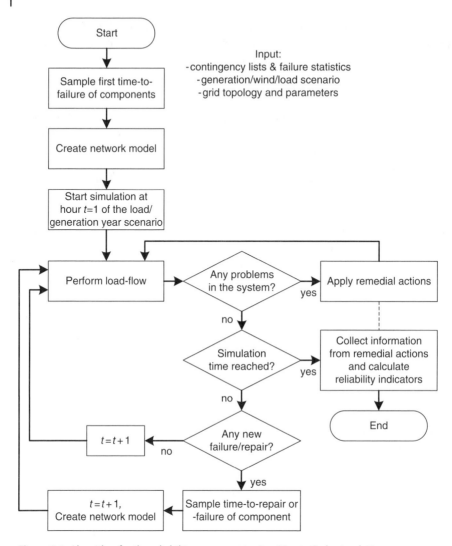

Figure 3.4 Algorithm for the reliability assessment using Monte Carlo simulation.

same input information as the state enumeration, albeit in a somewhat different way. First, from the failure frequencies specified in the contingency lists, the first time to failure (TTF) is sampled assuming an exponential distribution [4, 5]. The component with the smallest TTF will fail first and using its repair time, the time at which this component will be repaired is calculated. Until the component is repaired, a contingency exists in the network and the load flow is calculated for this period using the generation/wind/export/load scenario. If there is an overload in the network, remedial actions are applied in a similar way to state enumeration. Also similarly to state enumeration, intermediate results are collected to calculate the reliability indicators. A somewhat more complicated situation exists when a second contingency occurs before the first contingency is repaired. In this case, there are multiple contingencies in the network, which can also be combinations of independent/dependent failures as specified in the contingency lists. This situation ends when one of the contingencies is repaired, after which the other contingency is still being repaired. In contrast to state enumeration, it is possible (but very unlikely) that higher-order failure combinations occur during the Monte Carlo simulation, because component failures are randomly sampled and multiple components can be sampled as 'failed'. The simulation stops when the given simulation time is reached.

3.4.4   Output: Reliability Indicators

The results of the reliability analysis are various reliability indicators. As described in Section 3.3, these reliability indicators can be directly related to security-of-supply, but can also be additional reliability indicators. Together, these reliability indicators give a more complete understanding of the reliability of the studied transmission network.

3.5   Application Example: EHV Underground Cables

The reliability analysis as described in the previous section was applied to a case study of the Dutch extra-high voltage (EHV) transmission network. Currently, underground cables are installed in the backbone of the Dutch 380 kV network in the Randstad380 project [6]. So far, 380 kV Underground Cables (UGCs) have been installed in the Randstad380 South ring and another 380 kV cable project (Randstad380 North) is expected to come into operation in the near future. After the operation of both projects, around 20 kilometres will be underground in an entire route of approximately 80 km across the Randstad region. Figure 3.5 shows the Dutch EHV (380/220 kV) transmission network.

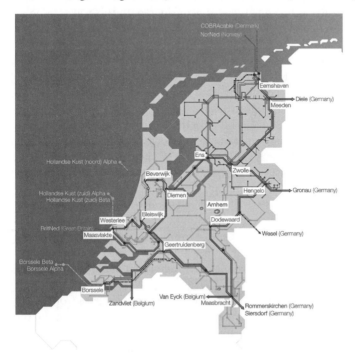

Figure 3.5 Dutch EHV transmission network (380/220 kV) [12].

As not much is known about the behaviour of EHV underground cables, various aspects like resonance behaviour, transient performance and reliability are being studied [7]–[10]. Regarding the reliability, previous research studied the reliability of cable components [7, 10] and the reliability of transmission links consisting of overhead lines and underground cables [9]. As underground cables will have an impact on the reliability of large transmission networks, the risk of further cabling of the EHV transmission network is also considered [2].

3.5.1 Input Parameters

The application example examines how the installation of more EHV underground cables (UGC) in the Dutch transmission grid affects the overall reliability level. Although the cable failure frequency is very close to that of overhead lines (OHL), the additional components of UGC (joints and terminations) reduce the reliability of the whole system [13]. Therefore, the unavailability of a connection is higher when it is partially or fully cabled than when it is completely an overhead line. Moreover, underground cables have a lower characteristic impedance. This is why it is important to study how the increased unavailability and the reduced impedance of the connection affect the overall reliability level.

As already described in Section 3.4, the reliability analysis consists of four major parts: contingency definition, load flow analysis, application of remedial actions if necessary and calculation of KPIs. A state enumeration is performed for the contingency types: independent single circuit failures, dependent double circuit failures, dependent triple circuit failures. All these contingencies refer to OHL and UGC failures. In addition, combinations of two of these contingency types are considered. In this way, 2nd-order contingencies can occur with two single circuit failures or with one double circuit failure and 6th-order contingencies are (only) reached by the combination of two triple circuit failures. Moreover, in this example three corrective actions are used with the following priority: national generator re-dispatch, cross-border re-dispatch and load curtailment. Taking these remedial actions into account, the state enumeration algorithm as shown in Figure 3.3 now becomes as shown in Figure 3.6.

In order to examine how further 380 kV cabling in the Dutch transmission network will impact the overall reliability level, simulations are performed by installing UGCs in three different connections (named Con1, Con2 and Con3). The PLC, the probability of overload and expected redispatch costs are calculated. The network topology and the required data are determined according to TenneT's scenario2020. It provides hourly data regarding generation (conventional/wind), load and export/import for the specific year. It is also important to describe the configuration of UGCs which is used in this study. As shown in Figure 3.7a, the UGCs consist of two circuits, and each circuit phase consists of two separate cables. In this configuration, a failure of a separate cable leads to a failure of one circuit, as operation with only one cable per circuit phase is undesirable in system operation.

In the case study, the cable length in the considered connections varies from 0% to 100% cabling with steps of 25%. The percentage refers to the transmission length of the connection and 0% means purely OHLs while 100% means a fully-cabled connection (as shown in Figure 3.7a). For example, in Figure 3.7b, 50% of the connection is cabled. The number of cable sections per connection can vary as well. For example, in Figure 3.7c

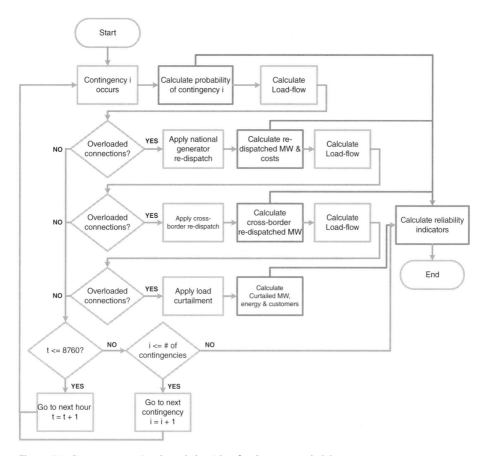

Figure 3.6 State enumeration-based algorithm for the case study [2].

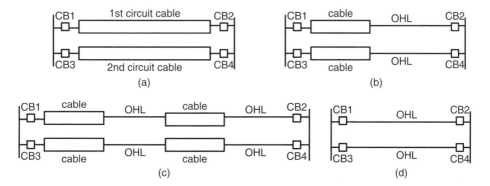

Figure 3.7 Configurations of underground cable connections.

there are two cable sections in the connection. Figure 3.7d shows the standard OHL connection configuration.

Due to the limited experience with EHV UGCs, there are no accurate values for their failure frequency and repair time. An earlier survey among European TSOs is used where a high and a low estimation of the failure frequencies are given, as shown in Tables 3.3 and 3.4 [7]. The repair time of UGCs is estimated as 730 h (1 month), which could be reduced to 336 h (2 weeks) if the repair process becomes more optimised in future [9]. For OHL, long experience has given more insight into failure statistics. The values are derived from actual failure statistics of the Dutch network [11].

3.5.2 Results of Analysis

The reliability analysis algorithm developed (Figure 3.6) is used to execute a number of simulation sets, and both categories of KPIs (directly linked to loss of load and not directly linked to loss of load) are calculated. Fig. 3.8 shows how the PLC changes when varying cable length is installed in the three considered connections. While one of these connections includes UGCs, the others are full OHLs. There are three curves, each devoted to a specific connection. The y-axis represents the PLC in h/y, while in the x-axis the cable length is presented as percentage of the connection length. The point 0% indicates that the three connections are OHLs, and this is the starting point for all curves.

The installation of UGC in Con1 has no impact on the PLC, while the other two curves illustrate an upward trend as cable length increases. However, while 50% cabling in Con2

Table 3.3 Failure frequency of EHV overhead lines and underground cables.

	Failure frequency	
OHL	0.00220 [/cctkm·y]	
	TSOs high	TSOs low
Cable	0.00120 [/cctkm·y]	0.00079 [/cctkm·y]
Joint	0.00035 [/comp·y]	0.00016 [/comp·y]
Termination	0.00168 [/comp·y]	0.00092 [/comp·y]

Table 3.4 Repair time of EHV overhead lines and underground cables.

	Repair time	
OHL	8 h	
	high	low
UGC	730 h	336 h

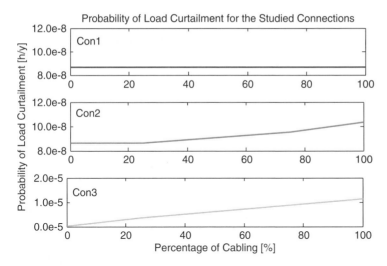

Figure 3.8 UGC in different connections – PLC.

causes a 5% increase in PLC, the same percentage of cabling in Con3 leads to a 70 times higher probability. In order to explain this phenomenon, the relative loadings of these connections were studied, and it was observed that the maximum loadings present similar behaviour with the amount of increase of the indicator, namely very small loading for Con1, larger for Con2 and even more considerable for Con3. It seems that the loading of the connection where UGCs are installed influences the impact on the reliability level.

Fig. 3.9 examines the probability of overload when UGCs are installed in each of the three connections. By comparing Fig. 3.8 with Fig. 3.9, it can be seen that the probability of overload in Con1 shows a very small upward trend as cable length increases (and is not constant like the PLC). The other two connections are characterised by an increasing

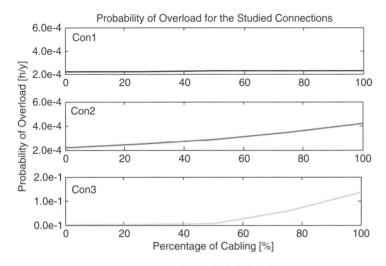

Figure 3.9 UGC in different connections – Probability of overload.

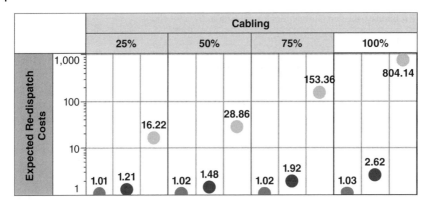

Figure 3.10 UGC in different connections – Expected redispatch costs. The leftmost point corresponds with Con1, the middle point corresponds with Con2, and the rightmost point corresponds with Con3.

trend as well. It is also clear that the values of the probability of overload are orders of magnitude larger than those of the PLC.

Fig. 3.10 illustrates the expected redispatch costs when UGCs are installed in each of the three connections. By considering a specific percentage of cabling, the three points of the connections are depicted next to each other. This occurs for each cabling percentage. The diagram is normalised (logarithmic), where the base 1 represents the starting point (0% cabling, three connections as OHL). The expected redispatch costs confirm the remarks made so far. They show a growing trend with rising cable length, and the installation of UGCs in Con3 seems the most expensive option (from redispatch costs point of view). Con3 is a heavily-loaded connection, and its failure appears several times in the contingencies which lead to generator redispatch. On the other hand, Con1 appears to be the most economical solution, since even at 100% cabling, the increase of the indicator is less than 10% from the starting point.

The results were verified using a Monte Carlo simulation as shown in Figure 3.11.

To sum up, the installation of UGC in specific connections might not influence the reliability indicators related to load curtailment, compared to the starting point. However, this does not mean that the level of reliability remains the same. It was shown in the results that in these cases, the reliability indicators which are not directly related to security-of-supply might demonstrate significant change leading to a lower reliability level. Therefore, it can be concluded that indicators not directly linked to load curtailment should be used as well.

3.6 Conclusions

This chapter discussed the probabilistic reliability analysis of transmission networks. It described how the planning and operation of power systems will change in future and what challenges can be expected. The most common probabilistic reliability indicators were presented and we discussed how these are related to deterministic criteria and risk categories and how these indicators can provide more insight into the reliability of power systems. The two probabilistic reliability analysis approaches, state enumeration

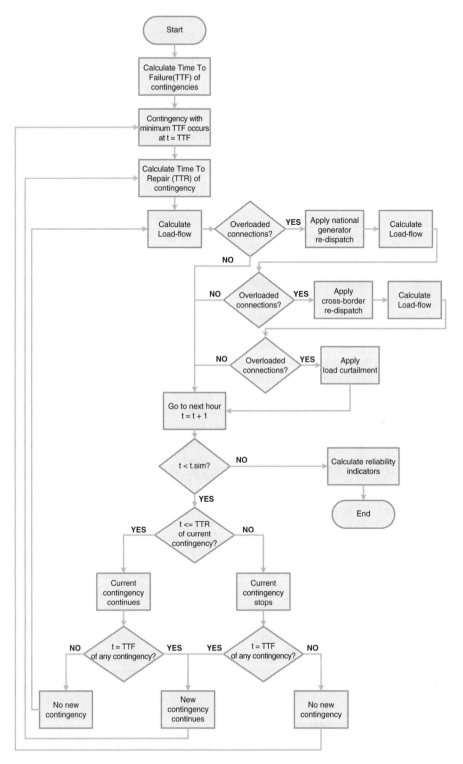

Figure 3.11 Monte Carlo simulation-based algorithm for the case study [2].

and Monte Carlo simulation, were explained. In a case study of the Dutch EHV transmission network, these approaches were applied to study the reliability impact of EHV underground cables.

The case study clearly showed that multiple reliability indicators provide more insight into the reliability of the transmission network than one single reliability indicator. In some cases, a reliability indicator such as the Probability of Load Curtailment does not show any difference whereas other reliability indicators such as the probability of redispatch can show a difference. For a TSO, both load curtailment and generation redispatch can be risks in the liberalised electricity market. Decisions on future developments, such as the installation of underground cables in the EHV transmission network, should be based on a combination of reliability indicators as presented in this chapter.

For future work, it is of interest to study further on how the results of probabilistic reliability analysis must be interpreted. Also, the development of fast probabilistic reliability analysis and decision making would be beneficial, such that probabilistic reliability analysis can effectively be applied in system operation as well. Clear and intuitive deterministic criteria have been used for a long time; it should be studied how probabilistic reliability analysis can best complement deterministic criteria and approaches.

References

1 S.R. Khuntia, B.W. Tuinema, J.L. Rueda, M.A.M.M. van der Meijden, "Time-horizons in the planning and operation of transmission networks: an overview", *IET Generation, Transmission & Distribution*, Vol. **10**, No. 4, Oct. 2016, pp 841–848.

2 N. Kandalepa, *Reliability Modeling in Transmission Networks – An explanatory Study for further EHV underground Cabling in the Netherlands*, MSc Thesis, Delft University of Technology, Delft, the Netherlands, 2015. Available online: repository.tudelft.nl.

3 M.Th. Schilling, A.M. Rei, M.B. Do Coutto Filho and J.C. Stacchini de Souza. On the Implicit Probabilistic Risk embedded in the Deterministic 'n-α' Type Criteria. *Proc. 7th International Conference on Probabilistic Methods Applied to Power Systems (PMAPS2002)*; Naples, Italy, 2002.

4 R. Billinton and R.N. Allan. *Reliability Evaluation of Power Systems*: Plenum Press, 2nd edition, New York, 1996.

5 W. Li. *Probabilistic Transmission System Planning*: Wiley Interscience – IEEE Press, Canada, 2011.

6 TenneT TSO, Ministries of EL&I and I&M, *Randstad380 project website*, Available online: http://www.randstad380kv.nl/, last accessed: November, 2014.

7 S. Meijer, J. Smit, X. Chen, W. Fischer, and L. Colla. Return of Experience of 380 kV XLPE Landcable Failures. *Jicable the 8th Int. Conf. on Insulated Power Cables*; Versailles, France, 2011.

8 L. Wu, *Impact on EHV/HV Underground Power Cables on Resonant Grid Behavior*, PhD Dissertation, Technical University Eindhoven, the Netherlands, 2014.

9 B.W. Tuinema, J.L. Rueda, L. van der Sluis, M.A.M.M. van der Meijden, "Reliability of transmission links consisting of overhead lines and underground cables," *IEEE Trans. Power Del.*, IEEE Early Access, 2015.

10 Cigré WG B1.10, "Update of Service Experience of HV Underground and Submarine Cable Systems", Cigré, Tech. Rep. TB379, ISBN 978-2-85873-066-7, Paris, Apr. 2009.

11 Tenne T TSO B.V., "NESTOR (Nederlandse Storingsregistratie) Database," TenneT TSO B.V., Arnhem, the Netherlands, 2007–2011.

12 Tenne T TSO B.V. Vision 2030. TenneT TSO B.V., Arnhem, the Netherlands, February 2008.

13 R. Bascom, C. Earle, K. M. Muriel, M. Nyambega, R. P. Rajan, M. S. Savage, et al., "Utility's strategic application of short underground transmission cable segments enhances power system", in T&D Conference and Exposition, 2014 IEEE PES, 2014, pp. 1–5.

14 TenneT TSO B.V. Grid map, Available: http://www.tennet.eu/company/publications/technical-publications/, 2017.

4

An Enhanced WAMS-based Power System Oscillation Analysis Approach

Qing Liu[1], Hassan Bevrani[2], and Yasunori Mitani[3]

[1] *Researcher, Department of Electrical and Electronics Engineering, Kyushu Institute of Technology, Kitakyushu, Japan*
[2] *Professor, Smart/Micro Grids Research Center, University of Kurdistan, Sanandaj, Iran*
[3] *Professor, Department of Electrical and Electronics Engineering, Kyushu Institute of Technology, Kitakyushu, Japan*

Modern interconnected wide area power systems around the world are faced with serious challenging issues in global monitoring, stability, and control mainly due to increasing size, changing structure, emerging new uncertainties, environmental issues, and rapid growth in distributed generation. The present chapter introduces an enhanced data processing method based on the Hilbert-Huang Transform (HHT) technology for studying power system oscillation dynamics.

The method discussed could be useful in developing a new oscillation monitoring system using real-time wide area measurement system performed by phasor measurement units (PMUs). The present methodology analyzes power system oscillation characteristics and estimates the damping of oscillatory modes from ambient data. The proposed method gives an indication for damping of transient oscillations following disturbances. It employs a system identification procedure that is carried out in a real-time environment. To demonstrate the effectiveness of the developed methodology, PMU-based measured data from the Japanese power system is used.

4.1 Introduction

The wide area measurement system (WAMS) is an important issue in modern electrical power system operation and control; and it is becoming more significant today due to the increasing size, changing structure, introduction of renewable energy sources, distributed smart/micro grids, environmental constraints, and complexity of power systems. The WAMS with phasor measurement units (PMUs) provides key technologies for monitoring, state estimation, oscillation analysis, system protections, and control of widely spread power systems [1].

Estimation of electromechanical mode properties can be carried out from PMU measurements. Different methods to do this have been developed over the last two decades; these methods utilize different types of dynamic responses which can be captured by PMUs. The Fast Fourier Transformation (FFT) is a popular tool for spectral analysis and has been already used for estimating parameters of various oscillation modes

Dynamic Vulnerability Assessment and Intelligent Control for Sustainable Power Systems, First Edition.
Edited by José Luis Rueda-Torres and Francisco González-Longatt.
© 2018 John Wiley & Sons Ltd. Published 2018 by John Wiley & Sons Ltd.

[2, 3]. However, FFT results are suspect in the event of nonstationary components in the original distorted signal, due to computing in a finite time window. Wavelet analysis with high resolution is known as a temporal frequency analysis method, which also shows the discrete transform in the time domain and spectral analysis in the time-frequency domain. Although the wavelet transform was introduced to solve the problem by presentation of frequency and energy content in the time domain, it still suffers from the convolution of a priori basis functions with the original signal [4].

The HHT technique is a new type of nonlinear and nonstationary signal processing method. It consists of an empirical mode decomposition (EMD) followed by the Hilbert transform and Hilbert spectrum analysis (HSA). Norden Huang developed the EMD combined with the Hilbert transform. It is used to decompose signals into nearly orthogonal basis functions with different instantaneous attributes. The decomposition provides a timescale separation that allows the extraction of different oscillation components [5]. It was initially proposed for geophysics applications, but has been widely applied in biomedical engineering, image processing, and structural safety problems. It has also been applied in power systems in the areas of power quality, sub-synchronous resonance, and to analyze inter-area oscillations. Some applications of HHT to WAMS-based power system oscillation analysis are summarized in [6–8].

Compared with other methods, the HHT has the advantage of analyzing low frequency oscillation signals, because a power system response following system disturbances contains both linear and nonlinear phenomena. Unlike the traditional methods such as the FFT and Wavelet Transform (WLT), the HHT processes the nonlinear and nonstationary signals.

On other hand, the HHT method is adaptive, which means that it can be adaptively extracted from the signal decomposed by the EMD. It is based on an adaptive mechanism, and the frequency is defined through the Hilbert transform. Consequently, the "base" of the FFT is the trigonometric functions, while the "base" of the WLT requires a pre-selection. As the name suggests, the EMD is a data-driven technique, empirical in nature with limited analytical justification of data. Therefore, the HHT provides an adaptive approach.

Moreover, the HHT is suitable for analysis of mutation signals. Concerning the Heisenberg uncertainty principle constraint [9, 10], many traditional algorithms produce a frequency window by using a constant time window. This property reduces precision in both the time and frequency domains for these algorithms. Nevertheless, there is no uncertainty principle limitation on time or frequency resolution from the convolution pairs based on a priori bases.

For these reasons, it can be concluded that applying the HHT method to analysis of power system oscillation signals is an effective choice. However, it still has some challenging issues that limit its application in power system stability studies; these need to be carefully resolved.

This chapter is organized as follows: the standard HHT method and its limitations are briefly reviewed in Section 2. Sections 3 and 4 address the enhanced HHT method, parameter identification, and evaluation. The application of the enhanced HHT method to a wide variety of problems in power system monitoring is demonstrated in Section 5, and finally, the chapter is concluded.

4.2 HHT Method

Before addressing the modified HHT method, a brief exposition of the fundamental HHT is introduced. The original EMD algorithm as provided in [9, 10] is referred to as "standard EMD" in this chapter.

4.2.1 EMD

EMD is performed to extract the intrinsic mode function (IMF) from a low frequency oscillation signal. The EMD can emphasize the local character of the original signal. Each oscillatory mode can be represented by an IMF with the following definition:

- In the whole dataset, the number of extremes and the number of zero-crossings must either be equal or differ at most by one.
- At any point, the mean value of the envelopes defined by the local maxima and local minima is zero.

With the above definition for the IMF, one can then decompose any function as follows:

a) Take a low frequency signal $s(t)$, and identify all the local extremes.
b) Connect all local maxima by a cubic spline line to produce the upper envelope $e_M(t)$. Repeat the procedure for the local minima to produce the lower envelope $e_m(t)$. The upper and lower envelopes should cover all the data between them. Their mean is designated as $m(t) = (e_M(t) + e_m(t))/2$, and the difference between the data is $c(t) = s(t) - m(t)$.
c) If $c(t)$ satisfies the definition of an IMF, it can be considered as one of the components of IMF h_1; then, compute the residue, $r(t) = s(t) - h_1$. If $c(t)$ does not satisfy the definition of an IMF, the procedure should be repeated until it satisfies the definition.
d) If $r(t)$ is not monotonic, then repeat steps from a) to c) replacing $s(t)$ with $r(t)$, to obtain the next IMF.

Subtract the decomposed components from the original signal, and then repeat the sifting processes for the residue components. Finally, the original signal is decomposed to the sum of several oscillations IMF and final residue component as follows [9]:

$$s(t) = \sum_{i=1}^{n} c_i(t) + r \qquad (4.1)$$

where $c_i(t)$ is the component of IMF and r is the residue component.

The above processing steps perform the EMD method to decompose the low frequency signal, which can be summarized as shown in Figure 4.1.

4.2.2 Hilbert Transform

For any continuous time signal $X(t)$, applying the Hilbert transform to each IMF component, the transformation can be expressed in the following form:

$$Y(t) = \frac{1}{\pi} \int_{-\infty}^{\infty} \frac{X(t)}{t - T} dt \qquad (4.2)$$

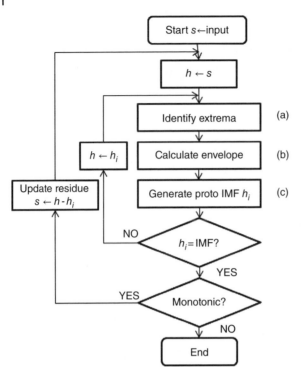

Figure 4.1 The EMD algorithm.

Multiple conjugate pairs of X(t) and Y(t) can be shown as:

$$Z(t) = X(t) + jY(t) = a(t)e^{j\theta(t)} \tag{4.3}$$

where a(t) is the instantaneous amplitude, $\theta(t)$ is the phase, and

$$a(t) = \sqrt{X^2(t) + Y^2(t)} \tag{4.4}$$

$$\theta(t) = tan^{-1}\frac{Y(t)}{X(t)} \tag{4.5}$$

The instantaneous frequency function is defined as follows:

$$f(t) = \frac{1}{2\pi}\frac{d\theta(t)}{dt} \tag{4.6}$$

$$\omega(t) = \frac{d\theta(t)}{dt} \tag{4.7}$$

For each IMF low frequency signal, an analysis function can be obtained through the Hilbert transformation:

$$Z_i(t) = c_i(t) + j\tilde{c}_i(t) = a_i(t)e^{j\theta_i(t)} \tag{4.8}$$

Then, instantaneous amplitude and frequency of each IMF can be easily obtained [9].

4.2.3 Hilbert Spectrum and Hilbert Marginal Spectrum

Eq. (4.8) enables us to represent the amplitude and the instantaneous frequency in a three-dimensional plot, in which the amplitude is the height in the time-frequency plane. This time-frequency distribution is defined as the Hilbert spectrum $H(\omega, t)$:

$$H(\omega, t) = Re \sum_{i=1}^{n} a_i(t)e^{-j \int \omega_i(t)dt} \tag{4.9}$$

With the defined Hilbert spectrum, the Hilbert marginal spectrum (HMS) of $h(\omega)$, can be defined as follows:

$$h(\omega) = \int_{0}^{T} H(\omega, t)dt \tag{4.10}$$

where T is the total data length.

The Hilbert spectrum offers a measure of amplitude contribution from each frequency and time value, while the marginal spectrum offers a measure of the total amplitude (or energy) contribution from each frequency value. The marginal spectrum represents the cumulative amplitude over the entire data span in a probabilistic sense. As pointed out in [11], the frequency in the marginal spectrum $h(\omega)$ has a totally different meaning from the Fourier spectral analysis. In the Fourier representation, the existence of energy at a frequency ω means there is a component of a sine or a cosine wave persisting through the time span of the data. Here, the existence of energy at the frequency ω only means that in the whole time span of the data, there is a higher likelihood for such a wave to appear locally. In fact, the Hilbert spectrum is a weighted non-normalized joint amplitude-frequency-time distribution. The weight assigned to each time-frequency cell is the local amplitude. Consequently, the frequency in the marginal spectrum indicates only the likelihood that an oscillation with such frequency exists. The exact occurrence time of that oscillation is given in the full Hilbert spectrum.

Therefore, the local marginal spectrum of each IMF component can be given by:

$$h_i(\omega) = \int_{0}^{T} H_i(\omega, t)dt \tag{4.11}$$

The local marginal $h_i(\omega)$ spectrum offers a measure of the total amplitude contribution from the frequency ω that we are particularly interested in.

4.2.4 HHT Issues

Although the HHT may become a promising method to extract the properties of nonlinear and nonstationary signals, like other signal analyses, the HHT also suffers from a number of shortcomings. In this section, the limitations of the standard EMD are highlighted.

In order to show the HHT characteristics, an example is given. Consider a simple *sine* signal as given in (4.12) as shown in Figure 4.2,

$$x = 2sin(2\pi * 20t) \tag{4.12}$$

Figure 4.3 and Figure 4.4 show the EMD result and time domain spectrum of the *sine* signal using the HHT method. Three IMF components and the residue signal are obtained as shown in Figure 4.3. As can be seen from the signal function, it should only get one IMF component. The occurrence of other two IMFs in Figure 4.3 is due to the problems caused by the EMD analysis process.

Figure 4.4 shows the change of three instantaneous frequency components over time. The first frequency component is around 20 Hz, which is similar to the frequency of

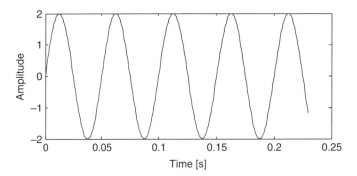

Figure 4.2 Test *sine* signal.

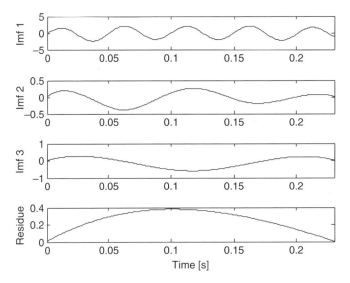

Figure 4.3 EMD analysis result.

the test sine signal. However, distortions appear at both endpoints; this phenomenon is called *the boundary end effect*. The reason for the appearance of the other two instantaneous frequency components, which are called pseudo-IMF components, is discussed in the next subsection.

Before proceeding further, an explanation for a frequency octave is provided. An octave is the frequency range between one frequency and its double or half-frequency. For instance, two frequencies are said to share an octave if their ratio lies between 0.5 and 2. Hence, 0.9 and 1.1 Hz share an octave, while 0.35 and 0.8 Hz lie in different octaves.

$$x = sin(2\pi * 0.4t) + 0.7sin(2\pi * 0.175t) \tag{4.13}$$

$$x = sin(2\pi * 0.5t) + 0.7sin(2\pi * 0.375t) \tag{4.14}$$

Figure 4.5 shows the signal of Eq. (4.13) and its EMD results. It can be seen that the standard application of EMD yields two mono-component IMFs corresponding to the two frequency components (0.4 and 0.175 Hz) in the given signal. Both of them

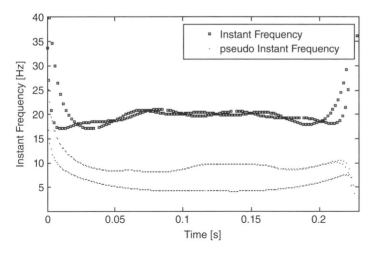

Figure 4.4 Time domain spectrum of HHT (Instantaneous frequency).

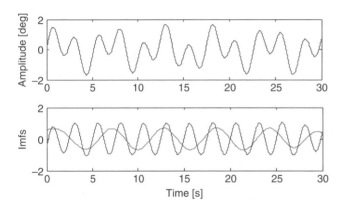

Figure 4.5 The signal in Eq. (4.13) (on top) and the first two IMFs (IMF 2 is plotted as dashed line).

clearly lie in different octaves. However, for the signal of Eq. (4.14), the standard application of EMD fails to yield the two components. The first IMF comes out to be identical to the original signal as shown in Figure 4.6, because the two frequencies (0.5 and 0.375 Hz) lie within the same octave. Thus, the implication is that if a signal contains two modes whose frequencies lie within an octave, standard application of EMD is unable to separate the modes.

From the above signal test, it can be seen that the available HHT algorithms typically include three key issues: (i) boundary end effect; (ii) mode mixing and pseudo-IMF component caused by EMD decomposition, especially for low frequency signals; and (iii) the possibility of parameter identification in power system oscillation analysis. These issues are briefly explained below.

4.2.4.1 The Boundary End Effect

The boundary end effect generated in the EMD process is one of the important factors that affect the quality of EMD. In general, a distortion known as the "Gibbs"

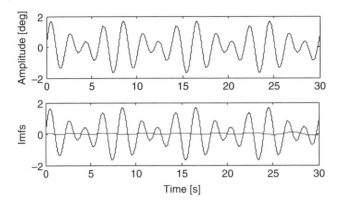

Figure 4.6 The signal of Eq. (4.14) (on top) and the first two IMFs (IMF 2 is plotted as dashed line).

phenomenon often appears at signal boundaries when processing signals concentrated in a limited period of time, which is known as the boundary end effect. The Gibbs phenomenon exists for many integral representations [12, 13] as well as for many series representations. The presence of this phenomenon is undesirable, since it is related to the behavior of the series approximating a discontinuous function f at a jump location t, implying a non-uniform approximation at t; so it is important to examine the ways to reduce or even avoid it. In [14] and [15], the Gibbs phenomenon has been shown to exist for Fourier interpolation. Since most multi-resolution analyses induce sampling expansions [16], the Gibbs phenomenon for wavelet sampling expansions has been examined in [17, 18].

In the EMD algorithm, every time "sieve" of a new IMF component has a certain relationship with the old IMF component. Therefore, the distortion at the endpoint will spread to the interior. As a result, the new IMF components obviously have distortion, especially for IMF lower frequency components. This will lead to seriously incorrect EMD decomposition results, and it can also bring endpoint distortion in the Hilbert transform process, which will affect the accuracy of the Hilbert spectrum analysis (HSA). Therefore, solving the boundary end effect issue for the HHT is highly important from both theoretical and practical points of view.

4.2.4.2 Mode Mixing and Pseudo-IMF Component

Mode mixing is defined as a single IMF either consisting of signals of widely disparate scales, or a signal of a similar scale residing in different IMF components [9]. The HHT method defines a unique phase for a real-valued signal at any time instant; it would be easy to compute its rate of change of phase and thereby its frequency. It is possible to define only one instantaneous frequency for a signal at any time instant. This poses a problem for multicomponent signals that have more than one frequency component existing at a given time. The instantaneous frequency obtained for such a signal would be meaningless unless the individual components are isolated before applying the Hilbert transformation. So, if the given signal is an intermittent signal, the mode mixing phenomenon will happen. Many researchers have worked to overcome this problem. One approach proposed is the Ensemble EMD (EEMD), which defines the true IMF components as the mean of an ensemble of trials, each consisting of the signal plus white noise of finite amplitude [19]. Masking signal-based EMD is also an effective method to enhance the discriminating capability of EMD [20, 22]. Besides mode mixing, there are some reasons for generating pseudo-IMF components, such as the boundary end effect, termination criterion for IMF components, and unreasonable sampling frequency selection. After solving the boundary end effect problem, the pseudo-IMF component can be eliminated to some extent.

4.2.4.3 Parameter Identification

It is well known that the EMD method is based on the local characteristic timescale signal. Any signal can be decomposed into the sum of IMF components adaptively in order to make instantaneous frequency have a real physical meaning. Afterward, each IMF component of the instantaneous amplitude and instantaneous frequency can be calculated by the Hilbert transform. In this chapter, the least squares method is employed to combine the HSA and to execute a power system low frequency oscillation parameter identification algorithm. Usually, 10–90% of the IMF component source is used for the identification algorithm. Concerning the poor quality of the IMF components, identification data is used to compress 20–80% of the source IMF components in order to avoid boundary end effects and unstable influence on the two ends caused by the Hilbert transform.

4.3 The Enhanced HHT Method

The HHT algorithm is divided into the EMD process and the HSA process. Removal of direct current (DC) and a band-pass filtering step are needed before applying the HHT algorithm. Since the low frequency oscillations (LFO) or generator rotor angle oscillations have a frequency between 0.1–2.0 Hz [23], the band frequency range is from 0.1 Hz to 2 Hz. The examined signals in the present chapter are: (i) stationary, nonstationary, and possibly nonlinear; (ii) low frequency in the range of 0.1–2 Hz; and (iii) individual component frequencies lying in different octaves.

The proposed improved HHT algorithm is summarized in Figure 4.7. Oscillation signal analysis in power systems is normally decomposed into low frequency oscillation analysis and sub-synchronous oscillation analysis. The IMF components decomposed by the EMD process include the spline interpolation algorithm (cubic spline interpolation function), the endpoint extension algorithm (alternatively direct continuation and mirror extension algorithm), and monotonic constraints (e.g., stop EMD screening process if extreme point is less than 3 and considered residual function is monotonous).

4.3.1 Data Pre-treatment Processing

The signal pre-treatment process includes DC removal and band-pass filtering. First, the DC part of the signal needs to be removed; second, in order to improve the efficiency of analysis, a band-pass filtering algorithm decomposes the useful frequency parts of the oscillating signal processing. Here, "useful" means that the frequency band is 0.1–2 Hz. The DC removal process and band-pass filtering is examined below.

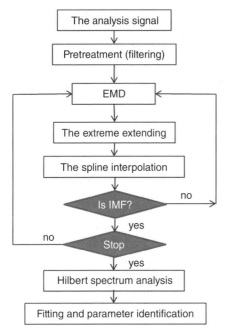

Figure 4.7 Oscillation mode extraction algorithm.

4.3.1.1 DC Removal Processing

The simplest way to removing the DC is to extract the $\bar{x}$ from the original signal x_i (t) $(i = 1, 2, \ldots, N)$; where $\bar{x}$ is average of x_i (t), and

$$\bar{x} = \sum_{i=0}^{N-1} x_i(t)/N \tag{4.15}$$

Hence, the processed signal x_i (t) can be written as

$$x(t) = x_i(t) - \bar{x} \tag{4.16}$$

4.3.1.2 Digital Band-Pass Filter Algorithm

Here, a Butterworth filter is used as a digital band-pass filter. Compared to the finite impulse response (FIR) filter, the infinite impulse response (IIR) filter offers the advantages of higher efficiency of amplitude-frequency characteristics, desirable accuracy, and low computational effort. Because of phase nonlinearity, it can be also used to analyse power system oscillation effectively.

The Butterworth filter is a type of signal processing filter that is designed to have as flat a frequency response as possible in the pass band. It is also referred to as a maximally flat magnitude filter. It has a magnitude response that is maximally flat in the pass band and monotonic overall. This smoothness comes at the price of decreased roll off steepness. Elliptic and Chebyshev filters generally provide steeper roll off for a given filter order.

The Butterworth filter algorithm can be summarized in five steps:

a) Find the low pass analog prototype poles, zeros, and gains using the Butterworth filter prototype function.
b) Convert the poles, zeros, and gain into the state-space form.

c) Use a state-space transformation to convert the low pass filter into a band-pass, high-pass, or band-stop filter with the given frequency constraints, if required.

d) Use bilinear transformation method (for digital filter design) to convert the analog filter into a digital filter through a bilinear transformation with frequency pre-warping. Careful frequency adjustment enables the analog and the digital filters to have the same frequency response magnitude at ω_n or at ω_1 and ω_2.

e) Convert the state-space filter block to its transfer function or zero-pole-gain form, as required.

For the sake of illustration, the following constructed linear/stationary oscillation signal is given as an example:

$$\delta = 5 \, sin(2\pi * 0.1t) + 2 \, sin(2\pi * 0.4t) + 1.5 \, sin(2\pi * 2t)$$
$$+ \, 0.5 \, sin(2\pi * 5t) + 0.2 \, sin(2\pi * 15t) + 15 \tag{4.17}$$

It contains five frequency components, 0.1 Hz, 0.4 Hz, 2 Hz, 5 Hz, and 15 Hz. After applying the DC preprocessing, the designed 4-order Butterworth band-pass filter (0.1–2 Hz) is used and the result is shown in Figure 4.8. The test data and filtered data are shown in the top and bottom of Figure 4.8, respectively. The FFT method is used to verify the performance of the filter as shown in Figures 4.9 and 4.10. It can be seen that the high frequency (5 Hz and 15 Hz) signals are effectively removed.

The reliability of the pre-treatment process to cope with nonlinearity/non-stationarity is also tested using the clipped signal. Here, the signal is distorted as given in Eq. (4.18) by clipping its components in some ranges:

$$\delta = 20 \, sin(2\pi * 0.3t) + 8 \, sin(2\pi * 1.1t) + 1.5 \, sin(2\pi * 4t) \tag{4.18}$$

The performance of the designed 4-order Butterworth band-pass filter (0.1–2 Hz) can be observed in Figure 4.11. The FFT spectrum is shown in Figure 4.12 in order to verify the performance of the designed filter. The 4 Hz component of the signal has been effectively removed.

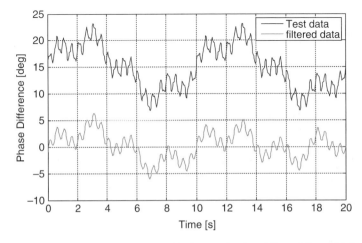

Figure 4.8 Pre-treatment process results: test (top) and filtered (bottom) data.

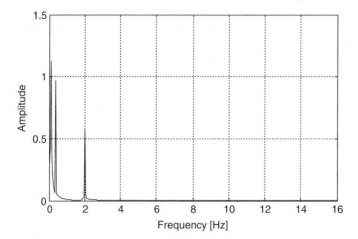

Figure 4.9 FFT spectrum of Butterworth filtered data.

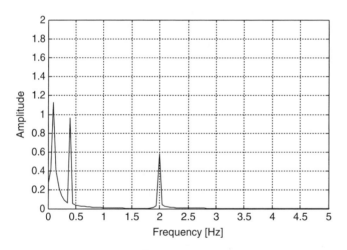

Figure 4.10 FFT spectrum of Butterworth filtered data (zoomed view).

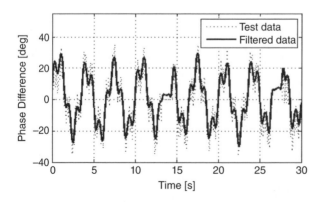

Figure 4.11 Pre-treatment process of nonlinear/nonstationary signal results.

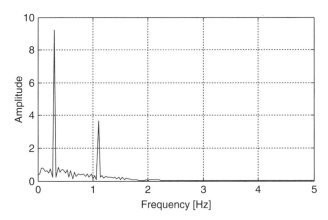

Figure 4.12 FFT spectrum of band-pass Butterworth filtered data.

4.3.2 Inhibiting the Boundary End Effect

The boundary end effect of the HHT algorithm is produced for two reasons: one is that, in the EMD process, a cubic spline interpolation exists to strike a mean envelope, in which there is an error at the mean envelope endpoint. The other reason is due to performing the Hilbert transform from the non-integer periodic sampling signal in the HAS, which generates distortion at the endpoint.

4.3.2.1 The Boundary End Effect Caused by the EMD Algorithm

Before introducing the iterated shift process, it is necessary to give a definition of the intrinsic mode function, which must satisfy two conditions. In other words, an IMF is a function that must satisfy two conditions according to the developed algorithm:

(i) The difference between the number of local extremes and the number of zero-crossings must be zero or one.
(ii) The running mean value of the defined envelopes by the local maxima and the local minima is zero.

Condition (i) is used to ensure that the signal has narrow band character, and condition (ii) is applied to assure that the instantaneous frequency does not include the unwanted fluctuations induced by an asymmetric wave.

Furthermore, in the case of using the cubic spline interpolation, the spline should conform to the following stipulations [24]:

(i) The piecewise function $S(x)$ is a third degree polynomial.
(ii) The piecewise function $S(x)$ interpolates all data points $S(x_i) = y_i$ for $i = 1, 2, \ldots, n\text{-}1$.
(iii) $S(x), S'(x), S''(x)$ are continuous on the interval $[x_1, x_n]$.

For such a signal, the interior extremes are easily identified. However, at both ends of the data, first or second order derivatives are required for the spline fitting. However, as the data curve does not provide any information on the envelope at the ends, the derivatives cannot be given unless the data is extended beyond the two ends.

To solve this problem, Huang used characteristic waves to get the maximum and minimum extremes [11]. The original signal only has the extreme in the data series, so the external extending points are not reliable. Different characteristic waves cause different results, and it is difficult to choose the proper waves for every iteration.

4.3.2.2 Inhibiting the Boundary End Effects Caused by the EMD

Usually, the extreme points or data sequence at both ends are increased/extended to inhibit the boundary end effect. However, this approach does not provide accurate extrapolating data sequences because of randomness of the real signals. So, it is required to adopt some extension rules to obtain a more accurate mean envelope.

The mirror extension method can be applied in order to solve the boundary end effect problems in the EMD. For the present study, consider the following mathematical expression as the analyzed signal:

$$x = 2 \, sin(2\pi * 0.4t) + 0.5 \, sin(2\pi * 2t) + 0.2 \, sin(2\pi * 15t) + 4 \tag{4.19}$$

The mirror extension method places two mirrors at the left and right end points which give symmetry about the extreme points. Extending the original signal by the mirrors provides a periodic signal sequence. The main steps of the mentioned method are taking the maximum and minimum values of the extended signal, seeking upper and lower envelopes via the cubic spline interpolation function, and finally seeking the average of the envelope.

The result of the EMD process applying a mirror extension method is shown in Figure 4.13. Comparing this with Figure 4.4, Figure 4.14 shows the boundary end effect problem of the EMD algorithm is properly inhibited by the mirror extension method. At the same time, pseudo-IMF components are also eliminated in this example.

4.3.2.3 The Boundary End Effect Caused by the Hilbert Transform

The error of instantaneous frequency and amplitude calculated in the Hilbert transform causes some impacts on the final results of the analysis, including the parameter identification quality. In what follows, the Hilbert transform of a sine signal given by (4.12) and its frequency spectrum is calculated to observe the boundary end effect resulting from the Hilbert transform.

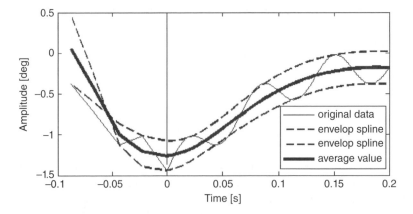

Figure 4.13 The mirror extension method.

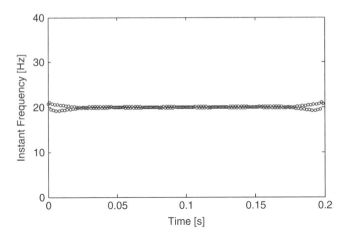

Figure 4.14 Inhibiting the boundary end effect problem of the EMD algorithm by the mirror extension method.

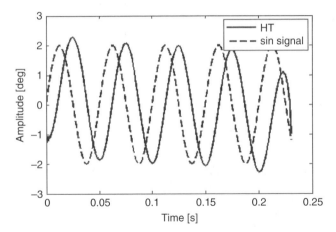

Figure 4.15 Non-integral periodic sine signal and its Hilbert transform (T=0.23).

Applying the Hilbert transform to the signal, the signal waveform and its Hilbert transformed wave are obtained as shown in Figure 4.15. It is known that there is $\pi/2$ phase difference between a sinusoidal signal and the Hilbert transform result. It is clear that the error generated by the Hilbert transform is reflected in the instantaneous frequency spectrum, which is shown in Figure 4.16. As can be seen, a serious distortion occurs at the endpoints.

To inhibit the endpoint distortions, the reasons of this distortion must be understood first. As shown in Figure 4.17, the Hilbert transformation process provides a bilateral spectrum of the real signal by applying Fourier transform. Then, filtering is used to remove one side of the spectrum, and the amplitude is doubled. Finally, applying the inverse Fourier transform to the doubled unilateral spectrum produces the final result.

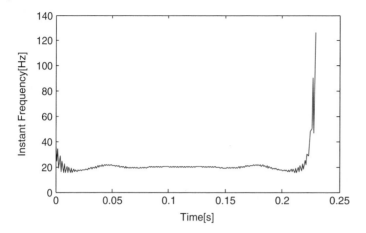

Figure 4.16 Instantaneous frequency of non-integral periodic sine signal.

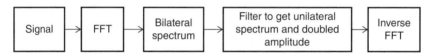

Figure 4.17 HT computing process.

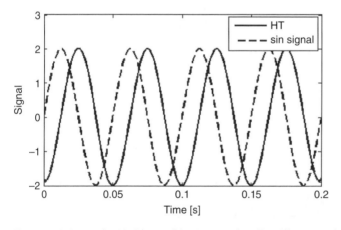

Figure 4.18 Integral periodic sampling sine signal and its Hilbert transform.

The core element throughout this process is the Fourier transform. The Fourier transform is used for periodic signals. Therefore, for a signal with non-periodic data series, the result will be mixed with error terms.

Figure 4.18 shows the results of Hilbert transform application to the integral periodic sampling of the sine signal (4.12). Comparing this with the non-integral periodic sampling data sequence in Figure 4.15, it can be seen that through the Hilbert transformation, a non-positive periodic sampling at the endpoint is distorted rather than the integral periodic sampling sequence.

Although it is known that the endpoint distortion can be inhibited by sampling integral periodic for the signal, it is very difficult to integrate periodic sampling of data in a real complex project. Therefore, it is necessary to use other methods to inhibit the boundary end effect produced by the Hilbert transformation.

4.3.2.4 Inhibiting the Boundary End Effect Caused by the HT

The boundary end effect produced by the Hilbert transform mainly appears at the ends of data series. The Hilbert transform results can be approximately obtained by shifting $\pm \pi/2$ phase to the original signal. So the ends of the original data can be extended according to certain rules, and then the Hilbert transform can be applied to the new extended data. Finally, after removing the Hilbert transform results at the extension data parts, the rest of the data is the Hilbert transform results of the original data. The most important part of this approach is the method selection for data extension. In this research, the autoregressive moving average (ARMA) model method is selected to establish extension data.

For some observed time series, a high-order autoregressive (AR) or moving average (MA) model is needed to model the underlying process. In this case, sometimes a combined ARMA model can be a more parsimonious choice.

An ARMA model expresses the conditional mean of y_t as a function of both past observations, $y_{t-1},...,y_{t-p}$, and past innovations, $\varepsilon_{t-1},...,\varepsilon_{t-q}$. The number of past observations (y_t) depends on p, which is the AR degree. The number of past innovations (y_t) depends on q, which is the MA degree. In general, these models are denoted by ARMA (p, q).

The form of the ARMA (p, q) model is [25]:

$$y_t = c + \phi_1 y_{t-1} + \cdots + \phi_p y_{t-p} + \varepsilon_t + \theta_1 \varepsilon_{t-1} + \cdots + \phi_q y_{t-q} \tag{4.20}$$

where ε_t is a white noise signal (uncorrelated innovation with mean zero).

Define the AR(p) and MA(q) as lag operator polynomials $\phi(L) = 1 - \phi_1 L - \cdots - \phi_p L^p$ and $\theta(L) = 1 + \theta_1 L + \cdots + \theta_q L^q$, respectively. The ARMA (p, q) model can be written as

$$\phi(L)y_t = c + \theta(L)\varepsilon_t \tag{4.21}$$

where $L_i y_t = y_{t-i}$. The signs of the coefficients in the AR lag operator polynomial $\phi(L)$ are opposite to the right hand side of Eq. (4.20).

The basic consideration to extend the ends of the original data according to the ARMA method is given in an algorithm that is summarized in Figure 4.19. Based on this algorithm, firstly the ARMA model to extend the original data is established. Then the Hilbert transform is performed, and finally the Hilbert transform (HT) results of the original signal are intercepted.

Comparing this with Figure 4.16, the distortions at both ends of the instantaneous frequency spectrum of non-periodic sampling signals are effectively inhibited after applying this algorithm, as shown in Figure 4.20.

Figure 4.19 Inhibiting the boundary end effect algorithm.

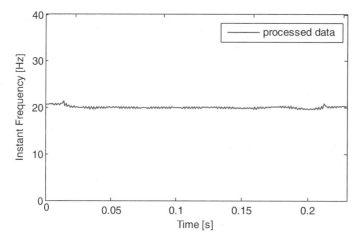

Figure 4.20 Processed instantaneous frequency spectrum.

4.3.3 Parameter Identification

In power systems, the oscillation mode component can be written as:

$$P(t) = Ae^{-\lambda t}cos(\omega t + \psi) = Ae^{-\lambda t}cos(2\pi ft + \psi) \tag{4.22}$$

or

$$P(t) = Ae^{-\xi\omega_0 t}\,cos(\omega_0\sqrt{1 - \xi^2}t + \psi) = Ae^{-\xi\omega_0 t}\,cos(\omega_d t + \psi) \tag{4.23}$$

where A is initial amplitude, ψ is initial phase, λ is damping coefficient, ω is oscillation angular frequency, f is oscillation frequency, ω_0 is un-damped angular frequency, ω_d is damped angular frequency, and ξ is damping ratio.

Comparing Eqs. (4.22) and (4.23) gives:

$$\begin{cases} a(t) = Ae^{-\lambda t} = Ae^{-\xi\omega_0 t} \\ \theta(t) = 2\pi ft + \psi = \omega_d t + \psi \end{cases} \tag{4.24}$$

For identification of the frequency oscillation signal, it is necessary to accurately extract the amplitude of the oscillation component and damping coefficient, and then calculate the damping ratio, which depends on the oscillation frequency and damping factor.

The frequency oscillation signal exists in the decomposed IMF components as an oscillation mode component. Therefore, parameter identification is needed. In order to obtain more accurate identification results, two issues should be considered. The least squares method is used to improve the accuracy of the damping coefficient and oscillation amplitude. It is also necessary to make some changes in the analytical signal form as shown in Eq. (4.25). Damping coefficient (λ) identification uses least squares through the logarithmic curve of instantaneous amplitude function of a single IMF component.

$$\begin{cases} lna(t) = -\lambda t + lnA = -\xi\omega_0 t + lnA \\ \omega(t) = \dfrac{d\theta}{dt} = 2\pi f = \omega_d = \omega_0\sqrt{1 - \xi^2} \end{cases} \tag{4.25}$$

On the other hand, reducing the IMF component data is also noteworthy. Since the data for parameter identification is obtained from the IMF component, the boundary end effect caused by the EMD is suppressed, but is not fully eliminated. At the same time, the endpoint distortion coming from the Hilbert transform is also not completely eliminated. Therefore, IMF can improve identification accuracy by removing small amounts of data at the both ends. In this work, 10–90% of the IMF component source is used for the identification algorithm.

4.4 Enhanced HHT Method Evaluation

4.4.1 Case I

The effectiveness of the proposed algorithm is used by application to the given oscillation signal in (4.17). This signal contains power system frequency components including inter-area oscillation (0.1–1 Hz), local oscillation (1–2 Hz) and noise oscillation (over 2 Hz).

This section also addresses the effect of the sampling time applied by the enhanced EMD algorithm. The results are shown in Figs. 4.21 and 4.22, with sampling time of 0.01 s and 0.033 s, respectively. Both figures include three oscillation modes and one residue. The frequencies of IMF components from high to low are 2 Hz, 0.4 Hz, and around 0.1 Hz. Comparing two results with different sampling times, it is easily to see that serious distortion occurred in the low frequency component with sampling time of 0.033 s (Figure 4.22), especially at the lowest frequency of 0.1 Hz. The reason for this distortion is the Gibbs phenomenon in the Butterworth filter processing, which is introduced in Section 4.3.1.

It is well known that low frequency oscillation in power systems is much more threatening than high frequency oscillation. In order to inhibit the distortion caused by Butterworth filter processing, the signal is analyzed without pre-treatment. Figure 4.23 for

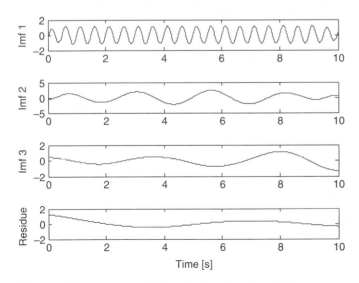

Figure 4.21 Pre-treatment EMD results (sampling time 0.01s).

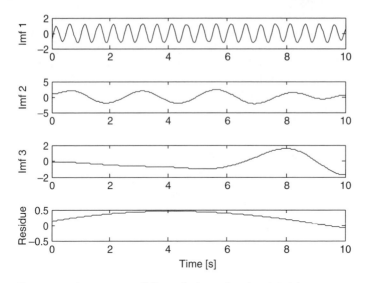

Figure 4.22 Pre-treatment EMD results (sampling time 0.033s).

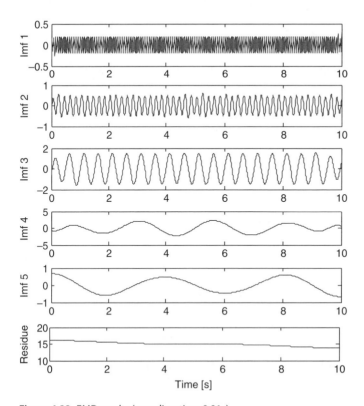

Figure 4.23 EMD results (sampling time 0.01s).

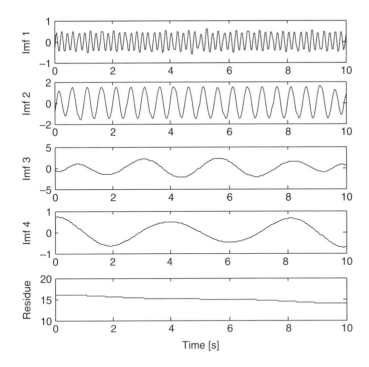

Figure 4.24 EMD results (sampling time 0.033s).

sampling time of 0.01 s and Figure 4.24 with sampling time of 0.033 s shows the EMD result without pre-treatment.

Figure 4.23 includes five oscillation modes and one residue at a sampling time of 0.01 s. The frequencies of IMF components from high to low are 15 Hz, 5 Hz, 2 Hz, 0.4 Hz, and around 0.1 Hz. The residue DC component is 15. Figure 4.24 includes four oscillation modes and one residue at a sampling time of 0.033 s. The frequencies of IMF components are 5 Hz, 2 Hz, 0.4 Hz, and around 0.1 Hz. The residue DC component is 15. It can be found that regardless of the sampling time, there are no serious distortions in the low frequency component. However, the high frequency component cannot be observed in the low sampling time cases.

From Figs. 4.21 and 4.22, it can be seen that the pre-treatment process can help to cut down noise performance and high frequency components. In the low sampling time cases, the pre-treatment process has worked, but without a high accuracy in the low frequency components (see Figs. 4.22 and 4.24).

In conclusion, these two comparisons indicate that by applying pre-treatment, the effects on the results are removed in high sampling time cases, while in the low sampling time situation, high frequency elements cannot be observed but low frequency distortion has happened. To observe low frequency components correctly at high sampling times, a pre-treatment process is needed. On the other hand, to observe low frequency elements clearly in low sampling time cases, the pre-treatment process should not be applied.

4.4.2 Case II

In this case, a constructed bimodal oscillation signal including damping coefficients is defined as follows:

$$x = 3e^{-0.15t} sin(2\pi * 2t) + 6e^{-0.2t} sin(2\pi t) + e^{-0.1t} sin(2\pi * 0.4t) + 15 \qquad (4.26)$$

This oscillation signal, shown in Figure 4.25, contains three decay elements with damping coefficients of -0.15, -0.2, and -0.1 and frequencies of 2 Hz, 1 Hz, and 0.4 Hz; and a DC component of 15. Here, we can get the IMFs by applying the improved EMD algorithm as shown in Figure 4.26. Table 4.1 shows the identified parameters for each IMF component. It can be seen that the improved HHT method can estimate the frequencies, damping coefficient, and damping ratio of the three modes effectively. Moreover, the estimation results are not affected by the sampling time. From the

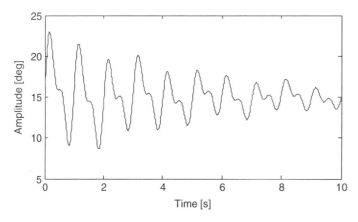

Figure 4.25 Bimodal test oscillation signal.

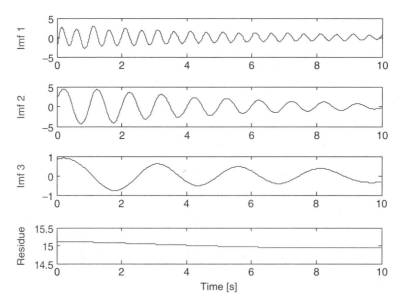

Figure 4.26 EMD results of bimodal test signal given in (4.26).

Table 4.1 Parameter identification results.

Mode	Frequency		Damping Coefficient		Damping Ratio	
Sampling Time(s)	0.01	0.033	0.01	0.033	0.01	0.033
1	2.0800	2.0766	−0.1490	−0.1516	0.0118	0.0120
2	0.9936	0.9951	−0.1948	−0.1978	0.0310	0.0315
3	0.3917	0.3903	−0.1032	−0.1000	0.0419	0.0407

results, it can be concluded that the improved HHT algorithm effectively decomposes the low frequency oscillation signal mode. The correct oscillation parameters can be obtained from the decomposed modal signal without considering the sampling time. This is an effective method for low frequency oscillation signal analysis as long as it can be correctly extracted.

4.4.3 Case III

The reliability of the proposed algorithm to deal with nonlinear/nonstationary signals is investigated by application to the oscillation signal given in Eq. (4.18), distorted by clipping its components in some ranges. This signal contains power system frequency components including inter-area oscillation (0.4 Hz), local oscillation (1.1 Hz) and noise oscillation (4 Hz). The decomposition performance with enhanced EMD algorithm can be observed in Figures 4.27 and 4.30 with a sampling time of 0.033 s. It can be seen that IMF 1 and IMF 2 in Figure 4.27 and IMF 1, IMF 2, and IMF 3 in Figure 4.28 preserve the

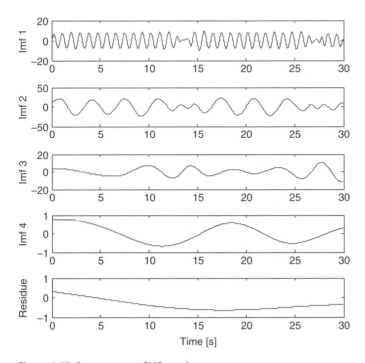

Figure 4.27 Pre-treatment EMD results.

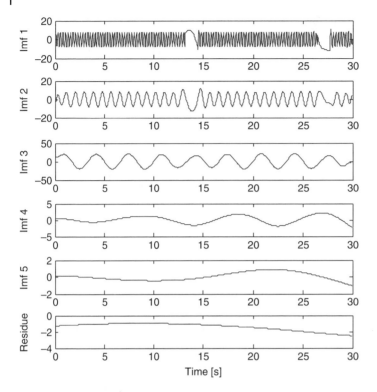

Figure 4.28 EMD results.

frequency of each corresponding component, while at the same time trying to capture the shape of the distorted signals, respectively. Both decompositions yield another two insignificant IMFs plus a residue. In order to show the difference analysis performance, Figure 4.29 shows the first two IMFs of the distorted signal obtained with pre-treatment EMD compared with original test signal, Figure 4.30 demonstrates that the IMF 2 and

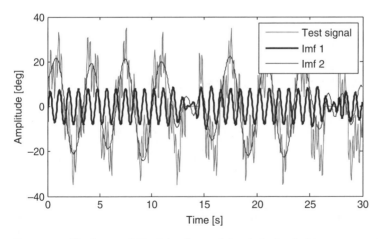

Figure 4.29 The first two IMFs of the distorted signal obtained with pre-treatment EMD.

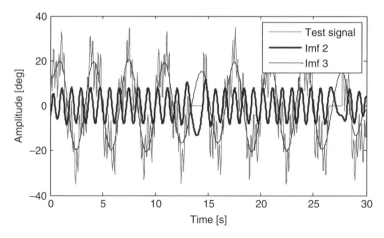

Figure 4.30 IMF 2 and IMF 3 of the distorted signal obtained without pre-treatment EMD.

IMF 3 of the distorted signal are obtained without pre-treatment of the EMD and the original test signal.

From the quantities shown in Tables 4.2 and 4.3, it can be seen that the HHT without pre-treatment is more reliable than the pre-treatment one in dealing with nonlinearities (here, the clipped signals). While, the higher order IMFs and the residue from HHT are insignificant in terms of magnitudes with respect to IMF 1 and IMF 2 through the pre-treatment process and to IMF 1, IMF 2, and IMF 3 with EMD-based filtering, which represent the components of the signal. The extra components yielded by EMD-based filtering have lower values than the applied pre-treatment. Moreover, the damping coefficient and ratio computation show that the EMD-based filtering is more accurate than the pre-treatment process.

Table 4.2 Parameter identification results with pre-treatment.

Mode	Frequency	Damping Coefficient	Damping Ratio
1	1.1260	−0.0109	0.0016
2	0.3327	−0.0256	0.00129
3	0.1638	−0.0078	−0.0497
4	0.0980	0.0528	−0.0900

Table 4.3 Parameter identification results with EMD-based filtering.

Mode	Frequency	Damping Coefficient	Damping Ratio
1	3.917	0.0038	−1.6185e-4
2	1.0916	−0.0017	2.5716e-4
3	0.2976	−0.0013	7.008 e-4
4	0.1311	−0.0206	0.025
5	0.0645	0.0313	−0.0672

4.5 Application to Real Wide Area Measurements

Generally, according to power system network structure and generator distribution, only one or two oscillation modes join together to dominate overall system dynamics. Furthermore, it is more common that for a studied system there is only one dominant oscillation mode which significantly influences overall system dynamics. The dominant oscillation mode can be separated from original measurements, and the problem is reduced to a parameter estimation problem of a single oscillation mode [1].

The Japan power network is used as a case study. Due to the longitudinal structure of this network, there are some significant low frequency oscillation modes across the whole system. Recently, a joint research project has been presented between some universities in Japan to develop an online wide area measurement of power system dynamics by using synchronized phasor measurement technique [3].

To establish a real WAMS, 13 PMUs are installed in universities/institutes in different geographical locations of Japan [3]. The PMUs are synchronized by the global positioning system (GPS) signal. This project was started to develop a WAMS covering the whole power system in Japan as a collaborative research initiative called *Campus WAMS* as shown in Figure 4.31. The PMUs measured voltage phasors in the monitoring location (laboratory) of the assigned university campuses over 24-hour schedules.

The Campus WAMS was developed to cover the typical power supply areas of the entire Japanese national-wide power grid. At least one PMU is installed in the supply network of each power company. Currently, the Campus WAMS project encompasses 13 PMUs (10 PMUs are installed in the supply area of the Western Japan 60-Hz system and 3 PMUs are located in the supply area of the Eastern Japan 50-Hz system). The locations of these PMUs are marked with filled circles associated with the respective city name in Figure 4.31. All PMUs are of the same model (NCT2000 Type-A manufactured by Toshiba Corp.), which is shown in Figure 4.32 [26].

Figure 4.31 PMU locations for the Campus WAMS project (Japan).

Figure 4.32 Phasor measurement unit (Toshiba NCT2000).

The main objective of this section is to present a monitoring and estimation scheme for power system inter-area oscillation based on wide area phasor measurements from the Campus WAMS. This scheme only focuses on the estimation of single dominant inter-area oscillation mode and includes the following steps: (i) estimating the center frequency of the single mode of interest, (ii) extracting the oscillation data from the original phasor measurements, (iii) analyzing the oscillation shape, (iv) identifying the oscillation model, and (v) estimating the oscillation parameters. Without considering any additional disturbance intentionally imposed on the system, a simplified second order oscillation model can be derived and identified based on the extracted oscillation data corresponding to the targeted single oscillation mode. The EMD-based filter technique is employed to separate the targeted single oscillation mode from other existing modes. The Hilbert marginal spectrum (HMS) analysis technique introduced in Section 4.2.3 is adopted to determine this center frequency from the original phasor measurements of power system output variables. Here, the center frequency means the real oscillation frequency of the targeted inter-area mode.

Figure 4.33 shows the waveforms of phase difference between Miyazaki University and Tokushima University stations. The twelve IMFs and one residue of the phase difference obtained using the EMD are listed in Figure 4.34. The frequencies of the

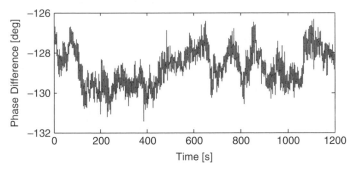

Figure 4.33 Waveforms of phase difference between Miyazaki University and Nagoya Institute of Technology University stations.

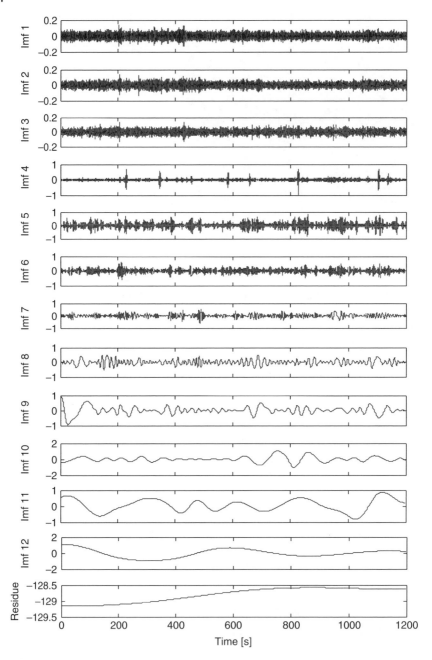

Figure 4.34 EMD results.

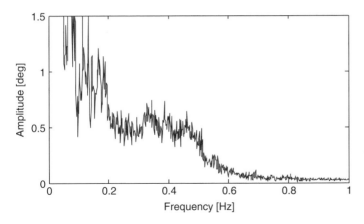

Figure 4.35 Hilbert marginal spectrum.

IMFs are ordered from high to low. The HMS technique is adopted to determine the real oscillation frequency of the targeted inter-area mode, which is demonstrated in Figure 4.35. The oscillation mode can be extracted in the frequency range of 0.3–0.5 Hz. Then, the oscillation shape of a short period of data from the extracted oscillation mode (FFT filtered) [3, 26] and obtained IMF 5 (EMD filtered) are compared with the original phasor difference (Figure 4. 36). It can be seen that the EMD filter can estimate the local characteristic more accurate than the FFT filter. The maximum amplitude value is extracted from the amplitude of the oscillation mode frequency range (Figure 4.37).

Figure 4.38 shows the time domain data from the extracted oscillation mode at the maximum point, which is used to estimate the power system oscillation parameters. In this case, the total computation time is less than two seconds. The resulting estimated parameter values for power system frequency oscillation are angular frequency (3.4344 rad/s), damping coefficient (-0.06891), and damping ratio (0.0820). Finally, the

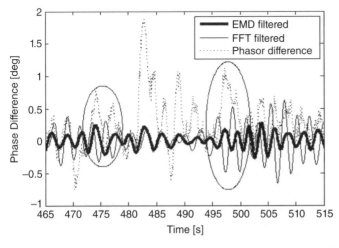

Figure 4.36 Short time phasor difference data from the extracted oscillation mode (FFT filtered) and obtained IMF 5 (EMD filtered).

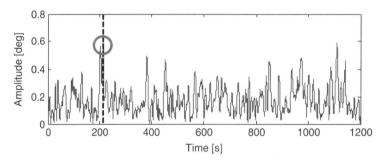

Figure 4.37 Amplitude of the extracted oscillation mode.

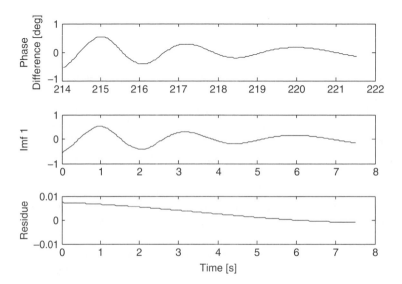

Figure 4.38 Short time data from the extracted oscillation mode and the obtained IMF.

FFT filter-based process calculates the same oscillation mode characteristics, resulting in an angular frequency of 2.9762 rad/s and damping ratio of 0.0543.

Summary

In recent years, detecting and analyzing the low frequency oscillation phenomenon in power systems has become an attractive research topic. As a result, many experts and scholars around the world have done valuable research work in the field of power system time-frequency analysis and have developed numerous processing methods for the oscillation signals.

In the present chapter, an integrated scheme based on an enhanced HHT algorithm is introduced for monitoring and detecting low frequency oscillations. Several key signal processing techniques are implemented to improve the HHT method. The relevant central frequency of the relevant mode is determined automatically and accurately, and this is then used to determine the oscillation mode parameters. The extracted frequency mode, damping, and oscillation shape can be properly detected by the proposed oscillation monitoring scheme.

References

1 Bevrani H., Watanabe M., and Mitani Y., *Power System Monitoring and Control*, Wiley-IEEE Press, New York, USA, 2014

2 Tsukui R., Beaumont P., Tanaka T., and Sekiguchi K., "Intranet-based protection and control", *IEEE Com-put Applications in Power*, vol. **14**, no. 2, pp.14–17, 2001

3 Hashiguchi T., Yoshimoto M., Mitani Y., Saeki O., and Tsuji K., "Oscillation mode analysis in power systems based on data acquired by distributed phasor measurement units", Proceedings of the 2003 IEEE International Symposium On Circuits And systems, 2003

4 Walter, G.G., *Wavelets and Other Orthogonal Systems With Applications*, CRC Press, 1994

5 IEEE Publications, *"Identification of Electromechanical Modes in Power Systems"*, IEEE Power & Energy Society, Special publication TP462, June 2012.

6 Messina, A.R., Vittal V., Ruiz-Vega D., and Enríquez-Harper G., "Interpretation and visualization of wide-area PMU measurements using Hilbert analysis", *IEEE Trans. on Power Systems*, vol. **21**, no. 4, November 2006, pp. 1763–1771.

7 Liu Q., Watanabe M., and Mitani Y., "Global oscillation mode analysis using phasor measurement units-based real data", *International Journal of Electrical Power & Energy Systems*, vol. **67**: pp. 393–400, 2015

8 Browne T.J., Vittal V., Heydt G.T., and Messina A.R., "A comparative asse ssment of two techniques for modal identification from power system measurements", *IEEE Trans. on Power Systems*, vol. **23**, no. 3, 2008, pp. 1408–1415.

9 Huang N.E., and Shen, S.S.P. (eds), *"The Hilbert–Huang Transform and Its Applications"*, 2nd Edition, World Scientific Publishing Company pp.1–15, 2005

10 Heisenberg W., "Über quantentheoretishe Umdeutung kinematisher und mechanischer Beziehungen", *Zeitschrift für Physik*, **33**, 879–893, 1925 (received July 29, 1925). English translation in: B. L. van der Waerden, *Sources of Quantum Mechanics*, NorthHolland publishing company, Amsterdam, Holland, 1967.

11 Huang N.E., Shen Z., Long S.R., et al, "The empirical mode decomposition and the Hilbert spectrum for nonlinear and non-stationary time series analysis", *Proceedings of the Royal Society London*, Series A, **454**(1998) 903–995

12 Jerri A., "The Gibbs Phenomenon in Fourier Analysis", *Splines and Wavelet Approximations*, Kluwer Academic Publishers, 1997

13 Karanikas C., "Gibbs phenomenon in wavelet analysis", *Result. Math.*, vol. **34**, pp. 330–341, 1998

14 Helmberg G., "The Gibbs phenomenon for Fourier interpolation", *J. Approx. Theory*, vol. **78**, pp. 41–63, 1994

15 Helmberg G. and Wagner P., "Manipulating Gibbs' Phenomenon for Fourier interpolation", *J. Approx. Theory*, vol. **89**, pp. 308–320, 1997

16 Atreas N.D., and Karanikas C., "Gibbs Phenomenon on sampling series based on Shannon's and Meyer's wavelet Analysis", *Fourier Analysis and Applications*, vol. **5**, no. 6, pp. 575–588, 1999

17 Shim H.T., Volkmer, "On the Gibbs Phenomenon on wavelet expansions", *J. Approx. Theory.*, vol. **84**, no. 1, pp. 74–95, 1996

18 Walter, G.G., and Shim, H.T., "Gibbs Phenomenon for sampling series and what to do about it", *Fourier Analysis and Applications*, vol. **4**, no. 3, pp. 357–375, 1998

19 Wu Z., and Huang, N.E., "Ensemble Empirical Mode Decomposition: a noise assisted data analysis method", p. 1, 2005

20 Deering R., "Fine -scale analysis of speech using empirical mode decomposition: Insight and applications," PhD Thesis, Department of Mathematics, Duke University, USA, 2006.

21 Deering R., and Kaiser, J.F., "The use of a masking signal to improve empirical mode decomposition," 2005 IEEE International Conference on Acoustics, Speech and Signal Processing, March 2005.

22 Laila D.S., Messina A.R., and Pal B.C., "A refined Hilbert–Huang transform with application to interarea oscillation monitoring" *IEEE Trans. on Power Syst. EMS*, vol. **24**, no. 2, pp. 610–620, May 2009,

23 Kundur P., "*Power System Stability and Control*", (McGraw-Hill: New York, 1994)

24 Hazewinkel M. (ed.), "Spline interpolation", *Encyclopedia of Mathematics*, Springer, ISBN 978-1-55608-010-4, 2001

25 MathWorks, *Matlab*, http://www.mathworks.com/products/matlab

26 Mitani Y., Saeki O., Hojo M., and Ukai H., "Online monitoring system for Japan Western 60 Hz power system based on multiple synchronized phasor measurements", Papers of Technical Meeting on Power Engineering and Power System Engineering, IEE Japan, PE-02-60, PSE-02-70, 2002 (in Japanese)

5

Pattern Recognition-Based Approach for Dynamic Vulnerability Status Prediction

Jaime C. Cepeda[1], José Luis Rueda-Torres[2], Delia G. Colomé[3], and István Erlich[4]

[1] Head of Research and Development, and University Professor, Technical Development Department, and Electrical Energy Department, Operador Nacional de Electricidad CENACE, Escuela Politécnica Nacional EPN, Quito, Ecuador
[2] Assistant professor of Intelligent Electrical Power Systems, Department of Electrical Sustainable Energy, Delft University of Technology, The Netherlands
[3] Professor of Electrical Power Systems and Control, Institute of Electrical Energy, Universidad Nacional de San Juan, Argentina
[4] Chair Professor of Department of Electrical Engineering and Information Technologies, Head of the Institute of Electrical Power Systems, University Duisburg-Essen, Duisburg, Germany

5.1 Introduction

Electric power systems have lately been operated dangerously close to their physical limits because of lack of investment, the use of congested transmission branches, and other technical reasons. Under these conditions, certain sudden perturbations can cause cascading events that may lead the system to blackouts. It is necessary to ensure that these perturbations do not affect security, so the requirement emerges of developing wide area protection schemes that allow guaranteeing service continuity. However, most actual wide area protection schemes are usually set to operate when specific pre-established operational conditions are reached, so they are unable to work under unconsidered contingencies that could entail cascading events. In these situations, the control of the system and the protection triggering should be adjusted depending on real-time event progress. This requirement emphasizes the need for advanced schemes to perform real-time adaptive control actions, with the aim of continuously guaranteeing system security while reducing the risk of power system blackouts [1]. Such a scheme depends upon timely and reliable provision of critical information in real time (via adequate measurement equipment and sophisticated communication networks) to quickly ascertain the system vulnerability condition (through appropriate tools to quickly analyze huge volumes of data) that subsequently allows the performance of remedial countermeasures (i.e., a Self-Healing Grid) [2]. A fundamental task of this smart structure is the real-time vulnerability assessment (VA), since this has the function of detecting the necessity of performing global control actions. Traditionally, VA approaches have been carried out to lead the system to a more secure steady-state operating condition by performing preventive control. However, in recent years, emerging

Dynamic Vulnerability Assessment and Intelligent Control for Sustainable Power Systems, First Edition.
Edited by José Luis Rueda-Torres and Francisco González-Longatt.
© 2018 John Wiley & Sons Ltd. Published 2018 by John Wiley & Sons Ltd.

technologies, such as Phasor Measurement Units (PMUs) and Wide Area Monitoring, Protection and Control Systems (WAMPAC), have enabled developing modern VA methods that permit the execution of corrective control actions [3–5]. This new framework is possible due to the great potential of PMUs to allow post-contingency dynamic vulnerability assessment (DVA) to be performed [3]. In view of this, a few PMU-based methods have been recently proposed in order to estimate the post-contingency vulnerability status by classifying real-time signal processing results obtained via application of different signal processing tools, including data mining techniques, as was presented in Chapter 1.

As highlighted in Chapter 1, the proposals discussed have not achieved good accuracy requirements due to their considerable overlap zone between stable and unstable cases. Thus, more appropriate mathematical tools should be developed in order to better adapt to power system dynamic variables and so increase the accuracy of results. In view of this, a novel approach to estimate post-contingency dynamic vulnerability regions (DVRs), taking into account three short-term stability phenomena (i.e., transient, voltage, and frequency stability—TVFS) is firstly deployed in this chapter. This proposal applies Monte Carlo (MC) simulation to recreate a wide variety of possible post-contingency dynamic data of some electrical variables, which could be available directly from PMUs in a real system (e.g., voltage phasors or frequencies). From this information, a pattern recognition method, based on empirical orthogonal functions (EOF), is used to approximately pinpoint the DVR spatial locations. Afterward, a comprehensive approach for predicting the power system's post-contingency dynamic vulnerability status is presented. This TVFS vulnerability status prediction considers the MC-based DVRs together with a support vector classifier (SVC), whose optimal parameters are determined through an optimization problem solved by the swarm version of the mean-variance mapping optimization ($MVMO^S$).

5.2 Post-contingency Dynamic Vulnerability Regions

As stated in Chapter 1, most existing VA methods are based on steady-state (Static Security Assessment—SSA) or dynamic (Dynamic Security Assessment—DSA) simulations of N-x critical contingencies. The aim of these methods is to determine whether the post-contingency states are within a "safe region" [4, 6] and accordingly to decide the most effective preventive control actions.

For this purpose, a Dynamic Security Region (DSR) might be established in order to determine the feasible operating region of an Electric Power System [6]. The process consists in determining the boundary of the DSR, which can be approximated by hyper-planes [6], and specifying the actual operating state relative position (or "security margin") with respect to the security boundary [4].

A fast direct method is presented in [6] in order to compute the DSR hyper-plane boundary for transient stability assessment (TSA) and to enhance the system stability margin by performing appropriate preventive strategies. Also, an approach to achieve on-line VA by tracking the boundary of the security region is outlined in [7]. These approaches have been focused on analyzing only one electrical phenomenon (commonly TSA), with the goal of leading the system to a more secure steady-state operating condition (by performing preventive control).

Figure 5.1 Dynamic vulnerability region concept.

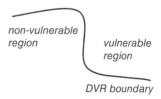

non-vulnerable
region

vulnerable
region

DVR boundary

The concept of DSR has been used in order to define a method for security assessment. In this sense, DSA is defined as "the determination of whether the actual operating state is within the DSR." The DSR can be defined in the injection space, depending on the analysis of TSA [6]. This DSR can be approximated by hyper-planes, and the computation of their boundaries allows specifying the "security margin" with respect to them [6]. The aim of this assessment is to enhance the system stability margin through performing appropriate preventive control actions.

Using the same concept as DSR, a DVR is defined in this chapter in order to perform post-contingency DVA with the aim of conducting corrective control actions. This DVR can also be specified by hyper-planes as in (5.1):

$$f(\mathbf{x}) = c \tag{5.1}$$

where $\mathbf{x}$ is an n-dimensional vector, and c is a constant that represents the value of the DVR boundary [7].

Figure 5.1 illustrates the DVR concept, where the DVR boundary delimits the "vulnerable" and the "non-vulnerable" regions.

The DVR and the hyper-plane boundaries of a dynamic system can be determined analytically or numerically. However, due to the huge complexity of bulk power systems, determining their DVRs analytically is not possible [7]. In this case, numerical methods offer the possibility of considering the complex power system physical model through simulation of several dynamic operating states. Monte Carlo methods are appropriate for analyzing the complexities in large-scale power systems with high accuracy, at the expense of greater computational effort [8].

By applying the introduced DVR concept, this chapter describes a novel approach to determining the post-contingency TVFS vulnerability status of an Electric Power System. This proposal allows assessing the second task related to real-time post-contingency DVA, which is the prediction of the system's tendency to change its conditions to a critical state.

5.3   Recognition of Post-contingency DVRs

With the aim of numerically determining the power system DVR and its hyper-plane boundaries, Monte Carlo-type (MC) simulation is performed to iteratively evaluate the post-contingency system time domain responses. In order to perform MC-based simulations, several input data are required. These data depend on the objective of the simulation, and they are usually represented by their corresponding probability distribution functions (PDFs). Since the proposal constitutes a wide area real-time post-contingency situational awareness methodology, capable of being updated even daily by the independent system operator (ISO) in the control center, the proposed uncertainty analysis encompasses only a specific short-term planning horizon. Thus, the probabilistic models of power system random variables, as well as the grid topology, should be structured within this short-term planning horizon so as to reflect the behavior of the system as realistically as possible.

In this connection, the basic data to be considered in the MC-based simulation, and which have to be previously prepared by the system operator, are: (i) short-term forecasting of nodal loads, (ii) short-term unit commitment, (iii) short-term system topology, and (iv) random generation of N-1 contingencies. Based on these system operating definitions, MC-type simulation is performed to iteratively evaluate the system responses, which would resemble those signals recorded in real time by smart metering, for instance by PMUs located throughout the system, with the ultimate goal of structuring a performance database.

Therefore, MC-based simulation is proposed as a method to obtain post-contingency data of some electrical variables, which would be available directly from PMUs in a real system (i.e., voltage phasors or frequencies), considering several possible operating conditions and contingencies including the most severe events that could lead the system to potential insecure conditions, and subsequent cascading events. Therefore, dynamic N-1 contingency simulations are performed via the MC-process, as depicted in Figure 5.2. First, the input variables to be considered in the MC simulation are randomly generated from the appropriate PDFs: that is, the load in each bus, the type of contingency (e.g., short circuit, generation outage, or load event), the faulted element (line, generator, or load), and the short circuit location, or the amount of load to be changed, or the branch outage in the case of static contingency analysis. Then, optimal power flow (OPF) is performed for every trial set of input variables in order to define feasible pre-contingency steady-state scenarios. Afterward, N-1 dynamic contingency time domain simulations are performed in order to obtain the post-contingency dynamic data that have led to the post-contingency dynamic vulnerability system status. These dynamic probabilistic attributes are then analyzed using a time series data mining technique based on EOFs, in order to recognize the system DVRs based on the patterns associated to the three short-term TVFS phenomena. This procedure is schematically summarized in Figure 5.2.

The ultimate goal of the methodology presented in Figure 5.2 is to map the spatial distribution of the DVRs, as well as to determine the EOFs that constitute the best orthogonal functions for extracting the predominant patterns buried in the post-contingency signals. These EOFs will be then used in real time in order to transform the actual measured signal into its corresponding pattern vector (by means of the projection of the data in the direction of the main EOFs), which has led to the information of the system post-contingency vulnerability status.

5.3.1 N-1 Contingency Monte Carlo Simulation

The DVRs are proposed in order to take advantage of the great potential of PMUs to allow performing post-contingency DVA that can be used to trigger corrective control actions in real time. Conventionally, these corrective actions are set to occur when specific pre-established operational conditions are reached, and they are unable to work under unconsidered contingencies that could begin cascading events.

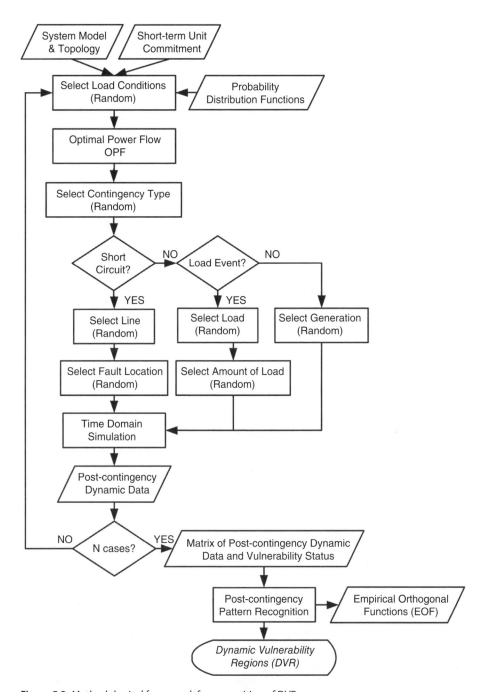

Figure 5.2 Methodological framework for recognition of DVRs.

Monte Carlo simulation allows obtaining the dynamic PMU post-contingency variables, which are used to determine the DVRs for the bulk power system under analysis. Figure 5.2 depicts the Monte Carlo simulation procedure in detail.

In order to adequately apply Monte Carlo simulation and obtain realistic results of the post-contingency dynamic response, some considerations and modeling requirements are taken into account:

- The DVRs can be defined for either short-term or long-term phenomena. This chapter focus on the three phenomena defined as short-term stability, which comprises transient stability, short-term voltage stability, and short-term frequency stability (TVFS) [9].
- Since the accuracy of the DVR boundaries depends on the accuracy of the models, the dynamic components (generators, motors, loads) and relevant control systems (such as excitation control system and speed governor systems) must be modeled with enough detail [8].
- The simulated events are based on N-1 contingencies, and the vulnerability criterion consists in the possibility of this kind of disturbance driving the system to further undesirable events (i.e., N-2 contingencies), which are considered as the beginning of a cascading event. For this purpose, three types of local protection devices are modeled in this chapter: out-of-step relays (OSR), under and over voltage relays (VR), and under and over frequency relays (FR). The tripping of one or more of these protection relays is used as the indicator of an imminent cascading event, and so it becomes a sensor of system vulnerability status. Thus, after each MC iteration, a vulnerability status flag gets the value of "1" (i.e., status of vulnerable) if one or more of the above mentioned local protections were tripped during the time domain simulation. Otherwise, in the case that none of the local protections was tripped, this vulnerability status flag acquires the value of "0" (i.e., status of non-vulnerable). This vulnerability status indicator is stored together with the post-contingency dynamic data. At the end of the MC simulation, a vulnerability status vector (**vs**) will contain all the vulnerability status flags corresponding to each iteration.

5.3.2 Post-contingency Pattern Recognition Method

The required mathematical tool for achieving the tasks involved in the post-contingency DVA has to be capable of predicting the post-contingency system security status and of specifying the actual dynamic state relative position with respect to the DVR boundaries, with reduced computational effort and quick-time response. Pattern recognition-based methods have the potential to effectively achieve this goal.

Pattern recognition is concerned with the automatic discovery of similar characteristics in data by applying computer algorithms and using them to take some action such as classifying the data into different categories [10], called classes [11].

Dynamic electrical signals can exhibit certain regularities (patterns) signalling a possibly vulnerable condition. However, these patterns are not necessarily directly evident in the electrical signal, though it may contain hidden information which can be uncovered by a proper pattern recognition tool [3].

This chapter applies a novel methodology for recognizing patterns in post-contingency bus features in order to numerically determine the DVRs of a specific Electric Power System. The method uses a time series data mining technique based on EOFs. The data consist of measurements of some dynamic post-disturbance PMU variables (voltage phasors or frequencies) at m spatial locations (buses where PMUs are installed) at r different times of n Monte Carlo cases. The measurements are arranged so that p time measures are treated as variables ($p = m \times r$) and every Monte Carlo simulation plays the role of observation, constituting an ($n \times p$) time series data matrix (**F**).

In the case of voltage angles, and since transient stability depends on angle separation, it is recommended not to use the voltage angles directly but to employ normalized angles obtained via the previous application of (5.2):

$$x_i(t) = x_i(t) - x_{i0} - \overline{X} \tag{5.2}$$

where $x_i(t)$ represents the voltage angle on bus i at time t; x_{i0} is its initial value prior to the perturbation; and $\overline{X}$ is the average value over the number of buses N_b at time t:

$$\overline{X} = \frac{\displaystyle\sum_{i=1}^{N_b} x_i(t)}{N_b} \tag{5.3}$$

Once matrix **F** has been structured, the method for obtaining EOFs is applied to this matrix. EOFs are the result of applying singular value decomposition (SVD) to time series data [12, 13]. Some authors consider that Principal Component Analysis (PCA) and EOF are synonymous [13]; nevertheless, this chapter uses a different interpretation of these two techniques. Whereas PCA is a data mining method that allows reducing the dimensionality of the data, EOF is a time series data mining technique that allows decomposing a discrete function of time f(t) (such as voltage angle, voltage magnitude, or frequency) into a sum of a set of discrete pattern functions, namely EOFs. Thus, EOF transformation is used in order to extract the most predominant individual components of a compound signal waveform (similar to Fourier analysis), which allows revealing the main patterns buried in the signal.

The main approaches related to EOFs have been developed for use in the analysis of spatio-temporal atmospheric science data, whereas their application in other scientific fields continues to be scarce. The data concerned consist of measurements of specific variables (such as sea level pressure, temperature, etc.) at n spatial locations at p different times [14]. As mentioned before, this chapter employs a variation of this definition, where the n spatial locations are replaced by n different post-contingency power system states (obtained from MC simulation), and the p different times consist of PMU instant values of post-disturbance dynamic variables (voltage phasors or frequencies) measured on m buses, at r different instants that depend on the selected time window (i.e., $p = m \times r$).

Therefore, an ($n \times p$) data matrix of discrete functions (**F**) is structured, where the post-contingency measurements at different power system states (n) are treated as observations, and the PMU samples belonging to a pre-specified time window (p time points) play the role of variables. Since the different power system states result from the application of MC-based simulations, n is conceptually greater than p ($n > p$), and so **F** is a rectangular matrix:

$$\mathbf{F} = \begin{pmatrix} f_1(t) \\ \vdots \\ f_n(t) \end{pmatrix} = \begin{pmatrix} x_{11} & \cdots & x_{1p} \\ \vdots & \ddots & \vdots \\ x_{n1} & \cdots & x_{np} \end{pmatrix} \tag{5.4}$$

where $\mathbf{f}_k$ is the k-th discrete function of time obtained in the k-th MC-repetition that consists of p samples.

Formally, the SVD of the real rectangular matrix $\mathbf{F}$ of dimensions $(n \times p)$ is a factorization of the form [15]

$$\mathbf{F}_{np} = \mathbf{U}_{nn}\boldsymbol{\Lambda}_{np}^{1/2}\mathbf{V}'_{pp} \tag{5.5}$$

where $\mathbf{U}$ is an orthogonal matrix whose columns are the orthonormal eigenvectors of $\mathbf{FF}'$, $\mathbf{V}'$ is the transpose of an orthogonal matrix whose columns are the orthonormal eigenvectors of $\mathbf{F}'\mathbf{F}$, and $\boldsymbol{\Lambda}^{1/2}$ is a diagonal matrix containing the square roots of eigenvalues from $\mathbf{U}$ or $\mathbf{V}$ in descending order, which are known as the singular values of $\mathbf{F}$.

Taking into account that $n > p$, this matrix decomposition can be written, as a finite summation, as follows:

$$\mathbf{F} = \sum_{i=1}^{p} \lambda_i^{1/2}\mathbf{u}_i\mathbf{v}'_i \tag{5.6}$$

where $\mathbf{u}_i$ and $\mathbf{v}_i$ are the i-th column eigenvectors belonging to $\mathbf{U}$ and $\mathbf{V}$ respectively, and $\lambda_i^{1/2}$ is the i-th singular value of $\mathbf{F}$.

From (5.6), and after some computations, each element of $\mathbf{F}$ (each discrete function) can be written as follows:

$$\mathbf{f}_k = \lambda_1^{1/2}u_{k1}\mathbf{v}_1 + \lambda_2^{1/2}u_{k2}\mathbf{v}_2 + \ldots + \lambda_p^{1/2}u_{kp}\mathbf{v}_p \tag{5.7}$$

It is worth mentioning that the expression shown by (5.7) actually represents the decomposition of the discrete function of time $\mathbf{f}_k$ into a sum of a set of discrete functions $(\mathbf{v}_j)$ which are orthogonal in nature (since they are the orthonormal eigenvectors of $\mathbf{F}'\mathbf{F}$), weighted by real coefficients resulting from the product of the j-th singular value of $\mathbf{F}$ by the j-th element of the eigenvector $\mathbf{u}_k$. Thus, $\mathbf{v}_j$ represents the j-th EOF and its coefficient $a_{kj} = \lambda_j^{1/2}u_{kj}$ is called the EOF score.

Based on a generalization of (5.7), it is possible to reconstruct the complete matrix $\mathbf{F}$ (i.e., the original data) using the EOFs and their corresponding EOF scores as given in (5.8):

$$\mathbf{F} = \sum_{i=1}^{p} \mathbf{a}_i\mathbf{v}'_i \tag{5.8}$$

where $\mathbf{a}_i$ is the i-th vector whose elements are all the a_{ij} EOF scores.

Comparing (5.6) and (5.8), it is easily concluded that $\mathbf{a}_i = \lambda_i^{1/2}\mathbf{u}_i$. Then, all a_{ij} EOF scores can be calculated by using its matrix form as follows:

$$\mathbf{A}_{np} = \mathbf{U}_{nn}\boldsymbol{\Lambda}_{np}^{1/2} \tag{5.9}$$

where $\mathbf{A}$ is the EOF score matrix.

From (5.5) and (5.9), it can be determined that

$$\mathbf{F}_{np} = \mathbf{A}_{np}\mathbf{V}'_{pp} \tag{5.10}$$

From the last equation, and based on the fact that $\mathbf{V}$ is an orthogonal matrix (whose main feature is that its transpose is equal to its inverse), the EOF score matrix $\mathbf{A}$ can be computed by (5.11):

$$\mathbf{A}_{np} = \mathbf{F}_{np}\mathbf{V}_{pp} \tag{5.11}$$

where matrix $\mathbf{V}$ contains the corresponding EOFs of $\mathbf{F}$ (i.e., the eigenvectors of $\mathbf{F}'\mathbf{F}$).

The sum of the singular values of $\mathbf{F}$ ($\lambda_i^{1/2}$) is equivalent to the total variance of the data matrix, and each i-th singular value offers a measurement of the explained variability (EV_i) given by EOF_i as defined by (5.12). Thus, the number of the chosen EOFs depends on the desired explained variability:

$$EV_i = \frac{\lambda_i}{\sum_{j=1}^{P} \lambda_j} \times 100 \tag{5.12}$$

It is worth mentioning that the main advantage of EOFs is their ability to determine the orthogonal functions that better adapt to the set of dynamic functions. This feature enables the mining of the patterns in the signal, and allows EOF to beat other signal processing tools such as Fourier analysis, which always employs the same pre-defined trigonometric pattern functions (i.e., sine and cosine), which are not always suited to representing specific dynamic behavior [16].

Once the EOFs (coefficients and scores) are computed, the corresponding EOF scores represent the projection of the data in the direction of the EOFs and form vectors of real numbers that represent the post-contingency dynamic behavior of the system. These pattern vectors permit mapping of the DVRs represented in the coordinate system formed by the EOFs.

Additionally, since some features in greater numeric ranges can dominate those of smaller numeric ranges, a numerical normalization is recommended before mapping the DVRs. Then, the pattern vectors have to be scaled in order to avoid this drawback [17]. In this chapter, a linear scaling in the range of [0, 1] is proposed.

Each pattern vector has a specific associated "class label" depending on the resulting vulnerability status vector (**vs**). These class labels might correspond to a non-vulnerable case (label 0) or a vulnerable case (label 1). Using the resulting pattern vectors and their associated vulnerability status class labels, the DVRs can be numerically mapped in the coordinate system formed by the main EOFs. Figure 5.3 presents the scheme of the proposed post-contingency pattern recognition method.

5.3.3 Definition of Data-Time Windows

In order to adequately show the system response for the different stability phenomena, several time windows (TWs) have to be defined. These time windows are established depending on the Monte Carlo statistics of the relay tripping times, influenced also by the WAMPAC communication time delay (t_{delay}), which represent the delays caused by measurement, communication, processing, and tripping. It is noteworthy that TW corresponds to the length of each time window, whose beginning (t_0) always accords with the instant of the fault clearing (t_{cl}).

Since three different types of relays are considered, three different tripping time statistical analyses will be carried out. First, the minimum MC tripping time (t_{min}) has to be determined. This time represents the maximum admissible delay for the actuation of any corrective control action, which has to be also affected by the WAMPAC communication time delay:

$$t_{min} = \min_{i=1\ldots n} \{t_{OSR_i}, t_{VR_i}, t_{FR_i}\} - t_{delay} \tag{5.13}$$

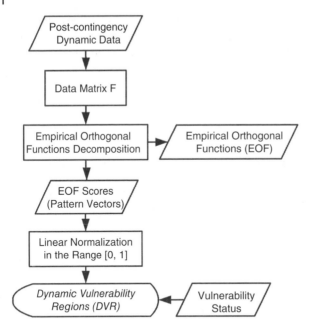

Figure 5.3 Post-contingency pattern recognition method.

where t_{OSRi}, t_{VRi}, and t_{FRi} are the tripping times of OSRs, VRs, and FRs of the n MC repetitions, respectively. Typically, OSRs present the fastest tripping time due to the fast time-frame evolution of TS.

Since the post-contingency data comprise the samples taken immediately after the fault is cleared, the first time window ($\mathrm{TW_1}$) is defined by the difference between $t_{\min}$ and the clearing time (t_{cl}):

$$\mathrm{TW}_1 \leq t_{\min} - t_{\mathrm{cl}} \tag{5.14}$$

The rest of the time windows are defined based on the statistical concept of confidence interval related to Chebyshev's inequality [18], which requires that at least 89% of the data lie within three standard deviations (3σ):

$$\mathrm{TW}_k \approx 3 \cdot \mathrm{std}\{t_{\mathrm{OSR/VR/FR}}\} + \mathrm{TW}_{k-1} \tag{5.15}$$

where $\mathrm{std}\{\cdot\}$ represents the standard deviation (σ) of the relay tripping time that most intersects the corresponding time window TW_k.

5.3.4 Identification of Post-contingency DVRs—Case Study

For illustrative purposes, this section shows the recognition of the post-contingency DVRs of the IEEE New England 39-bus, 60 Hz [19] and 345 kV test system, slightly modified in order to satisfy the N-1 security criterion and to include dynamic load models, automatic voltage regulators (AVR), governors (GOV), and the three mentioned types of relays: OSR, VR, and FR. Additionally, six PMUs have already been located at buses 2, 9, 11, 17, 22, and 28, based on the results presented in [20]. The test system single-line diagram is depicted in Figure 5.4, which also includes the PMU location.

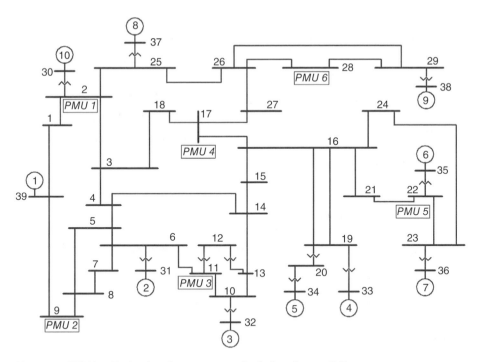

Figure 5.4 IEEE New England 39-Bus test system single-line diagram [19].

The placement PMUs are considered to have a one-cycle (16.67 ms) updating period. Also, it is assumed that the WAMPAC communication time delay is $t_{delay} = 250$ ms (i.e., a conservative value derived from the reference data presented in Chapter 1).

Simulations were carried out in DIgSILENT PowerFactory and consisted of several contingencies where the causes of vulnerability could be transient instability, short-term voltage instability, or short-term frequency instability (TVFS). These contingencies are generated by applying the Monte Carlo method.

Two types of events are considered: three phase short circuits and generator outage. The short circuits are applied at different locations of the transmission lines, depending on the Monte Carlo simulation. The disturbances are applied at 0.12 s, followed by the opening of the corresponding transmission line at 0.2 s (i.e., 80 ms fault clearing time t_{cl}). Likewise, the generator to be tripped is also chosen by the Monte Carlo method, and this type of contingency is applied at 0.2 s. Several operating states have been considered by varying the loads of PQ buses, depending on three different daily load curves. Then, optimal power flow (OPF) is performed in order to establish each operating state, using the MATPOWER package. After that, time domain simulations have been performed using the DIgSILENT PowerFactory software, so that the post-contingency PMU dynamic data could be obtained.

Both voltage components (magnitude and angle), and bus frequencies are considered as potential input variables. A total number of 10 000 cases have been simulated, from which 7600 are stable or "non-vulnerable" and 2400 are unstable or "vulnerable": 1308 are transient unstable, 682 are frequency unstable, and 410 are voltage unstable. It is worth to mention that 84 cases belonging to the 7600 TVFS non-vulnerable instances

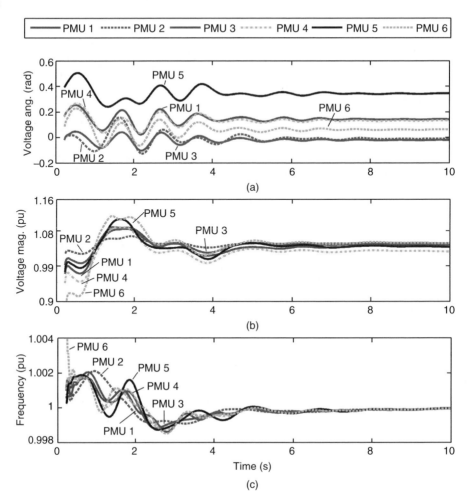

Figure 5.5 39-bus-system: non-vulnerable case

correspond to oscillatory unstable cases. Nevertheless, since oscillatory phenomena are treated in a different way in this book, these affairs will be separately analyzed in Chapter 6.

For illustrative purposes, the post-contingency dynamic responses of two different MC iterations are presented in Figures 5.5 and 5.6. These figures depict the simulation results of one non-vulnerable case and one voltage unstable case (vulnerable case). It is possible to observe in the figures the particular signal dynamic behavior for each case, which expose the existence of certain patterns revealing the actual system security level as well as the future system vulnerability status tendency.

By using the MC simulation results, consisting of the post-contingency PMU dynamic data, the relay tripping times, and the vulnerability status vector, the DVRs are identified for this specific power system. First, several time windows are defined based on the Monte Carlo statistics of the relay tripping times. Figure 5.7 presents the 39-bus-system relay tripping time histograms, where the reference frame corresponds to the time simulation.

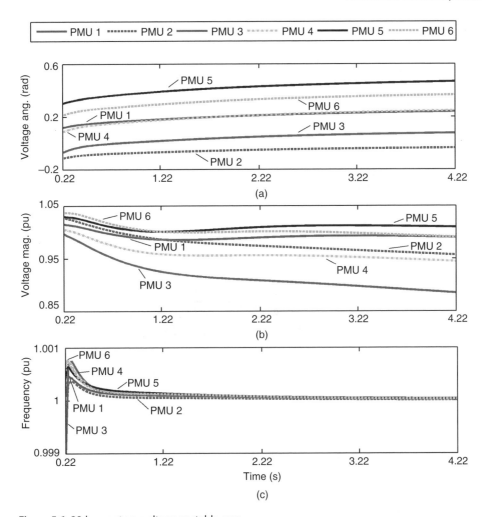

Figure 5.6 39-bus-system: voltage unstable case.

The out-of-step relay time has a mean of 1.2252 s, a standard deviation of 0.3746 s, and a minimum value of 0.8342 s. Thus, vulnerability assessment has to be done in less than 0.5842 s ($t_{min} = 0.8342$ s $- t_{delay}$). For this reason, an adequate data window (TW$_1$) for TS phenomenon would be 300 ms ($t_{min} - t_{cl} = 0.3842$ s) starting from fault clearing.

In this test system, the tripping of voltage relays presents a mean of 4.1275 s, a standard deviation of 1.6872 s, and a minimum value of 3.22 s, whereas the frequency relay tripping time has a mean of 10.6829 s, a standard deviation of 2.4921 s, and a minimum value of 5.987 s. Using these values, and (5.15), the rest of the time windows are determined. This time window definition is summarized in Table 5.1.

After TW definition, the corresponding EOF coefficients and scores (i.e., pattern vectors) are determined using the resulting MC dynamic data. After this, the EOF scores are computed. For instance, the first four TW$_3$ voltage-magnitude-based EOFs are presented in Figure 5.8. Similarly, EOFs are determined for each electric variable at every TW.

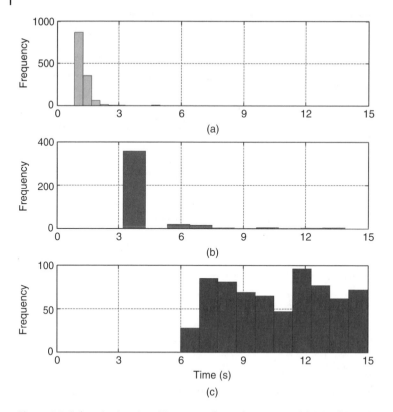

Figure 5.7 Relay tripping time Histograms for 39-bus-system: (a) OSR, (b) VR, (c) FR.

Table 5.1 Time window definition for 39-bus-system.

Time Window	std{$t_{OSR/VR/FR}$} (s)	$3 \times std\{t_{OSR/VR/FR}\} + TW_{k-1}$ (s)	TW (s)
TW_1	–	–	0.30
TW_2	0.3746	1.4238	1.50
TW_3	0.3746	2.6238	2.70
TW_4	0.3746	3.8238	3.90
TW_5	1.6872	8.9616	9.00

Once the EOF scores are computed, the corresponding DVRs can be mapped. For instance, Figure 5.9 presents the three dimensional distribution of the pattern vectors obtained from the voltage magnitudes corresponding to TW_3. In the figure, the ellipsoidal area (enclosing the "vulnerable" pattern vectors represented by circles) represents the vulnerable region; whereas the white area (relating to the "non-vulnerable" diamond pattern vectors) corresponds to the non-vulnerable region.

These areas have been empirically delimited, bordering the obtained pattern shapes which depend on the pattern vector spatial locations. The plot presents the DVRs corresponding to the first three EOFs depicted in Figure 5.8.

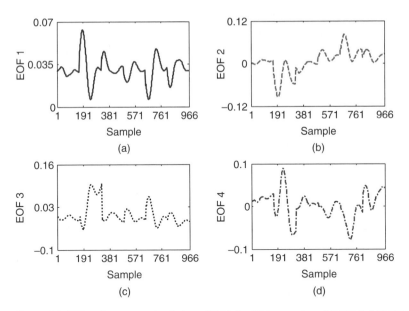

Figure 5.8 TW$_3$ voltage-magnitude-based EOFs for 39-bus-system: (a) EOF 1, (b) EOF 2, (c) EOF 3, (d) EOF 4.

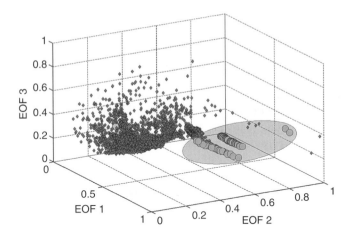

Figure 5.9 TW$_3$ voltage-magnitude-based DVRs for 39-bus-system.

5.4 Real-Time Vulnerability Status Prediction

The DVRs will be used to specify the actual dynamic state relative position with respect to their boundaries, which might be established using an intelligent classifier. For this purpose, a support vector classifier (SVC) is proposed in this chapter since this classifier computes an optimal hyper-plane that separates the vector space depending on the spatial location of each class [11, 21]; this agrees with the hyper-plane-based definition of DVRs laid out in Section 5.2. Additionally, the SVC has the property of being more robust in avoiding over-fitting problems [8], which is also desirable in order to get enough robustness and generalization in classifying an actual case in real time. This classifier acquires decision functions that classify an input into one of the given classes through training using input-output (features-label) pair data. The optimal decision function is called the Optimal Hyper-plane (OH), and it is determined by a small subset of the training set which are called the Support Vectors (SV), using the concept of the VC (Vapnik–Chervonenskis) dimension as the theoretical basis [11]. Figure 5.10 shows an illustration of an SVC solution for a two-class data classification problem, where the SV and the OH have been determined. The classified vectors belong to one of two different groups, that is, "class 1" or "class 2," and they are represented in a two-dimensional plane whose axes are the first and second variables (x_1 and x_2) of the feature vectors ($\mathbf{x}$).

The SVC needs an a priori off-line learning stage, in which the classifier has to be trained using a training set of data. Hence, the data have to be split into training and testing sets. Each element in the training set contains one "target value" (class labels) and several "attributes" (features). The objective of the SVC is to yield a training data based model, which predicts the target values of the test data given only the test data features [17].

Given a training set of features-label pairs ($\mathbf{x}_i$, y_i), $i = 1, \ldots, l$ where $\mathbf{x}_i \in R^n$ and $y \in \{1, -1\}^l$, for a two-class classification problem, the support vector classifier requires the solution of the optimization problem shown in (5.16) [17]:

$$\min_{\mathbf{w},b,\xi} \quad \frac{1}{2}\mathbf{w}^T\mathbf{w} + C\sum_{i=1}^{l} \xi_i$$

$$\text{subject to} \quad y_i\left(\mathbf{w}^T\phi(\mathbf{x}_i) + b\right) \geq 1 - \xi_i, \tag{5.16}$$

$$\xi_i \geq 0$$

where $\mathbf{w}$ is an n-dimensional weight vector, b is a bias term, ξ_i is a slack variable associated with $\mathbf{x}_i$, C is the margin parameter, and $\phi(\mathbf{x}_i)$ is the mapping function from $\mathbf{x}$ to the feature space [11]. It is worth mentioning that $\mathbf{w}$, b, and ξ_i are determined via the SVC optimization process, whereas C is a parameter to be specified a priori.

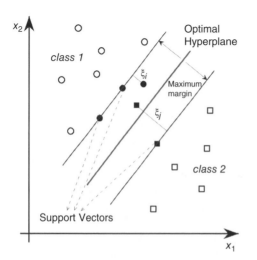

Figure 5.10 Support vectors and optimal separating hyper-plane of SVC.

The mapping function $\phi(\mathbf{x}_i)$ is usually defined as the so called "kernel function" $K(\mathbf{x}_i, \mathbf{x}_j)$, as in (5.17) [17]:

$$K(\mathbf{x}_i, \mathbf{x}_j) \equiv \phi(\mathbf{x}_i)^T \phi(\mathbf{x}_j) \tag{5.17}$$

There are several types of kernel function, such as linear, polynomial, and radial basis function (RBF), among others. Figure 5.10 presents, for instance, an OH determined using a linear kernel function.

In this chapter, an RBF kernel (5.18) is used because it is capable of handling possible nonlinear relations between labels and features [17]:

$$K(\mathbf{x}_i, \mathbf{x}_j) = e^{-\gamma \|\mathbf{x}_i - \mathbf{x}_j\|^2}, \gamma > 0 \tag{5.18}$$

Before training the SVC, it is necessary to identify the best parameters C of (5.16), and γ of (5.18) [17], as well as W_m that represents a weight factor used to change the penalty of class m, which is useful for training classifiers using unbalanced input data [21]. For this purpose, an optimization problem is solved via the swarm version of the mean-variance mapping optimization (MVMO^S), for which the details of implementation can be obtained from [22].

Based on the requirement of dynamic vulnerability status prediction, a two-class classifier is structured that sorts the power system vulnerability status into "vulnerable" or "non-vulnerable" using the pattern vectors obtained from the pattern recognition method as inputs of SVC. The classifier needs an a priori off-line learning stage, in which SVC has to be trained using the previously recognized DVRs (i.e., pattern vectors and vulnerability status). For real-time implementation, the classifier uses the post-contingency voltage phasors (angles and magnitudes) and frequencies, obtained from the PMUs, transformed by the EOFs stored in the control center processor as table-based reference functions. Then, the classifier will classify the system's tendency to become "vulnerable" or "non-vulnerable."

Figure 5.11 presents a flowchart of the proposed real-time vulnerability status classification algorithm, including the off-line training stage.

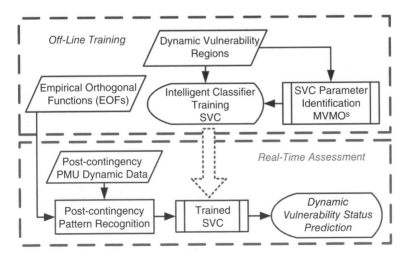

Figure 5.11 Post-contingency vulnerability status prediction methodology.

5.4.1 Support Vector Classifier (SVC) Training

In dynamic vulnerability status classification, the target values resemble the MC vulnerability status "class label." These class labels might correspond to non-vulnerable cases (label 0) or vulnerable cases (label 1), depending on the triggering status of the different relays. Hence, a two-class classifier has to be used. The attributes, instead, are represented by the post-contingency pattern vectors that better represent the DVRs belonging to each time window. In this connection, and based on the fact that some signals better show the evolution of specific phenomena than others, a procedure for improving the feature extraction and selection has to be followed in order to choose the appropriate pattern vectors before training the SVC. Then, it is necessary to choose an adequate number of EOFs that allows maintaining as much of the variation presented in the original variables as possible. This analysis constitutes the proposed *feature extraction stage*. For this purpose, the i-th EOF desired explained variability (EV_i), represented by (5.12), is used as a weight measure. After several tests, it has been established that the number of EOFs chosen will be that which permits obtaining an EVi of more than 97%. In the case where the EVi requirement is satisfied already with the first EOF, two EOFs will be used in order to ensure the mapping of at least two-dimensional DVRs. On the other hand, based on the premise that some electrical variables better show the evolution of specific phenomena than others, it is necessary to select those signals that allow the best DVR representation for each TW. The selection will permit obtaining the best classification accuracy. This analysis can be seen as the *feature selection stage*. Therefore, a decision tree (DT) method for feature selection, originally introduced in [23], has been adapted to this problem, based on the classification accuracy obtained considering a combinatorial analysis of the potential electrical variables. Additional details of this feature extraction and feature selection method can be found in [22].

Once the optimal SCV parameters have been identified and the best features have been determined, the SVC is trained for each data-time window. The objective is to obtain a robust enough SVC that allows predicting the vulnerability status by using as input the best pattern vectors obtained from the EOFs. For this purpose, 10-fold cross validation (CV) is used to analyze the robustness of the classifier training as explained in [22]. For illustrative purposes, the performance of the trained SVC for each of the five TW determined in the previously presented case study is presented in Table 5.2. This table also includes the performance of other classifiers, such as: decision tree classifier (DTC), pattern recognition network (PRN: a type of feed-forward network),

Table 5.2 Classification performance of 39-bus-system.

Classifier	mean{CA_i} for Time Window (%)				
	TW_1	TW_2	TW_3	TW_4	TW_5
DA	97.440	99.966	99.494	98.034	97.178
DTC	98.200	99.931	99.736	99.436	99.291
PRN	98.760	99.897	99.770	99.029	98.993
PNN	98.930	99.977	99.770	99.137	99.055
SVC	99.290	100.00	99.885	99.880	99.727

discriminant analysis (DA), and probabilistic neural networks (PNN: a type of radial basis network). The performance of the classification is evaluated, in this comparison, by means of the mean of each CV-fold classification accuracy (CA_i) for each TW [22]. The Library for Support Vector Machines (LIBSVM) [21] is used for designing, training, and testing the SVC models.

From the results, it is possible to observe that the SVC outperforms all other classifiers in terms of classification accuracy. In addition, the SVC is the only classifier that permits obtaining more than 99% of CA for all TWs. These results allow us to form the conclusion that the SVC presents an excellent performance when adequate parameters and features are used. Then, suitable parameters permit exploiting the SVC advantage of being more robust to over-fitting problems, which is reflected in a more robust 10-fold CV accuracy. This pre-requisite has been optimally satisfied via the proposed $MVMO^S$-based parameter identification method.

5.4.2 SVC Real-Time Implementation

For real-time implementation, the off-line trained SVCs will be in charge of classifying the post-contingency dynamic vulnerability status of the power system, using the actual post-contingency PMU voltage phasors and frequencies as data. First, the dynamic signals must be transformed to the corresponding EOF scores. For this purpose, the measured data have to be multiplied by the EOFs determined in the off-line training and stored in the control center processor. Then, the obtained EOF scores will be the input data to the trained SVCs, which will automatically classify the system status into "vulnerable" or "non-vulnerable." This vulnerability status prediction will be then used by the "performance-indicator-based real-time vulnerability assessment" module (for more details, see Chapter 6) that is responsible for computing several vulnerability indices, which are the final indicators of the system vulnerability condition. Since this classification uses patterns hidden in the input signal as input data, it is capable of predicting the post-contingency dynamic vulnerability status before the system reaches a critical state. Then, the methodology presented in this chapter allows the assessment of the second aspect regarding the vulnerability concept, whose causes concern TVFS issues. Figure 5.12 illustrates how the proposed TVFS vulnerability status prediction method might be implemented in a control center.

In order to validate the complete TVFS vulnerability status prediction presented in this chapter, the power system protection concepts of dependability (Dep (5.19): ability for tripping when there exists the necessity to trip) and security (Sec (5.20): ability not to trip when there is no necessity to trip) are used. It is worth mentioning that security is more critical than dependability because a miss-classification might suggest wrong corrective control actions:

$$Dep(\%) = \frac{N^\circ \text{ vulnerable samples correctly classified}}{Total\, N^\circ \text{ vulnerable samples in data set}} \times 100 \tag{5.19}$$

$$Sec(\%) = \frac{N^\circ \text{ non} - \text{vulnerable samples correctly classified}}{Total\, N^\circ \text{ non} - \text{vulnerable samples in data set}} \times 100 \tag{5.20}$$

Table 5.3 shows a summary of the number of vulnerable and non-vulnerable cases corresponding to the case study of the 39-bus system, where the percentages of

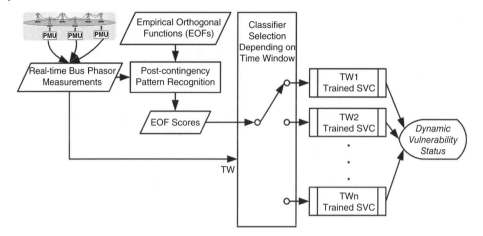

Figure 5.12 SVC real-time implementation in a control center.

Table 5.3 Vulnerability status prediction performance for 39-bus-system.

Feature	Value
Non-vulnerable cases correctly classified	7587
Vulnerable cases correctly classified	2390
Non-vulnerable cases classified as vulnerable	13
Vulnerable cases classified as non-vulnerable	10
Complete classification accuracy (%)	99.770
Security (%)	99.829
Dependability (%)	99.583

complete classification accuracy, security, and dependability are also presented. Results highlight the excellent performance of the methodology regarding security and dependability, offering more than 99% confidence in both measures. The small percentage of errors is due to the existence of overlapping zones between vulnerable and non-vulnerable regions.

Considering the necessity of ensuring very quick response for WAMPAC applications, the methodology presented in this chapter must deliver in a very short elapsed time. Thus, in order to verify the speed of the real-time computations of the proposed method, a complete classification test has been made using only one of the cases analyzed in the previous simulations, which would represent a real-time event. Simulations have been run in Windows 7 Ultimate-Intel (R) Core (TM) i3-2350M-2.30GHz-6GB RAM. In this instance, the resulted SVC elapsed time was 0.68 ms, and the entire process presented in Figure 5.12 takes only 1.47 ms. This very quick response of the proposed methodology matches the very short time delay requirement of this type of real-time application.

5.5 Concluding Remarks

This chapter presents a novel post-contingency pattern recognition method for predicting the dynamic vulnerability status of an Electric Power System. The methodology begins with the determination of post-contingency DVRs using MC simulation and EOFs, which allow finding the best pattern functions for representing the particular dynamic signal behavior. This proposal considers three different short-term instability phenomena as the potential causes of vulnerability (TVFS), for which several time windows have been defined. These DVRs are then used to specify the actual dynamic state relative position with respect to their boundaries, which is established using an intelligent classifier together with an adequate feature extraction and selection scheme. In this connection, SVC is used due to its property of being particularly robust to over-fitting problems when adequate parameters and features are selected. From the results, it is possible to observe that the post-contingency TVFS vulnerability status prediction method permits assessment of the second aspect regarding the vulnerability concept by analyzing the system's tendency to reach an unstable condition with high classification accuracy and low time-consumption requirements. In addition, this mathematical tool has been capable of predicting system response using only small windows of post-contingency data obtained directly from PMUs located in some high voltage transmission buses. Thus, it is not necessary to calculate other physical variables (e.g., rotor angles or machine rotor speeds), which are needed in the traditional TSA. The obtained classification accuracy (more than 99%) outperforms the few methods reported by the literature that also aim to estimate system vulnerability status via the application of other types of signal processing techniques (such as Short Time Fourier Transform (STFT) [24] or Local Correlation Network Pattern (LCNP) [25], as discussed in Chapter 1), in which the considerably large overlap zone provokes confusion between vulnerable and non-vulnerable regions. The presented approach, in contrast, allows a better adaptation of the system dynamic variables of interest by using empirical orthogonal functions based on the fact that this technique permits finding the best pattern functions for representing the particular signal dynamic behavior. In this manner, an improvement in the accuracy of classification results is obtained compared to the application of other signal processing tools, such as STFT or LCNP. The TVFS vulnerability status prediction presented in this chapter is not alone sufficient to achieve recognition of the specific symptom of system stress that is the cause of the vulnerability, which might help in orienting the adequate corrective control action. Hence, the need to use some complementary decision techniques (e.g., Key Performance Indicators (KPIs)) that allow an increase in security confidence and permit recognizing the causal symptoms of vulnerability; these will be discussed in Chapter 6 of this book. This complementary analysis, based on performance indices, will help toward better decision making.

References

1 K. Moslehi, and R. Kumar, "Smart Grid – A Reliability Perspective", IEEE PES Conference on Innovative Smart Grid Technologies, January 19–20, 2010, Washington, DC.

2 M. Amin, *"Toward Self-Healing Infrastructure Systems"*, Electric Power Research Institute (EPRI), IEEE, 2000.

3 J. C. Cepeda, D. G. Colomé, and N. J. Castrillón, *"Dynamic Vulnerability Assessment due to Transient Instability based on Data Mining Analysis for Smart Grid Applications"*, IEEE PES ISGT-LA Conference, Medellín, Colombia, October 2011.

4 S. C. Savulescu, et al, *Real-Time Stability Assessment in Modern Power System Control Centers*, IEEE Press Series on Power Engineering, Mohamed E. El-Hawary, Series Editor, a John Wiley & Sons, Inc., Publication, 2009.

5 Z. Huang, P. Zhang, et al, "Vulnerability Assessment for Cascading Failures in Electric Power Systems", *Task Force on Understanding, Prediction, Mitigation and Restoration of Cascading Failures, IEEE PES Computer and Analytical Methods Subcommittee*, IEEE Power and Energy Society Power Systems Conference and Exposition 2009, Seattle, WA.

6 Y. Zeng, P. Zhang, M. Wang, H. Jia, Y. Yu, and S. T. Lee, "Development of a New Tool for Dynamic Security Assessment Using Dynamic Security Region", 2006 International Conference on Power System Technology.

7 M. A. El-Sharkawi, *"Vulnerability Assessment and Control of Power System"*, Transmission and Distribution Conference and Exhibition 2002: Asia Pacific., Seattle, WA, USA, Vol. 1, pp. 656–660, October 2002.

8 Z. Dong, and P. Zhang, *Emerging Techniques in Power System Analysis*, Springer, 2010.

9 P. Kundur, J. Paserba, V. Ajjarapu, et al, "Definition and Classification of Power System Stability", *IEEE/CIGRE Joint Task Force on Stability: Terms and Definitions. IEEE Transactions on Power Systems*, Vol. 19, pp. 1387–1401, August 2004.

10 C. Bishop, *Pattern Recognition and Machine Learning*, Springer, 2006.

11 S. Abe, *Support Vector Machines for Pattern Classification*, Second Edition, Springer, 2010.

12 E. N. Lorenz, *"Empirical Orthogonal Functions and Statistical Weather Prediction"*, Statistical Forecasting Project Rep. 1, MIT Department of Meteorology, 49 pp, December 1956.

13 H. Björnsson, and S. Venegas, "A Manual for EOF and SVD Analyses of Climatic Data", C2GCR Report No. 97-1, McGill University, February 1997, [Online]. Available at: http://www.geog.mcgill.ca/gec3/wp-content/uploads/2009/03/Report-no.-1997-1.pdf.

14 I. Jollife, *Principal Component Analysis*, 2nd. Edition, Springer, 2002.

15 D. Peña, *Análisis de Datos Multivariantes*, Editorial McGraw-Hill, España.

16 J. Cepeda, and G. Colomé, *"Benefits of Empirical Orthogonal Functions in Pattern Recognition applied to Vulnerability Assessment"*, IEEE Transmission and Distribution Latin America (T&D-LA) 2014, Medellín, Colombia, Septiembre 2014.

17 C-W. Hsu, C-C. Chang, and C-J. Lin, "A Practical Guide to Support Vector Classification", April 15, 2010, [Online]. Available: http://www.csie.ntu.edu.tw/~cjlin.

18 J. Han, and M. Kamber, *Data Mining: Concepts and Techniques*, second edition, Elsevier, Morgan Kaufmann Publishers, 2006.

19 M. A. Pai, *Energy Function Analysis for Power System Stability*, Kluwer Academic Publishers, 1989.

20 J. Cepeda, J. Rueda, I. Erlich, and G. Colomé, *"Probabilistic Approach-based PMU placement for Real-time Power System Vulnerability Assessment"*, 2012 IEEE PES

Conference on Innovative Smart Grid Technologies Europe (ISGT-EU), Berlin, Germany, October 2012.

21 C-C. Chang, and C-J. Lin, "LIBSVM: A Library for Support Vector Machines", 2001. [Online]. Software Available at: http://www.csie.ntu.edu.tw/~cjlin/libsvm.

22 J. Cepeda, J. Rueda, G. Colomé, and I. Erlich, "Data-Mining-Based Approach for Predicting the Power System Post-contingency Dynamic Vulnerability Status", *International Transactions on Electrical Energy Systems*, Vol. 25, Issue 10, pp. 2515–2546, October 2015.

23 S. P. Teeuwsen, *Oscillatory Stability Assessment of Power Systems using Computational Intelligence*, Doktors der Ingenieurwissenschaften Thesis, Universität Duisburg-Essen, Germany, March 2005.

24 I. Kamwa, J. Beland, and D. Mcnabb, "PMU-Based Vulnerability Assessment using Wide-Area Severity Indices and Tracking Modal Analysis", IEEE Power Systems Conference and Exposition, pp. 139–149, Atlanta, November, 2006.

25 P. Zhang, Y. D. Zhao, et al, *Program on Technology Innovation: Application of Data Mining Method to Vulnerability Assessment*, Electric Power Research Institute (EPRI), Final Report, July 2007.

6

Performance Indicator-Based Real-Time Vulnerability Assessment

Jaime C. Cepeda[1], José Luis Rueda-Torres[2], Delia G. Colomé[3], and István Erlich[4]

[1] *Head of Research and Development, and University Professor Technical Development Department, and Electrical Energy Department Operador Nacional de Electricidad CENACE, and Escuela Politécnica Nacional EPN, Quito, Ecuador*
[2] *Assistant professor of Intelligent Electrical Power Systems, Department of Electrical Sustainable Energy, Delft University of Technology, The Netherlands*
[3] *Professor of Electrical Power Systems and Control, Institute of Electrical Energy, Universidad Nacional de San Juan, Argentina*
[4] *Chair Professor of Department of Electrical Engineering and Information Technologies, Head of the Institute of Electrical Power Systems, University Duisburg-Essen, Duisburg, Germany*

6.1 Introduction

Electric power systems usually support perturbations which, depending on their severity, eventually could initiate cascading events that might cause total or partial blackouts [1]. In this context, vulnerability assessment needs to be done in order to determine the necessity of triggering different types of protection mechanisms (i.e., System Integrity Protection Schemes (SIPS) or Special Protection Schemes (SPS)) designed to attenuate the risk of these possible blackouts [2, 3]. However, the conditions that lead a system to blackouts are not easy to identify because the process of system collapse depends on multiple interactions [4]. Therefore, vulnerability assessment should be done considering these interactions that, in fact, relate several electrical phenomena that can be defined as vulnerability symptoms [5].

Vulnerability assessment is carried out by checking the system performance under the severest contingencies with the purpose of detecting the conditions that could initiate cascading failures and might provoke system collapses [6]. A vulnerable system can be defined as a system that operates with a "reduced level of security that renders it vulnerable to the cumulative effects of a series of moderate disturbances" [6]. The concept of vulnerability involves the system security level (static and dynamic security) [7] and the tendency to change its conditions into a critical state, known as the "Verge of Collapse State" [5]. A Vulnerable Area is a specific section of the system where vulnerability begins to develop. The occurrence of an abnormal contingency in the vulnerable areas and highly stressed operating conditions define a system in the Verge of Collapse State. Vulnerable areas are characterized by, in total, five different symptoms of system stress: transient stability, poorly-damped power oscillations, voltage instability, frequency deviations outside limits, and overloads, as analyzed in Chapter 1.

Dynamic Vulnerability Assessment and Intelligent Control for Sustainable Power Systems, First Edition.
Edited by José Luis Rueda-Torres and Francisco González-Longatt.
© 2018 John Wiley & Sons Ltd. Published 2018 by John Wiley & Sons Ltd.

Many methods have been proposed for vulnerability assessment [4, 6]. Conventional methods are based on steady state or dynamic simulations of contingencies, that is, static security assessment (SSA) or dynamic security assessment (DSA) [4, 6–9], which can also be associated to a specific probability of occurrence [4, 6]. These methods require long processing times, which restricts online applications. Modern high-performance computing (HPC) techniques have improved calculation times, permitting online applications [8], but does not allow real-time uses. Emerging technologies such as Phasor Measurement Units (PMUs) and Wide Area Monitoring Systems (WAMS), which provide voltage and current phasor measurements with updating periods of one cycle, have allowed the development of modern approaches for real-time dynamic vulnerability assessment (DVA) [9–11]. Other emerging methods permit identification of indicators or patterns that show system vulnerability, which use artificial intelligence [6] or modern data mining tools [10, 11]. Some of these approaches permit assessment of the post-contingency system security level, but they do not consider the fact that vulnerability begins to develop in specific areas of the system, which instead is tackled in [12] using some performance indicators. Moreover, most of these approaches have been focused on analyzing only one symptom of vulnerability, while the approach that considers several phenomena only allows prediction of power system post-contingency dynamic vulnerability status [11] but does not consider the valuation of the security level via performance indicators.

This chapter presents a comprehensive real-time vulnerability assessment methodology oriented to determine vulnerable areas of an electric power system and to assess vulnerability of each one of them, considering both aspects involved in the vulnerability concept, that is: (i) assessment of the system's tendency to reach a critical state, and (ii) evaluation of the actual security level. In the present chapter, the methodology for identifying vulnerable areas in real time, based on electrical coherency, is firstly introduced. Then, the evaluation of system security level due to the three short-term stability phenomena (i.e., transient, voltage, and frequency stability (TVFS)) is presented. After this, the assessment for oscillatory stability and overloads, that is, the slowest phenomena, is defined. Finally, both tasks involved in the concept of vulnerability are joined and the complete vulnerability assessment is presented. All of these vulnerability symptoms are evaluated via different key performance indicators (KPIs) that also combine the results of the post-contingency vulnerability status prediction based on Dynamic Vulnerability Regions (DVR) presented in Chapter 5 by using arrays of logic gates.

6.2 Overview of the Proposed Vulnerability Assessment Methodology

The present chapter presents a comprehensive methodology to assess post-contingency system dynamic vulnerability in real time by using PMU measures as input data. This premise is based on the fact that after a contingency occurs, power system variables (i.e., voltage phasors, current phasors, and frequencies) begin to change, and these variables can be measured, in real time, by PMUs adequately located throughout the system. Using these data as inputs, the proposed real-time DVA methodology is capable of giving early warning alerts about possible system blackouts. The framework of the proposal

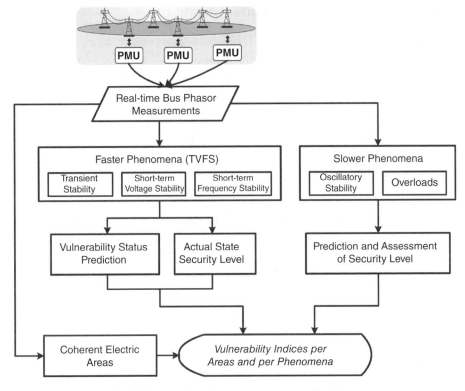

Figure 6.1 Framework of the proposed real-time assessment methodology.

is schematically sketched in Figure 6.1. As can be seen, the results of the complete assessment are several KPIs that evaluate the power system vulnerability condition.

The procedure depicted in Figure 6.1 describes the steps to be followed in real time to achieve the proposed post-contingency vulnerability assessment (VA). The scheme begins with the PMU data acquisition at each updating period. These data are then received and pre-processed in the control center in order to filter possible noise and debug potential outliers (this chapter assumes that this pre-processing step has already been accomplished).

The approach allows carrying out VA through identifying vulnerable areas and ranking them according to their grade of vulnerability, which is evaluated via several KPIs, taking into account several signals of system stress that have been split into "faster" and "slower" phenomena depending on their evolution timeframe. For this purpose, both tasks involved in the vulnerability concept are assessed through the prediction of the future vulnerability status and the evaluation of the actual state security level. Afterwards, several performance indices are computed that can be seen as a measure of the post-contingency system vulnerability per each area.

In real time, the power system is firstly partitioned into coherent electric areas via an adequate mathematical tool that analyzes coherency from the PMU dynamic data. Then, five vulnerability indices (indicators) that reflect the performance of the system as regards the symptoms of system vulnerability are computed for each electric area. These indicators cover the assessment of both tasks involved in the vulnerability concept.

6.3 Real-Time Area Coherency Identification

As noted in [13], interconnected power systems are usually partitioned into areas for different reasons such as reduction of required measurements for each coherent area, improvement of corrective control actions, or facilitation of system modeling in the case of huge or complex interconnected systems. The objective of system partitioning in this chapter is to detect, in real time, the vulnerable areas in which the possibility of a cascading event beginning exists.

The bus where a PMU is located represents the "centroid" of a coherent electrical area [14], which can be seen as the "associated PMU coherent area." Since the coherent electrical areas can change depending on the operating state and the actual contingency, it is necessary to analyze the possibility of existing coherency between the associated PMU coherent areas. Then, the aim of the real-time area coherency identification is to join two or more associated PMU coherent areas into "zones," if coherency between them is detected.

Hence, the recursive clustering approach for decomposing large power systems into coherent electrical areas presented in [14], and initially introduced in [13], is used. This coherency calculation method is adapted to be used in real time. In this connection, the three dissimilarity matrices for voltage angle, voltage magnitude, and frequency ($\mathbf{D}_\theta$, $\mathbf{D}_V$, and $\mathbf{D}_F$) [14] are built using the data measured by the PMUs located in the power system. Then, the size of these matrices is ($N_{PMU} \times N_{PMU}$), with N_{PMU} being the number of PMUs placed in the system.

Since this clustering algorithm is an iterative method, it improves its results according to the number of data items used to find the clusters, which are actualized when each phasor measurement arrives to the control center. Because of this fact, the implemented algorithm will be run at each PMU updating period, and the coherent groups might change during the event evolution.

$$\mathbf{D} = \begin{bmatrix} 0 & \delta_{12} & \cdots & \delta_{1N_{PMU}} \\ \delta_{21} & 0 & \cdots & \delta_{2N_{PMU}} \\ . & . & \cdots & . \\ \delta_{N_{PMU}1} & \delta_{N_{PMU}2} & \cdots & 0 \end{bmatrix} \tag{6.1}$$

Once the power system is partitioned into real-time coherent electrical areas, several performance indicators associated with each vulnerability symptom are calculated for each electrical area, as explained in the next sections.

One important aspect to be considered before applying the suggested algorithm is to specify the boundaries of the associated PMU coherent areas a priori, which will be discussed in the following subsection.

6.3.1 Associated PMU Coherent Areas

Since fast coherency may change depending on the operating state and the contingency magnitude, the number and the structure of the coherent electrical areas might be different. Monte Carlo (MC)-based dynamic simulations and clustering analysis, similar to those applied for placing the PMUs in [14], can be carried out in order to analyze the different possibilities of coherent areas for different operating states and contingencies. In this connection, once PMUs are placed (via the method depicted in [14]), a probabilistic analysis of coherent areas has to be done in order to determine the most representative group of coherent buses associated to each PMU (i.e., the probabilistic areas associated to PMUs), which will permit structuring a database for real-time use.

Thus, a probabilistic analysis of the results of the fast dynamic coherency Monte Carlo analysis described in [14], via histograms, would reveal those buses belonging to each associated PMU area that present the highest frequency. The real-time area coherency identification is then used to join these associated PMU areas if there is high coherency behavior during the development of the actual contingency.

For illustrative purposes, the identification of the probabilistic areas associated to PMUs, based on MC-based simulations is carried out to the IEEE New England 39-bus, 60 Hz and 345 kV test system, previously presented in Figure 5.4 of Chapter 5, including the PMUs already located [14]. Figure 6.2 presents histograms of the 39-bus-system probabilistic coherency area results, showing the buses that belong to the same area as

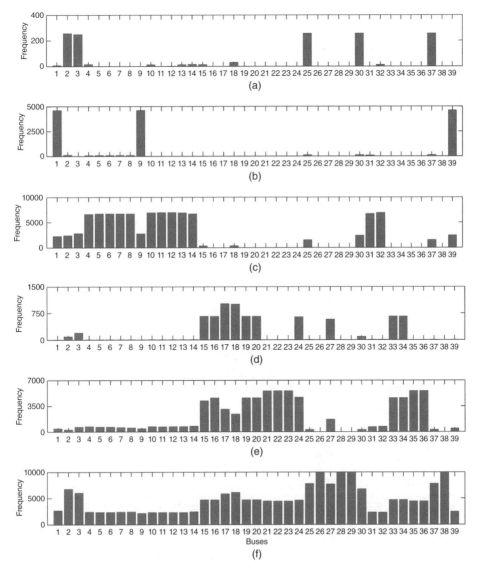

Figure 6.2 39-bus-system associated PMU areas: (a) associated PMU 1 area, (b) associated PMU 2 area, (c) associated PMU 3 area, (d) associated PMU 4 area, (e) associated PMU 5 area, (f) associated PMU 6 area.

each PMU already located (i.e., areas associated to each PMU). From the histograms, the buses that present the highest frequencies are selected to compose each associated PMU coherent area, considering that each PMU can belong to only one area as the basic premise. Table 6.1 shows the 39-bus-system associated PMU area resulted database.

The results are also illustrated in Figure 6.3, which present the buses that belong to each associated PMU area over the system single-line diagram. The different coherency distribution in each MC iteration is reflected in the overlap that exists between coherent areas, which can be seen in the figure.

Table 6.1 39-bus-system associated PMU area database.

PMU	Associated PMU coherent Areas/Buses
PMU 1	Area 1 = {1, 2, 3, 4, 18, 25, 30, 37}
PMU 2	Area 2 = {1, 6, 7, 8, 9, 31, 39}
PMU 3	Area 3 = {3, 4, 5, 6, 7, 8, 10, 11, 12, 13, 14, 31, 32}
PMU 4	Area 4 = {3, 15, 16, 17, 18, 19, 20, 24, 27, 33, 34}
PMU 5	Area 5 = {15, 16, 19, 20, 21, 22, 23, 24, 33, 34, 35, 36}
PMU 6	Area 6 = {25, 26, 27, 28, 29, 37, 38}

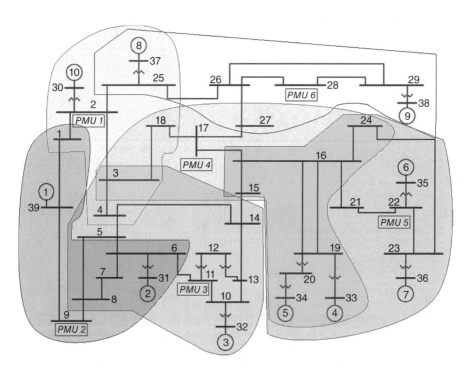

Figure 6.3 39-bus-system probabilistic associated PMU areas.

6.4 TVFS Vulnerability Performance Indicators

The evaluation of each of the symptoms of a stressed system mentioned above has to be done in order to properly assess vulnerability. In this section, three vulnerability indices, which resemble the system security level as regards transient stability, short-term voltage stability, and short-term frequency stability issues (TVFS), are presented.

These indices are then joined with the results obtained from the TVFS vulnerability status prediction depicted in Chapter 5 in order to perform both tasks involved in the vulnerability concept (i.e., actual security assessment and prediction of future vulnerability status).

For illustrative purposes, dynamic simulations consisting in several contingencies where the causes of vulnerability could be transient instability, short-term voltage instability, or short-term frequency instability (TVFS) have been carried out on the 39 bus test system. These contingencies are generated by applying the Monte Carlo method. Two types of events are considered: three phase short circuits and generation outage. The short circuits are applied at different locations of the transmission lines, depending on the Monte Carlo simulation. The disturbances are applied at 0.12 s, followed by the opening of the corresponding transmission line at 0.2 s (i.e., fault clearing time t_{cl}). Likewise, the generation to be tripped is also chosen by the Monte Carlo method, and this type of contingency is applied at 0.2 s.

Several operating states have been considered by varying the load of the PQ buses, depending on three different daily load curves. Then, optimal power flow (OPF) is performed in order to establish each operating state, using the MATPOWER package. After that, time domain simulations have been performed using the DIgSILENT PowerFactory software, so that the post-contingency PMU dynamic data could be obtained.

For instance, post-contingency dynamic responses of four different MC iterations are presented in Figures 6.4–6.7. These figures depict the simulation results of one non-vulnerable case, one transient unstable case, one voltage unstable case, and one frequency unstable case, respectively. It is possible to observe in the figures the particular dynamic signal behavior for each one of the cases, which exposes the existence of certain patterns revealing the actual system security level, as well as the future system-vulnerability-status tendency. These cases will be used as illustrative examples throughout this chapter.

6.4.1 Transient Stability Index (TSI)

The real-time transient stability index (TSI) presented in [10] can be used as a KPI for transient stability. This index is determined via the prediction of area-based center-of-inertia-(COI)-referred rotor angles using PMU measurements as input data, and is capable of giving a quick quantification of the actual TS level. The method uses an intelligent COI-referred rotor angle regressor based on Monte Carlo simulation, a support vector regressor (SVR), and PMU data as input. This regressor is used for obtaining the COI-referred rotor angles directly from PMU measurement in real time.

For real-time implementation, the previously off-line trained SVR [10] will be in charge of estimating the COI-referred rotor angles for each area associated to the PMUs, using the actual post-contingency PMU voltage phasors as inputs. Once the associated PMU COI-referred rotor angles are estimated, the recursive clustering

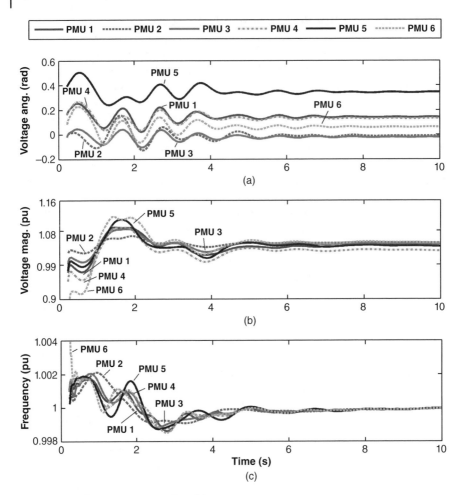

Figure 6.4 39-bus-system non-vulnerable case.

approach for determining coherent electrical areas is first applied to these angles in order to determine the boundaries of the real-time coherent areas, as explained in section 6.3. After the definition of real-time areas, new area-COI-referred rotor angles will be computed, representing the real-time coherency. The expressions that permit computation of these new COI-referred rotor angles are presented by (6.2) and (6.3):

$$\delta_k^{COI_{System}} = \frac{1}{M_{k_T}} \sum_{j=1}^{v} M_j \delta_j^{COI_{System}} \tag{6.2}$$

$$M_{k_T} = \sum_{j=1}^{v} M_j \tag{6.3}$$

where $\delta_j^{COI_{System}}$ is the estimated COI-referred rotor angle of the associated PMU j-th area, v is the number of associated PMU areas that belong to the real-time area k determined by the real-time recursive coherency method, and M_j is the total inertia moment of the

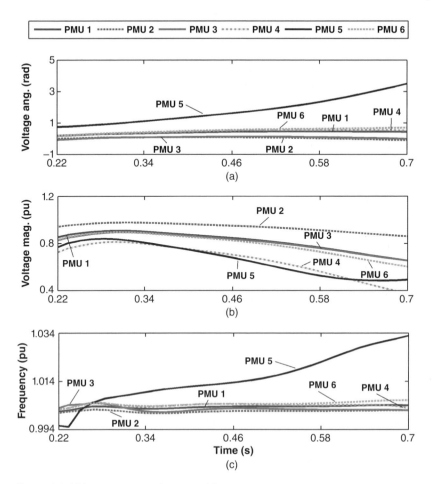

Figure 6.5 39-bus-system transient unstable case.

j-th associated PMU area, whereas M_{kT} and $\delta_k^{COI_{System}}$ denote total inertia moment and the new COI-referred-rotor angle of the k-th real-time coherent area.

Finally, using the $\delta_k^{COI_{System}}$ computed value and the TSA criterion presented in [10], it is possible to establish a TSI in the range of [0, 1] for each k-th area as shown by (6.4). This TSI_k will be computed after each set of PMU data arrives to the control center and might be used as an indicator of the area being imminently out-of-step:

$$TSI_k = \begin{cases} 0 & \text{if } \left| \delta_k^{COI_{System}} \right| < \delta_{lim} \\ \dfrac{\left| \delta_k^{COI_{System}} \right| - \delta_{lim}}{\pi - \delta_{lim}} & \text{if } \delta_{lim} \leq \left| \delta_k^{COI_{System}} \right| \leq \pi \\ 1 & \text{if } \left| \delta_k^{COI_{System}} \right| > \pi \end{cases} \quad (6.4)$$

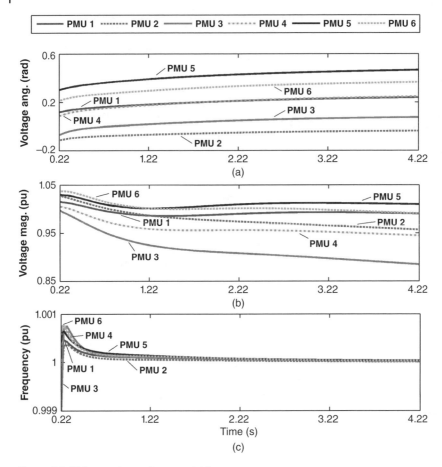

Figure 6.6 39-bus-system voltage unstable case.

where TSI_k, and $\delta_k^{\text{COI}_{\text{System}}}$ are the TS index and COI-referred rotor angle in radians of the k-th real-time area. Since TSI is conceived to give early warning, it is not necessary to compute it when COI-referred rotor angles are within a normal operating range. Then, the maximum admissible steady state COI-referred rotor angle (i.e., the maximum allowed angle determined by steady state constraints—δ_{lim}) is used as the lower limit of the TSI function.

As an illustration, the COI-referred rotor angles for the unstable case presented in Figure 6.5 are estimated [10]. Figure 6.8 shows a comparison between the computed (solid lines) and the estimated (dotted lines) associated PMU COI-referred rotor angles for this case.

6.4.2 Voltage Deviation Index (VDI)

In the context of vulnerability, it is a fact that bus voltages are good indicators of the health of the system. So, a voltage deviation index (VDI) can be used in order to assess vulnerability. This index has to be calculated when each phasor measurement arrives at the control center.

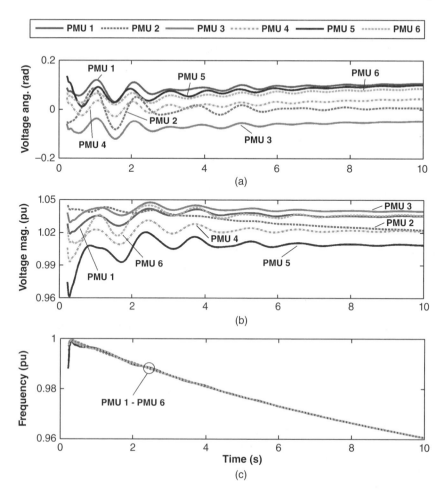

Figure 6.7 39-bus-system frequency unstable case.

Since the defined vulnerability assessment timeframe covers only the short-term voltage stability phenomena, the proposed index is intended to reflect the power system voltage performance as regards the voltage dynamic period.

During this dynamic period, the under and over-voltage relays (VR) might operate when specific vulnerable conditions are reached. In this connection, the defined VDI resembles the operating characteristics of this type of relay, which is triggered when the voltage drops below a threshold value V_{lower} (or exceeds a threshold value V_{upper}) during more than a pre-defined period of time (tv_{max}). Figure 6.9 schematizes the definition of an under-voltage VR triggering operation.

Using the previously defined VR characteristics, it is possible to define a VDI for each bus where PMUs are located, as follows:

$$\text{VDI}_i = \begin{cases} 0 & \text{if } V_{lower} \leq V \leq V_{upper} \\ \min\left\{1, \frac{t_n}{tv_{max}}\right\} & \text{if } V < V_{lower} \vee V > V_{upper} \end{cases} \tag{6.5}$$

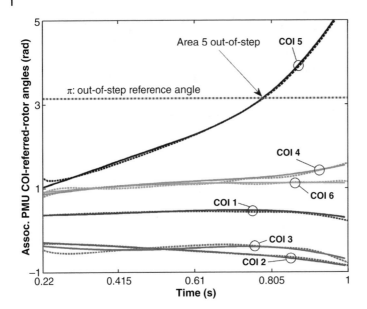

Figure 6.8 Associated PMU COI-referred rotor angles for transient unstable case of Figure 6.5.

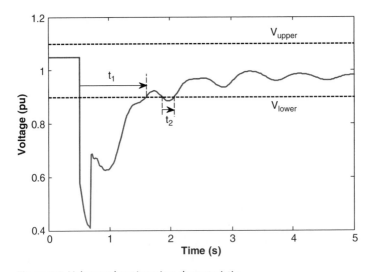

Figure 6.9 Voltage relay triggering characteristic.

where V is the instant i-th PMU voltage, V_{lower} and V_{upper} are the voltage threshold values, tv_{max} is the maximum pre-defined period of time before the VR triggers, and t_n is the period of time in which $V < V_{lower}$ or $V > V_{upper}$.

Then, the VDI for each real-time electric area k is determined by the following expression:

$$VDI_k = \max\{VDI_i | PMU\, i \in Electric\, Area\, k\} \tag{6.6}$$

6.4.3 Frequency Deviation Index (FDI)

The frequency deviation from its rated value is a clear indicator of the dynamic effect produced by the contingency. As noted in [8], the higher the frequency deviation, the bigger the effect produced by the contingency.

Using this concept, an approach for assessing vulnerability in each area is used based on the fact that frequency can also be measured in PMUs, and these data might be used in control centers with similar updating periods and phasors.

In order to assess the effect of the frequency behavior, the frequency measured in each PMU is assessed, since the frequency is not the same in all system buses during dynamic conditions.

Typical frequency threshold value settings are 57–58.5 Hz for a 60 Hz system and 48–48.5 Hz for a 50 Hz system (Δ_{fmax} from 2.5% to 5%) [15]. Similarly, over frequency protection of generators usually has a threshold value of 61.7 Hz (Δ_{fmax} around 2.8% for a 60 Hz system) [16].

Using this information, a Frequency Deviation Index (FDI) for each PMU can be calculated as follows:

$$FDI_i = \min\left\{1, \frac{|\Delta f_i|}{\Delta f_{max}}\right\} \qquad (6.7)$$

where $|\Delta f_i|$ is the frequency variation measured in PMU i, and Δf_{max} is the maximum admissible system frequency deviation. Notice that Δf_{max} changes depending on the frequency phenomenon. That is, if $\Delta f_i < 0 \Rightarrow \Delta f_{max}$, this corresponds to an under frequency limit; otherwise, if $\Delta f_i > 0 \Rightarrow \Delta f_{max}$ corresponds to an over frequency limit.

Then, using the previous index per PMU, the FDI for each real-time electrical area k is determined as follows:

$$FDI_k = \max\{FDI_i | PMU\,i \in \text{Electrical Area}\,k\} \qquad (6.8)$$

6.4.4 Assessment of TVFS Security Level for the Illustrative Examples

In this section, the computation of each TVFS performance indicators is illustrated in order to show the real-time application of the stage involved with the evaluation of the system security level. For instance, the TSI, VDI, and FDI are computed for each one of the unstable cases presented in Figures 6.5, 6.6, and 6.7, respectively. First, the real-time coherency analysis is performed in order to partition the system into specific coherent zones (by clustering the associated PMU electric areas).

The TVFS indices per zone for each one of these unstable cases are presented in Figure 6.10 (Zone A = {Area 1, Area 4, Area 6}, Zone B = {Area 2, Area 3}, and Zone C = {Area 5}), Figure 6.11 (Zone A = {Area 1, Area 2, Area 3, Area 6}, Zone B = {Area 4}, Zone C = {Area 5}), and Figure 6.12 (Zone A = {Area 1, Area 2, Area 3, Area 6}, Zone B = {Area 4}, Zone C = {Area 5}).

Note in the figures how each indicator reaches the value of "1" at 0.72 s, 3.5 s, and 15 s, respectively. At this moment, the assessment of TVFS security level gives an alert of the system collapse. Nevertheless, at the instant in which the alert is given, the system is already collapsing. Thus, the need for an early warning cannot be achieved simply by the evaluation of the actual security level by means of performance indicators. In this

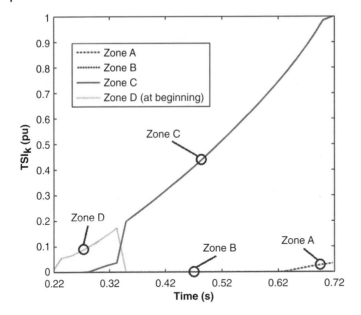

Figure 6.10 TSI for transient unstable case of Figure 6.5.

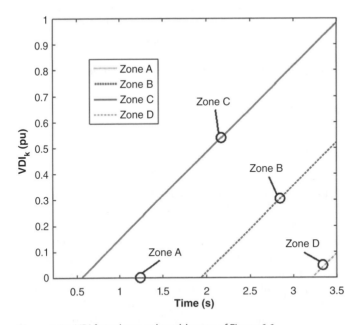

Figure 6.11 VDI for voltage vulnerable case of Figure 6.6.

connection, the use of a prediction mechanism is essential in order to be able to apply any corrective action before the system collapses, which will be incorporated by the inclusion of the vulnerability status prediction results, presented in Chapter 5. This aspect will be analyzed in the next subsection.

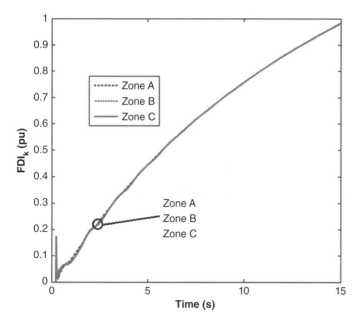

Figure 6.12 FDI for frequency vulnerable case of Figure 6.7.

6.4.5 Complete TVFS Real-Time Vulnerability Assessment

In this subsection, the previously defined TVFS indices (i.e., TSI, VDI, and FDI) are combined with the results obtained from the TVFS vulnerability status prediction depicted in Chapter 5 with the aim of performing both tasks involved in the vulnerability concept, that is: assessment of the actual security level, and of the system's tendency to reach a critical state.

For this purpose, a "penalty" is included into the indices depending on the result obtained from the vulnerability status prediction, taking into account the time window where the vulnerability status has been predicted and adequate arrays of logic gates. This penalty has the aim of forcing the corresponding TVFS index to the value of "1" when the vulnerability status pattern recognition results in an alert about the system's tendency to change its conditions to a critical state (i.e., when the vulnerability status prediction is set to "1"). In order to increase security, this penalty is applied only if the corresponding TVFS index has exceeded a minimum pre-specified value.

In order to determine the time windows where two or more different phenomena might occur, the statistical results of the relay tripping times and the pre-defined time windows (TWs) are used (according to explanation given in Chapter 5).

From the analysis for the 39 bus test system (see Chapter 5), TW_1 and TW_2 allow predicting the transient unstable cases; TW_3 includes the information of voltage vulnerable cases; while TW_4 and TW_5 contain the patterns belonging to vulnerability caused by voltage or frequency phenomena. This analysis is summarized in Figure 6.13, whereas the corresponding defined logic schemes are presented in Figure 6.14.

After specifying the logic schemes, the TVFS performance indicators are again computed.

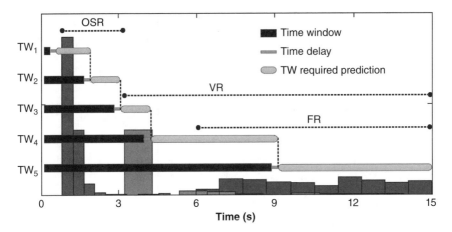

Figure 6.13 Time window analysis for structuring the 39-bus-system logic schemes.

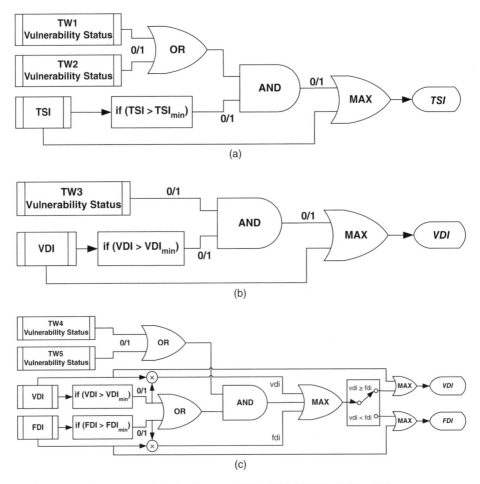

Figure 6.14 39-bus-test system logic schemes: a) $TW_1 \vee TW_2$, b) TW_3, c) $TW_4 \vee TW_5$.

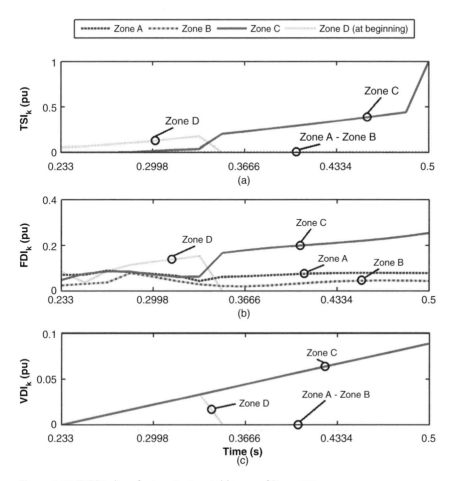

Figure 6.15 TVFS indices for transient unstable case of Figure 6.5.

The behavior of the TVFS indices for the transient unstable case is presented in Figure 6.15. Note how the real-time coherency identification method joins the pre-specified associated PMU coherent areas into a new set of coherent zones, which change during event evolution. At the beginning, four different zones are determined; but after seven samples (at around 0.34 s), the iterative algorithm improves the results, suggesting the formation of three coherent zones, which present the best final coherency distribution. Then, for this specific case, the system is partitioned into three coherent zones at the moment of vulnerability detection, that is: Zone A = {Area 1, Area 4, Area 6}, Zone B = {Area 2, Area 3}, and Zone C = {Area 5}. Figure 6.15 (a) shows how the TSI of Area C reaches the value of "1" at 0.5 s after the corresponding application of the penalty when the vulnerability status is set to "1" by the TW_1 status prediction classifier. In the figure, the slope change at around 0.48 s is a product of this penalty application. Since in this case the loss of synchronism occurs at 0.9854 s (tripping of out-of-step relay OSR), the operator will have around 480 ms to perform any corrective control action.

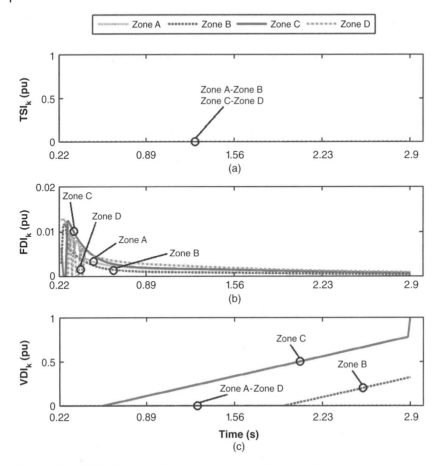

Figure 6.16 TVFS indices for voltage vulnerable case of Figure 6.6.

Likewise, Figure 6.16 shows the TVFS vulnerability indicators corresponding to the voltage vulnerable case presented in Figure 6.6. In this case, the real-time area coherency iterative algorithm orients the formation of three or four different coherent zones at the beginning, from which the final coherency distribution corresponds to four coherent zones. These zones are: Zone A = {Area 1}, Zone B = {Area 2}, Zone C = {Area 3}, and Zone D = {Area 4, Area 5, Area 6}.

The VDI of Zone C reaches the value of 1 at 2.9 s, when the vulnerability status prediction corresponding to TW3 warns about the voltage vulnerability risk. Considering that the voltage relay VR tripped at 3.42 s, the methodology allows alerting about the vulnerability risk 520 ms before it occurs.

Figure 6.17, in contrast, presents the TVFS vulnerability indices corresponding to the frequency vulnerable case of Figure 6.7. At this instant, the real-time area coherency identification detects the formation of three different coherent zones from the beginning. These zones are: Zone A = {Area 1, Area 2, Area 3, Area 6}, Zone B = {Area 4}, and Zone C = {Area 5}. The FDI of Zone A reaches the value of 1 at 9.2 s, when the vulnerability status prediction corresponding to TW_5 alerts about the frequency vulnerability

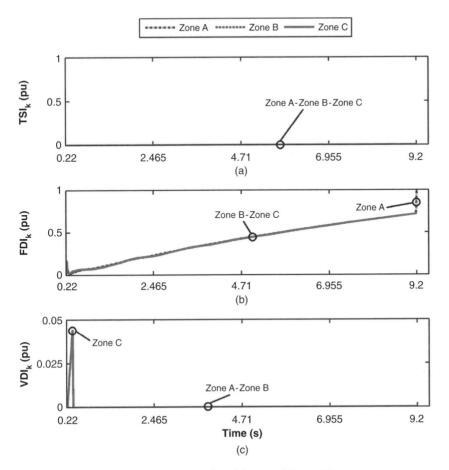

Figure 6.17 TVFS indices for frequency vulnerable case of Figure 6.7.

risk. In this case, the TVFS vulnerability assessment warns about vulnerability 3.71 s before it occurs due to the corresponding frequency relay (FR) tripping at 12.912 s.

6.5 Slower Phenomena Vulnerability Performance Indicators

Two vulnerability symptoms are characterized by timeframes that vary from seconds to several minutes, and even hours. These symptoms correspond to system stability regarding electromechanical oscillations and power system component overloads.

Since the timeframes of oscillatory stability and overloads are larger than the corresponding TVFS timeframes, these phenomena have been treated slightly different in this chapter, and will be described in the following subsections.

6.5.1 Oscillatory Index (OSI)

Oscillations may be identified in electrical signals after a perturbation has occurred. These electrical signals can be decomposed in their oscillatory modes using measurement-based modal identification, which is perhaps the most appropriate option to quickly estimate the parameters associated to system critical modes, and allows having a clear picture of the actual system damping performance. To date, there has been a significant development of different types of mode estimators, which can be grouped into ringdown (linear), mode-meter (ambient), and nonlinear/non-stationary analysis methods [17].

Ringdown methods (e.g., Prony analysis [18]) are based on the assumption that the signal under analysis resembles a sum of damped sinusoids. Such kinds of signal, which can be easily distinguished from ambient noise, arise following a large disturbance (e.g., adding/removing large loads, generator tripping, and severe short circuits). Mode meters are primarily tailored to ambient data, which results from random variations of small load switching around system steady state conditions. So far, mode meters have been subdivided into parametric, non-parametric, block processing, and recursive methods. On the other hand, nonlinear/non-stationary analysis methods, such as, for instance, the Hilbert–Huang Transform or Continuous Wavelet Transform (CWT) based methods [17, 19], or even other algorithms developed for this purpose (such as the WAProtector™ client-server application used by the Ecuadorian ISO—CENACE [20]), can be employed to track the time-varying attributes of modal parameters. The application of these tools is of particular relevance in today's power system operational context in order to enable the system operators to be aware of potential small-signal stability problems.

As is well-known, the damping ratio associated with dominant system modes can vary significantly depending on different factors, such as, for instance, grid strength, generator operating points, loading conditions and the number of power transfers [21]. Thus, continuous evaluation of the system damping performance should be performed in order to capture the operating conditions which could entail major damping concerns.

Since the fast-phenomena assessment presented in Section 6.4 might not detect oscillatory phenomena, it is necessary to continuously carry out oscillatory supervision by applying one of the mathematical tools listed above. Therefore, in order to complement the TVFS vulnerability assessment proposed in this chapter, Prony analysis is used to estimate the frequency, damping, and magnitude of dominant modes that are observable in a given real-time electrical signal.

Prony analysis is one of the most useful methods to analyze the frequency modes buried in an electrical signal due to its high efficiency for system dynamic component estimation [18]. However, it is necessary to avoid the effect of nonlinear relations and fast transients present in the signal, using for example a Discrete Wavelet Transform (DWT) [19]. Prony analysis also needs a specific data window depending on the frequency of the modes present in the signal. This problem can be overcome by defining the window length using the results of an off-line modal analysis that reveals the typical system mode frequency.

By applying Prony analysis, it is possible to analyze the damping of each mode in real time, in order to define the modes that have poor or negative damping, which are called the critical modes. Once critical modes have been identified for each area, an index that represents the possible damping problems in the system can be calculated.

For this purpose, it is necessary to define a threshold damping (ζ_{lim}) for which the system is considered in some level of oscillatory problem. The selected threshold damping depends on system expert knowledge and agreed operating policies.

For instance, Argentina's Wholesale Electricity Market Management Company (CAMMESA) specifies the damping thresholds for inter-area mode oscillations as 15% or higher for normal operating conditions, 10% or higher for critical (highly loaded) operating conditions or N-1 operating conditions, and 5% or higher for post-contingency operating conditions. For local mode oscillations, the damping threshold is 5% or higher for any operating conditions [22].

For illustrative purposes, the threshold damping has been specified as 5% for the test power system used in this chapter, based on the discussed post-contingency requirement. In addition, the chosen electrical signal is the branch injection active power, belonging to the buses where PMUs have been located, that presents the highest current probabilistic measure of observability (PO_{gm}) indices proposed in [14]. This injection power can be easily calculated using the voltage and current phasors measured by the PMU.

Using the previously specified data, an oscillatory index (OSI) for the poorest damped mode, obtained from application of Prony analysis to power measured at each PMU, can be defined by a function, as follows:

$$OSI_i = \begin{cases} 1 & \zeta_i \leq 0 \\ -\dfrac{1}{\zeta_{lim}}\zeta_i + 1 & 0 < \zeta_i < \zeta_{lim} \\ 0 & \zeta_i \geq \zeta_{lim} \end{cases} \tag{6.9}$$

where $\zeta_i = \min\{\zeta \text{ of } P_i \text{ modes}\}$ is the poorest damped mode obtained from application of Prony analysis to power measured P_i at each PMU i, and ζ_{lim} is the system threshold damping.

Then, the OSI for each electrical area k is determined as follows:

$$OSI_k = \max\{OSI_i | \text{PMU} i \in \text{Electrical Area} k\} \tag{6.10}$$

For illustrative purposes, the OSI is computed for the 39 bus test system. From a total number of 10 000 simulated MC cases, 84 instances were detected as oscillatory unstable. These cases are used to show the behavior of the proposed OSI. First, Prony analysis is applied to the post-contingency power injections at each PMU. After this, OSIs are computed for each PMU.

For instance, Figure 6.18 presents histograms of the determined dominant modes at PMU 3, for the 84 analyzed cases, respectively, where (a) corresponds to the frequency of the estimated inter-area modes, (b) represents the damping ratio of the estimated inter-area modes, (c) depicts the frequency of the estimated local modes, and (d) shows the damping ratio of the estimated local modes. Similar analysis is developed for each PMU.

Table 6.2 presents a summary of the estimated oscillatory modes that shows the distribution of the poorly-damped modes for each PMU in terms of the cumulative distribution function (i.e., $P(\zeta_i < 5\%)$). It can be noted that, in most of the cases, the inter-area modes are the least-damped modes. Additionally, note that this type of oscillatory mode is observable in all PMUs, showing more notoriety in PMUs 4, 5, and 6 as these PMUs show the largest $P(\zeta_i < 5\%)$ values (highlighted in Table 6.2).

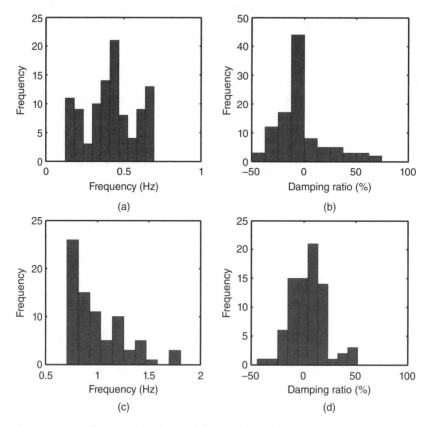

Figure 6.18 Oscillatory modes detected at PMU 3: (a, b) inter-area, (c, d) local.

Table 6.2 Distribution of the poorly-damped modes per each PMU.

PMU	Inter-area modes		Local modes	
	mean{f} (Hz)	$P(\zeta_i < 5\%)$ (%)	mean{f} (Hz)	$P(\zeta_i < 5\%)$ (%)
PMU 1	0.5074	78.16	1.2033	38.53
PMU 2	0.4140	83.53	1.0482	68.13
PMU 3	0.4213	78.43	0.9868	53.16
PMU 4	0.4706	91.03	1.1433	52.17
PMU 5	0.5547	93.59	1.1005	48.45
PMU 6	0.4735	86.42	1.1944	83.62

Also, there are cases presenting poorly-damped local modes. These types of modes are mainly observable in PMU 6 as shown by its high $P(\zeta_i < 5\%)$.

From the results, it is possible to appreciate the existence of cases where both critical modes (i.e., local and inter-area) appear poorly damped. This aspect suggests an overlap

Table 6.3 OSIs per PMU—Summary of number of cases.

PMU	OSI = 0	0 < OSI < 1	OSI = 1
PMU 1	11	8	65
PMU 2	10	7	67
PMU 3	5	3	76
PMU 4	5	3	76
PMU 5	5	5	74
PMU 6	8	1	75

between cases belonging to each critical mode. In these instances, it is not easy to discern which mode is responsible for the instability issue. Nevertheless, whichever mode is predominant, the instability problem has been detected by the modal identification technique, which is the main purpose of the vulnerability assessment. Thus, the modal damping results will be then used in order to compute the proposed oscillatory index (OSI). In this connection, once critical mode estimation has been carried out, the corresponding OSIs are computed, which allow grading the level of system vulnerability as regards oscillatory threats.

Table 6.3 shows a summary of the number of cases that correspond to three OSI ranges. In most of the cases, OSI gives adequate early warning as regards the actual occurrence of oscillatory issues, reaching the value of 1. It is worth mentioning that in 83 cases, at least 1 PMU provides an OSI equal to 1, which highlights the excellent performance of OSI in alerting to oscillatory instability. Moreover, the only case where OSIs do not reach the value of 1, the OSI of PMU 5 attains the value of 0.5006, which also gives an alert of system oscillatory risk.

6.5.2 Overload Index (OVI)

While much work has been directed toward to development of methods for assessing the causes of vulnerability regarding stability issues (TVFS and oscillatory stability), the possible overloads have often been treated as negligible in vulnerability assessment tasks. However, sometimes high electric post-contingency currents might provoke overloads which could increase the system vulnerability problem [8]. So, a system can be stable following a contingency, yet insecure due to post-fault system conditions resulting in equipment overloads [23]. Some approaches have tackled the problem of overload using off-line vulnerability assessment via estimation of the overloads by means of distribution factors (dc-DFs) [24]. These dc-DFs show the approximate line flow sensitivities (DC power flow assumptions) due to a system topology change.

Using the same concept of distribution factors, overloads might be estimated in real time. However, since dc-DFs are based on DC power flow linear approximations, dc-DF-based overload estimation would present significant errors, which are not admissible for real-time decision making applications. To overcome this drawback, a methodology to estimate the possible overloads in real time based on statistical ac-DFs (SDFs), which depend on the system response to different type of contingencies (i.e., power injection changes or branch outages) considering most of the probable operating scenarios (i.e., MC-based simulation), has been presented in [23]. Afterward, the proposed approach is combined with a knowledge-based intelligent classifier to structure the function of a real-time overload estimator using the pre-contingency operating data which can be obtained from PMUs and/or a SCADA/EMS. Finally, an overload index (OVI) is computed in real time using the results of the SDF-based overload estimation [23].

First, using the probabilistic models of input parameters based on a short-term operating scenario and via optimal power flow (OPF) computations, MC-based AC contingency analysis is performed to iteratively calculate ac-DFs. After this, SDFs are defined by the mean and standard deviation of ac-DFs, considering two types of filters that permit improving the accuracy of SDFs. For real-time implementation, these SDFs are then joined with an intelligent classifier (based on principal component analysis (PCA) and the support vector classifier (SVC)) in order to structure a table-based post-contingency overload estimation algorithm, which allows the computing of OVIs depending on the pre-contingency operating state and the actual contingency [23].

MC-based simulation is used as a method to obtain pre and post-contingency power flow results (i.e., contingency analysis), considering several possible operating conditions and N-1 contingencies (i.e., branch outages and variation in power injections). These considerations allow inclusion of the most severe events that could lead the system to potential overload conditions, and further N-2 contingencies, which are considered as the beginning of a cascading event.

The Injection Shift Factor (ISF) ac-$\psi^i_{e_k}$ of a branch e_k and the Line Outage Distribution Factor (LODF) ac-$\varsigma^{(eq)}_{ek}$ are calculated using the results of each MC simulation (i.e., results from AC power flows: ac-DFs), as shown in (6.11) and (6.12) [23]:

$$\text{ac-}\psi^i_{e_k} = \frac{f_{e_k} - f_{e_k\text{pre-c}}}{\Delta S_i} \tag{6.11}$$

$$\text{ac-}\varsigma^{(e_q)}_{e_k} = \frac{f_{e_k} - f_{e_k\text{pre-c}}}{f_{e_q\text{pre-c}}} \tag{6.12}$$

where k and q are branches, f_{ek} is the post-contingency apparent power flow, $f_{ekpre-c}$ is the pre-contingency apparent power flow, and ΔS_i is the change of apparent power injection in bus i.

Then, the SDFs are calculated based on the probabilistic attributes represented by the mean and standard deviation of all MC-based DFs (i.e., ISFs and LODFs, respectively):

$$\left. \text{SDF}^j_k \right|_{\text{ac-DF}=\text{ac-}\psi^{j=i}_{e_k} \text{ or ac-}\varsigma^{j=(e_q)}_{e_k}} = \begin{cases} \displaystyle\operatorname*{mean}_{m=1\ldots n} \left\{ \left. \text{ac-DF}^j_k \right|_m \right\} \\ \displaystyle\operatorname*{std}_{m=1\ldots n} \left\{ \left. \text{ac-DF}^j_k \right|_m \right\} \end{cases} \tag{6.13}$$

where ac-DF^j_k is the distribution factor (ISF or LODF) of branch k provoked by contingency j (i.e., variation of power injection or branch outage), and n is the number of scenarios (m).

Since ac-DFs directly depend on each operating scenario, the complete set of short-term possible scenarios have to be previously filtered in order to improve the accuracy of SDFs (i.e., more representative mean and lower standard deviation), with the aim of obtaining more robust results. In this connection, two types of scenario filters are proposed in [23]: (i) to establish representative clusters of scenarios depending on operating similitudes, and (ii) to reduce the sample of scenarios to those with the highest probability of overload.

Once SDFs have been calculated, a table for real-time implementation has to be structured. This table includes the computed SDFs, structuring a three-dimensional matrix (or table) whose dimensions represent: (i) the branch that shows the power flow sensitivity, (ii) the type and location of the contingency, and (iii) the clusters of scenarios. For real-time applications, a classifier has to be capable of orienting the selection of the best SDFs considering the pre-contingency operating state (that can be obtained from a SCADA/EMS) and the actual contingency (that can be quickly alerted by local IEDs) [23].

After a contingency occurs, re-distributed flows (post-contingency steady state power flow f_{ek}) are estimated in each branch using the corresponding SDFs selected from the table depending on the SVC classification results, as follows:

$$
f_{e_k} = \begin{cases} f_{e_k \text{ before}} + \Delta S_i \times SDF_k^j \Big|_{\text{ac-DF=ac-}\psi_{e_k}^{j=i}} & \text{if contingency is Power Injection} \\ & \text{Change in bus } i \\ f_{e_k \text{ before}} + f_{e_q \text{ before}} \times SDF_k^j \Big|_{\text{ac-DF=ac-}\varsigma_{e_k}^{j=(e_q)}} & \text{if contingency is Branch } e_q \text{ outage} \end{cases}
$$

(6.14)

Since there are, in general, three categories of power transfer limits: thermal, angle stability, and voltage, branch overload threshold values have to be pre-defined for each system [23]. Thus, it is necessary to pre-establish the power transfer limit (restricted by one or more of the three limits) corresponding to each element of the system. Then, an overload risk band, with upper and lower limits (S_{upper} and S_{lower}), can be structured in order to calculate an overload index (OVI) [23]. Therefore, once the re-distributed flows are estimated, an overload index (OVI) can be calculated for each branch using the function defined by (6.15):

$$
OVI_{e_k} = \begin{cases} 0 & f_{e_k} \leq S_{lower} \\ \dfrac{1}{S_{upper} - S_{lower}}(f_{e_k} - S_{lower}) & S_{lower} < f_{e_k} < S_{upper} \\ 1 & f_{e_k} \geq S_{upper} \end{cases}
$$

(6.15)

where S_{upper} and S_{lower} are the upper and lower limits of the overload risk band for each grid element.

Then, the OVI for each electrical area is determined as follows:

$$
OVI_k = \max\{OVI_{e_k} | \text{Branch } e_k \in \text{Electrical Area } k\}
$$

(6.16)

For illustrative purposes, a MC-based contingency analysis simulation of the 39 bus test system, consisting of 10 000 different operating states, has been performed. The output constitutes the pre and post-contingency AC power flow results, which are used to calculate the corresponding ac-DFs. Based on the MC results and the pre-defined overload risk bands, the most critical branches are established. In this case, twelve edges present risk of overload, and these are the critical branches for which SDFs have to be calculated. After analysis, six clusters of operating scenarios have been defined for which the SDFs are computed, getting a maximum std{ac-DF} of 0.1. In the cases where SDFs are not computed, the variation of power flow does not surpass the filter limit, so it does not provoke overloads; hence the corresponding SDFs are not included in the SDF-table for real-time assessment. In these cases, the OVIs are automatically set to zero. Table 6.4 shows some of the computed SDFs (i.e., mean{ac-DF}) that correspond to different branch outages.

Table 6.4 SDFs resulting from branch outages.

Outage Branch (e_q)	Branch (e_k)	Cluster / SDF (mean{ac-DF})					
		1	2	3	4	5	6
L15-16	L03-04	0.793	0.809	0.792	0.812	0.798	-
L01-02	L03-04	0.664	0.706	0.709	-	0.645	-
L22-23	L23-24	0.317	0.399	0.185	-	0.265	-
T06-31	L21-22	0.014	0.008	0.359	-	0.349	-
T06-31	L15-16	0.102	-	0.289	-	0.285	-
L01-02	L15-16	0.331	-	0.325	-	0.308	-
L04-14	L03-04	0.314	-	-	-	0.324	-
L08-09	L15-16	0.352	-	-	-	0.337	-
T10-32	L15-16	0.052	-	-	-	0.044	-
L16-24	L21-22	1.029	1.036	1.036	1.009	0.975	-
L16-21	L23-24	1.016	1.017	1.018	1.003	1.013	0.988
L21-22	L23-24	-	-	1.009	0.998	1.006	0.979

Table 6.5 OVIs for Branch Outages—Summary of number of cases.

Branch (e_k)	OVI = 0		$0 < OVI \leq 0.5$		$0.5 < OVI \leq 1$	
	Sim.	SDF est.	Sim.	SDF est.	Sim.	SDF est.
L03-04	234,994	234,993	6	7	0	0
L05-06	234,993	234,993	7	7	0	0
L05-08	234,986	234,983	14	17	0	0
L06-07	234,986	234,986	14	14	0	0
L07-08	235,000	235,000	0	0	0	0
L14-15	234,943	234,945	57	55	0	0
L15-16	233,952	233,959	972	965	76	76
L16-17	233,224	233,226	1293	1291	483	483
L16-21	234,458	234,479	542	521	0	0
L17-18	235,000	235,000	0	0	0	0
L21-22	231,681	231,682	2876	2875	443	443
L23-24	234,718	234,717	282	283	0	0

In real-time, after a contingency occurs, the SDFs will estimate the possible branch overload. Once estimation has been completed, the final step is to calculate the corresponding OVIs, which permit ranking the level of system vulnerability as regards overloads. These indices might orient the selection of corrective control actions (e.g., focused load shedding) if needed. Table 6.5 shows a summary of the number of cases that correspond to three OVI ranges. Most of the cases show excellent accuracy in estimation, mainly in the range (0.5, 1] where possible corrective control actions might be required.

6.6   Concluding Remarks

The vulnerability assessment methodology presented in this chapter allows assessing system vulnerability based on the fact that vulnerability begins to develop in specific areas of the system. In this connection, the grid is partitioned into coherent electrical areas via clustering techniques, and each one is ranked depending on several performance indices regarding vulnerability. The performance indices or indicators reflect the actual system stress level as regards five different post-contingency phenomena: transient instability, short-term voltage instability, frequency deviations outside limits (TVFS phenomena), poorly-damped power oscillations, and overloads.

Additionally, the second aspect regarding the vulnerability concept, that is the system's tendency to reach a critical state, has been included by means of a penalty into the TVFS indices (TSI, VDI, and FDI) via a scheme of logic gates.

Likewise, the possible overloads have been assessed in this chapter via an overload index (OVI). This methodology has been structured based on statistical distribution factors (SDFs), which permit the prediction of the post-contingency re-distribution of the power flow in each system branch.

On the other hand, since oscillatory stability is a slower phenomenon, its assessment does not strictly need a prediction of the future system condition. For this phenomenon, the application of modal identification techniques (such as Prony analysis) provides good enough results to giving early warning. This type of evaluation has to be performed as a complementary assessment of system vulnerability. In this connection, an oscillatory index (OSI) has been presented in this chapter.

The complete real-time post-contingency vulnerability assessment methodology depicted in this chapter gives five indicators of system vulnerability as regards different phenomena. Then, it allows identification of the phenomena involved in system vulnerability and the grid zone where the system stress is high. Therefore, this methodology might be easily used as a signal for triggering focused corrective control strategies that enable improvements in post-contingency system security.

References

1 U. Kerin, G. Bizjak, E. Lerch, O. Ruhle, and R. Krebs, "Faster than Real Time: Dynamic Security Assessment for Foresighted Control Actions", 2009 IEEE Bucharest Power Tech Conference, June 28–July 2, Bucharest, Romania.

2 J. D. McCalley, and Fu Weihui, "Reliability of Special Protection Systems", *IEEE Transactions on Power Systems*, IEEE Power & Energy Society, vol. 14, no 4, pp. 1400–1406, November 1999.

3 M. Amin, "Toward Self-Healing Infrastructure Systems", *Electric Power Research Institute (EPRI), IEEE, Computer*, vol. 33, no 8, pp. 44–53, 2000.

4 I. Dobson, P. Zhang, et al, "Initial Review of Methods for Cascading Failure Analysis in Electric Power Transmission Systems", IEEE PES CAMS Task Force on Understanding, Prediction, Mitigation and Restoration of Cascading Failures, IEEE Power Engineering Society General Meeting, Pittsburgh, PA USA July 2008.

5 D. McGillis, K. El-Arroudi, R. Brearley, and G. Joos, "The Process of System Collapse Based on Areas of Vulnerability", Large Engineering Systems Conference on Power Engineering, pp. 35–40, Halifax, NS, July 2006.

6 Z. Huang, P. Zhang, et al, "Vulnerability Assessment for Cascading Failures in Electric Power Systems", Task Force on Understanding, Prediction, Mitigation and Restoration of Cascading Failures, IEEE PES Computer and Analytical Methods Subcommittee, IEEE Power and Energy Society Power Systems Conference and Exposition 2009, Seattle, WA.

7 A. Fouad, Qin Zhou, and V. Vittal, "System Vulnerability As a Concept to Assess Power System Dynamic Security", *IEEE Transactions on Power Systems*, vol. 9, no 2, May 1994, pp. 1009–1015.

8 J. Gimenez, and P. Mercado, "Online Inference of the Dynamic Security Level of Power Systems Using Fuzzy Techniques", *IEEE Transactions on Power Systems*, vol. 22, no. 2, pp. 717–726,May 2007.

9 S. C. Savulescu, et al, *Real-Time Stability Assessment in Modern Power System Control Centers*, IEEE Press Series on Power Engineering, Mohamed E. El-Hawary, Series Editor, a John Wiley & Sons, Inc., Publication, 2009.

10 J. Cepeda, J. Rueda, G. Colomé, and D. Echeverría, "Real-time Transient Stability Assessment Based on Centre-of-Inertia Estimation from PMU Measurements", *IET Generation, Transmission & Distribution*, vol. 8, issue 8, pp. 1363–1376, August, 2014.

11 J. Cepeda, J. Rueda, G. Colomé, and I. Erlich, "Data-Mining-Based Approach for Predicting the Power System Post-contingency Dynamic Vulnerability Status", *International Transactions on Electrical Energy Systems*, vol. 25, issue 10, pp. 2515–2546, October 2015.

12 J. Cepeda, and G. Colomé, "Vulnerability Assessment of Electric Power Systems through identification and ranking of Vulnerable Areas", *International Journal of Emerging Electric Power Systems*, vol. 13, issue 1, May 2012.

13 I. Kamwa, A. K. Pradham, G. Joos, and S. R. Samantaray, "Fuzzy Partitioning of a Real Power System for Dynamic Vulnerability Assessment", *IEEE Transactions on Power Systems*, vol. 24, no. 3, pp. 1356–1365, August 2009.

14 J. Cepeda, J. Rueda, I. Erlich, and G. Colomé, "*Probabilistic Approach-based PMU placement for Real-time Power System Vulnerability Assessment*", 2012 IEEE PES Conference on Innovative Smart Grid Technologies Europe (ISGT-EU), Berlin, Germany, October 2012.

15 K. Seethalekshmi, S. N. Singh, and S. C. Srivastava, "*WAMS Assisted Frequency and Voltage Stability Based Adaptive Load Shedding Scheme*", IEEE PES General Meeting, Calgary, pp. 1–8, 2009.

16 IEEE C37.106, "Guide for Abnormal Frequency Protection for Power Generating Plants", IEEE Power Engineering Society, 2003.

17 Power System Dynamic Performance Committee, *Identification of Electromechanical Modes in Power Systems*, IEEE Task Force Report, Special Publication TP462, June 2012.

18 J. Hauer, C. Demeure, and L. Scharf, "Initial Results in Prony Analysis of Power System Response Signals", *IEEE Transactions on Power Systems*, vol. 5, no. 1, pp. 80–89, February 1990.

19 J. Rueda, C. Juárez, and I. Erlich, "Wavelet-based Analysis of Power System Low-Frequency Electromechanical Oscillations," *IEEE Transactions on Power Systems*, vol. 26, no. 3, pp. 118–133, Feb. 2012.

20 J. Cepeda, G. Argüello, P. Verdugo and A. De La Torre, "Real-time Monitoring of Steady-state and Oscillatory Stability Phenomena in the Ecuadorian Power System", IEEE PES Transmission & Distribution Conference and Exposition Latin America (T&D-LA), September 2014.

21 S. P. Teeuwsen, *Oscillatory Stability Assessment of Power Systems using Computational Intelligence*, Doktors der Ingenieurwissenschaften Thesis, Universität Duisburg-Essen, Germany, March 2005.

22 G. Amico, R. Molina, and V. Sinagra, "Rules for Determining Transmission Limits in the Power Grid of Argentina", Proceedings of The CIGRE XII Iberian–American Regional Meeting, Foz do Iguazú, Brazil 2007.

23 J. Cepeda, D. Ramírez, G. Colomé, "Probabilistic-based Overload Estimation for Real-Time Smart Grid Vulnerability Assessment", IEEE Transmission and Distribution Latin America (T&D-LA), Montevideo, Uruguay, September 2012.

24 J. Rossmaier, B. Chowdhury, "*Further Development of the Overload Risk Index, an Indicator of System Vulnerability*", North American Power Symposium (NAPS), Starkville, MS, USA, pp. 1–6, October 2009.

7

Challenges Ahead Risk-Based AC Optimal Power Flow Under Uncertainty for Smart Sustainable Power Systems

Florin Capitanescu

Researcher, Luxembourg Institute of Science and Technology (LIST), Environmental Research and Innovation (ERIN) Department, Belvaux, Luxembourg

7.1 Chapter Overview

In order to meet the ambitious long-term sustainability objectives set by governments, most electric power systems worldwide are undergoing significant structural and operational changes due to mainly the large-scale integration of intermittent renewable generation (wind, solar, etc.) and the increasing adoption of emerging flexibility-enabling technologies (storage batteries, HVDC links, etc.). This challenging energy transition toward smarter sustainable power systems requires in turn adapting decision making tools (e.g., for planning, operational planning, and real-time control) to cope with the prevailing operational environment.

This chapter focuses on optimal power flow (OPF) which, together with unit commitment and dynamic simulation-based stability control, are the main decision making problems to be solved in day-ahead operational planning. The chapter discusses the challenges and proposes solutions to leverage OPF computations for future smart sustainable power systems, fostering the development of the first generation of tools balancing risk-based security criteria and renewable generation forecast uncertainty. It describes the stages needed to reach this goal and details the contents of their corresponding building blocks. Illustrative examples using a small test system are used through the whole chapter to support the various concepts presented.

The chapter is organized as follows. Section 7.2 revisits the conventional (deterministic) OPF problem. Section 7.3 presents a formulation of the OPF problem that integrates a probabilistic, risk-based security criterion. Section 7.4 addresses the extension of the OPF framework to take into account renewable generation forecast uncertainty. Finally, Section 7.5 is devoted to advanced issues related to the current and future generation of OPF packages.

Dynamic Vulnerability Assessment and Intelligent Control for Sustainable Power Systems, First Edition.
Edited by José Luis Rueda-Torres and Francisco González-Longatt.
© 2018 John Wiley & Sons Ltd. Published 2018 by John Wiley & Sons Ltd.

7.2 Conventional (Deterministic) AC Optimal Power Flow (OPF)

7.2.1 Introduction

Initiated in the early 1960s [1], AC optimal power flow (OPF) calculations are nowadays fundamental in power systems planning, operational planning, operation, control, and electricity markets [2–7].

OPF problems search for optimal values for a set of decision variables, according to a certain objective, while ensuring power system *static security* (i.e., satisfying grid power flow equations and other operational constraints with respect to both the base case and a set of plausible contingencies). The consideration of contingency constraints (e.g., the N-1 security criterion, an operating rule to be enforced by transmission system operators) is vital for practical applications and represents by far the most important and challenging requirement for industrial OPF software [2–6]. However, although this OPF challenge has been expressed for over fifty years [8], most research endeavors ignore it. This presentation asserts that contingency constraints make it intrinsically part of the OPF problem, a concept that has also been named in the literature as *security-constrained optimal power flow* (SCOPF) or, as a notable special case, *security-constrained economic dispatch* (SCED).

OPF requirements, challenges, and solution methodologies have been the subject of an important number of survey papers stemming from teams of experts with different backgrounds (industry, academia, and software developers) [2–6], out of which the recent paper [2] is absolutely outstanding while [6] provides a critical review of latest developments in this area.

7.2.2 Abstract Mathematical Formulation of the OPF Problem

OPF is in its general form a non-convex, nonlinear, large-scale, static optimization problem with both continuous and discrete variables. Because OPF is very versatile in terms of optimization goals (e.g., secure economic dispatch, minimum cost/number of control means re-dispatch, minimum power losses, maximum reactive power reserves, etc.) and the means to fulfill them, it requires careful problem formulation. An abstract traditional OPF problem formulation is as follows [2, 3, 5]:

$$\min_{\mathbf{x}_0, \mathbf{u}_0, \mathbf{x}_k, \mathbf{u}_k} f_0(\mathbf{x}_0, \mathbf{u}_0) \tag{7.1}$$

$$\text{s.t. } \mathbf{g}_0(\mathbf{x}_0, \mathbf{u}_0) = \mathbf{0} \tag{7.2}$$

$$\mathbf{h}_0(\mathbf{x}_0, \mathbf{u}_0) \leq \mathbf{0} \tag{7.3}$$

$$\mathbf{g}_k(\mathbf{x}_k, \mathbf{u}_k) = \mathbf{0}, \qquad k = 1, \dots, K \tag{7.4}$$

$$\mathbf{h}_k(\mathbf{x}_k, \mathbf{u}_k) \leq \mathbf{0}, \qquad k = 1, \dots, K \tag{7.5}$$

$$|\mathbf{u}_k - \mathbf{u}_0| \leq \Delta\mathbf{u}_k, \qquad k = 1, \dots, K, \tag{7.6}$$

where subscripts 0 and k denote the base case (or pre-contingency state) and post-contingency state, respectively; K is the number of postulated contingencies; $\mathbf{x}_0$, $\mathbf{x}_k$ are the sets of state variables (i.e., real and imaginary parts of bus voltages) in the pre-contingency state and post-contingency state (i.e., after the corrective actions took place), respectively; $\mathbf{u}_0$ is the set of preventive controls (e.g., generator active powers, generator voltages, taps position of phase shifter and on load tap changing transformer, etc.); $\mathbf{u}_k$ is the set of corrective controls (e.g., generator active powers, generator voltages, etc.); f_0 is the objective function to be minimized (e.g., cost of generation dispatch, power losses, etc.); functions $\mathbf{g}_k$ model mainly the AC power flow network equations in state k; functions $\mathbf{h}_k$ model the operational limits (e.g., mainly on branch currents and voltage magnitudes) in state k; and $\Delta\mathbf{u}_k$ is the set of allowed variations of corrective actions (according to their ramping ability and alloted time for control).

The objective (7.1) considers the cost of base case operation. The constraints (7.2)–(7.3) and (7.4)–(7.5) enforce the feasibility of the base-case and post-contingency states, respectively. The constraints (7.6) impose ranges on corrective actions.

OPF can be used to cover the static power system security according to any of the following modes:

- *preventive security* [8]: only preventive actions are taken to ensure that in the event of a contingency, no operational limit (e.g., branch currents and voltage magnitudes) will be violated;
- *also corrective security* [9]: both preventive and corrective actions ensure that, if a contingency occurs, violated limits can be removed without loss of load.

7.2.3 OPF Solution via Interior-Point Method

For the sake of simplicity[1] it is assumed that (i) the OPF formulation (7.1)–(7.6) contains only continuous variables, leading to a nonlinear programming (NLP) problem, and (ii) the size of the problem allows the direct solution via an NLP method.

Several notable approaches proposed in the literature for solving NLP OPF problems are today the core of various commercial OPF software (e.g., chronologically [3]: sequential linear programming (SLP) [10], sequential quadratic programming (SQP) [11], Newton method [12], interior-point method (IPM) [13, 14], etc.)

Since the early 1990s, the IPM has gained increasingly significant popularity in the power systems community owing to three major advantages: (i) ease of handling inequality constraints by logarithmic barrier functions, (ii) speed of convergence, and (iii) a feasible initial point is not required [13–17].

The basics of IPM are briefly detailed in what follows.

7.2.3.1 Obtaining the Optimality Conditions In IPM

The abstract OPF formulation (7.1)–(7.6) can be compactly written as follows:

$$\min f(\mathbf{y}) \tag{7.7}$$

$$\text{s.t. } \mathbf{g}(\mathbf{y}) = \mathbf{0}; \quad \mathbf{h}(\mathbf{y}) \leq \mathbf{0} \tag{7.8}$$

where $f(\mathbf{y})$, $\mathbf{g}(\mathbf{y})$, and $\mathbf{h}(\mathbf{y})$ are twice continuously differentiable functions, $\mathbf{y}$ is a vector grouping both decision variables and state variables, the vectors $\mathbf{y}$, $\mathbf{g}$, and $\mathbf{h}$ having sizes m, p, and q, respectively.

The IPM requires four steps to obtain optimality conditions as follows.

---

1 A more comprehensive methodology for OPF solution is briefly presented in Section 7.5 (see Figure 7.3).

First step: slack variables are added to inequality constraints, transforming them into equality constraints and non-negativity conditions on slacks:

$$\min f(\mathbf{y}) \tag{7.9}$$

$$\text{s.t. } \mathbf{g}(\mathbf{y}) = \mathbf{0}; \quad -\mathbf{h}(\mathbf{y}) - \mathbf{s} = \mathbf{0}; \quad \mathbf{s} \geq \mathbf{0}, \tag{7.10}$$

where the vectors $\mathbf{y}$ and $\mathbf{s} = [s_1, \ldots, s_q]^T$ are called *primal variables*.

Second step: the non-negativity conditions are added to the objective function as logarithmic barrier terms, leading to an equality constrained optimization problem:

$$\min f(\mathbf{y}) - \mu \sum_{i=1}^{q} \ln s_i \tag{7.11}$$

$$\text{s.t. } \mathbf{g}(\mathbf{y}) = \mathbf{0}; \quad -\mathbf{h}(\mathbf{y}) - \mathbf{s} = \mathbf{0} \tag{7.12}$$

where μ is a positive scalar called the *barrier parameter*, whose key role in IPM will be explained later on.

Third step: the equality constrained optimization problem is transformed into an unconstrained one, by building the Lagrangian:

$$L_\mu(\mathbf{z}) = f(\mathbf{y}) - \mu \sum_{i=1}^{q} \ln s_i - \boldsymbol{\lambda}^T \mathbf{g}(\mathbf{y}) - \boldsymbol{\pi}^T [-\mathbf{h}(\mathbf{y}) - \mathbf{s}]$$

where the vectors of Lagrange multipliers $\boldsymbol{\lambda}$ and $\boldsymbol{\pi}$ are called *dual variables* and the vector $\mathbf{z}$ gathers all variables, that is, $\mathbf{z} = [\mathbf{s} \ \boldsymbol{\pi} \ \boldsymbol{\lambda} \ \mathbf{y}]^T$.

Fourth step: the *perturbed* Karush–Kuhn–Tucker (KKT) first order necessary optimality conditions of the problem are obtained by setting to zero the derivatives of the Lagrangian with respect to all variables [18]:

$$\begin{bmatrix} \nabla_s L_\mu(\mathbf{z}) \\ \nabla_\pi L_\mu(\mathbf{z}) \\ \nabla_\lambda L_\mu(\mathbf{z}) \\ \nabla_y L_\mu(\mathbf{z}) \end{bmatrix} = \begin{bmatrix} -\mu \mathbf{1} + \mathbf{S}\,\boldsymbol{\pi} \\ \mathbf{h}(\mathbf{y}) + \mathbf{s} \\ -\mathbf{g}(\mathbf{y}) \\ \nabla f(\mathbf{y}) - \mathbf{J_g}(\mathbf{y})^T \boldsymbol{\lambda} - \mathbf{J_h}(\mathbf{y})^T \boldsymbol{\pi} \end{bmatrix} = \mathbf{0} \tag{7.13}$$

where $\mathbf{S}$ is a diagonal matrix of slack variables, $\nabla f(\mathbf{y})$ is the gradient of f, $\mathbf{J_g}(\mathbf{y})$ is the Jacobian of $\mathbf{g}(\mathbf{y})$, and $\mathbf{J_h}(\mathbf{y})$ is the Jacobian of $\mathbf{h}(\mathbf{y})$.

Central to the IPM is the theorem [18], which proves that as μ tends to zero the solution $\mathbf{y}(\mu)$ converges to a local optimum of the original problem (7.7)–(7.8). IPM solution algorithms (e.g., pure primal dual [13–15], predictor-corrector [13–15], multiple centrality corrections [16, 17]) exploit this finding, solving the perturbed KKT optimality conditions (7.13) while gradually decreasing μ to zero as iterations progress. To facilitate understanding, the basic primal dual IPM algorithm is outlined hereafter.

7.2.3.2 The Basic Primal Dual Algorithm

1) Set iteration count $k = 0$. Chose $\mu^k > 0$. Initialize $\mathbf{z}^k$ such that slack variables and their corresponding dual variables are strictly positive $(\mathbf{s}^k, \boldsymbol{\pi}^k) > \mathbf{0}$.

2) Solve the linearized KKT conditions (7.13) for the Newton direction $\Delta \mathbf{z}^k$:

$$\mathbf{H}(\mathbf{z}^k) \begin{bmatrix} \Delta \mathbf{s}^k \\ \Delta \boldsymbol{\pi}^k \\ \Delta \boldsymbol{\lambda}^k \\ \Delta \mathbf{y}^k \end{bmatrix} = \begin{bmatrix} \mu^k \mathbf{1} - \mathbf{S}^k \boldsymbol{\pi}^k \\ -\mathbf{h}(\mathbf{y}^k) - \mathbf{s}^k \\ \mathbf{g}(\mathbf{y}^k) \\ -\nabla f(\mathbf{y}^k) + \mathbf{J_g}(\mathbf{y}^k)^T \boldsymbol{\lambda}^k + \mathbf{J_h}(\mathbf{y}^k)^T \boldsymbol{\pi}^k \end{bmatrix} \tag{7.14}$$

where $\mathbf{H}(\mathbf{z}^k) = \partial^2 L_\mu(\mathbf{z}^k)/\partial \mathbf{z}^2$ is the second derivatives Hessian matrix.

3) Compute the maximum step length $\alpha^k \in (0, 1]$ along the Newton direction $\Delta \mathbf{z}^k$ such that $(\mathbf{s}^{k+1}, \boldsymbol{\pi}^{k+1}) > \mathbf{0}$:

$$\alpha^k = \min \left\{ 1, \gamma \min_{\Delta s_i^k < 0} \frac{-s_i^k}{\Delta s_i^k}, \gamma \min_{\Delta \pi_i^k < 0} \frac{-\pi_i^k}{\Delta \pi_i^k} \right\} \tag{7.15}$$

where $\gamma \in (0, 1)$ is a safety factor[2] aiming to maintain the strict positiveness of slack variables and their corresponding dual variables.

Update the solution:

$$\mathbf{s}^{k+1} = \mathbf{s}^k + \alpha^k \Delta \mathbf{s}^k \qquad \boldsymbol{\pi}^{k+1} = \boldsymbol{\pi}^k + \alpha^k \Delta \boldsymbol{\pi}^k$$
$$\mathbf{y}^{k+1} = \mathbf{y}^k + \alpha^k \Delta \mathbf{y}^k \qquad \boldsymbol{\lambda}^{k+1} = \boldsymbol{\lambda}^k + \alpha^k \Delta \boldsymbol{\lambda}^k$$

4) Convergence test. A locally optimal solution is found and the optimization process terminates when change in primal feasibility, scaled dual feasibility, scaled complementarity gap and objective function from one iteration to the next meet some tolerances (typically $\epsilon_1 = 10^{-4}, \epsilon_2 = 10^{-6}$) [13–15]:

$$\max \left\{ \max_i \left\{ h_i(\mathbf{y}^k) \right\}, ||\mathbf{g}(\mathbf{y}^k)||_\infty \right\} \le \epsilon_1 \tag{7.16}$$

$$\frac{||\nabla f(\mathbf{y}^k) - \mathbf{J_g}(\mathbf{y}^k)^T \boldsymbol{\lambda} - \mathbf{J_h}(\mathbf{y}^k)^T \boldsymbol{\pi}^k||_\infty}{1 + ||\mathbf{y}^k||_2 + ||\boldsymbol{\lambda}^k||_2 + ||\boldsymbol{\pi}^k||_2} \le \epsilon_1 \tag{7.17}$$

$$\frac{\rho^k}{1 + ||\mathbf{y}^k||_2} \le \epsilon_2 \tag{7.18}$$

$$\frac{|f(\mathbf{y}^k) - f(\mathbf{y}^{k-1})|}{1 + |f(\mathbf{y}^k)|} \le \epsilon_2 \tag{7.19}$$

where $\rho^k = (\mathbf{s}^k)^T \boldsymbol{\pi}^k$ is the *complementarity gap*.

5) Update the barrier parameter:

$$\mu^{k+1} = \sigma \frac{\rho^k}{q}$$

where the *centering parameter* is usually set as $\sigma = 0.2$.

Set $k = k + 1$ and go back to step 2.

As with any approach, the IPM has also some drawbacks: (i) the heuristic to decrease the barrier parameter, (ii) the positivity condition on slack variables and their dual variables, which may shorten the Newton step length, slowing down convergence, and (iii) lack of efficient warm/hot start, which should generally further speed-up OPF solution methodologies (see Figure 7.3).

---

2 A typical value of the safety factor is $\gamma = 0.99995$.

7.2.4 Illustrative Example

7.2.4.1 Description of the Test System

The main characteristics of the abstract OPF formulation are best unveiled for educational purposes on a small system. Figure 7.1 sketches a 5-bus system, inspired from [19], whose bus and line data are provided in Tables 7.1 and 7.2, respectively.

In Table 7.1, all notations are self-explanatory, except of ΔP_G which denotes the ramping ability of a generator (both up and down) as a corrective control action for the assumed time horizon for removing post-contingency violated limits.

In Table 7.2, R_{ij}, X_{ij}, and B_{ij} are the resistance, the reactance, and the susceptance, respectively, of the line linking bus i and bus j. Note that the line current limit I_{ij}^{max} (in A) corresponds to 1100 MVA limit and hence to 11 pu current limit in 100 MVA base.

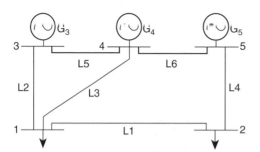

Figure 7.1 One-line diagram of the 5-bus system.

Table 7.1 5-bus system data and initial state.

Bus	P_L MW	Q_L MVar	P_G MW	Q_G MVar	V pu	V^{min} pu	V^{max} pu	P_G^{min} MW	P_G^{max} MW	Q_G^{min} MVar	Q_G^{max} MVar	ΔP_G MW
1	1100	400	–	–	0.954	0.92	1.05	–	–	–	–	–
2	500	200	–	–	0.950	0.92	1.05	–	–	–	–	–
3	–	–	700.0	69.5	1.0	0.92	1.05	150	1500	−500	750	200
4	–	–	600.0	304.9	1.0	0.92	1.05	150	1500	−500	750	200
5	–	–	333.8	146.9	1.0	0.92	1.05	150	1500	−500	750	200

Table 7.2 5-bus system: line data.

Line	Bus i	Bus j	V^{nom} kV	R_{ij} Ω	X_{ij} Ω	B_{ij} μS	I_{ij}^{max} A
L1	1	2	400	3.2	16	160	1587.7
L2	1	3	400	6.4	32	320	1587.7
L3	1	4	400	3.2	16	160	1587.7
L4	2	5	400	6.4	32	320	1587.7
L5	3	4	400	6.4	32	320	1587.7
L6	4	5	400	6.4	32	320	1587.7

The following generator cost curves, in (US) $/h$, are assumed:

$$C(P_{G_3}) = 100 + 25P_{G_3} + 0.01P_{G_3}^2,$$
$$C(P_{G_4}) = 100 + 30P_{G_4} + 0.01P_{G_4}^2,$$
$$C(P_{G_5}) = 100 + 60P_{G_5} + 0.01P_{G_5}^2.$$

For the sake of simplicity, current and voltage limits are considered the same in both pre-contingency and post-contingency states. Furthermore, only N-1 lines outages are considered in the contingencies set, hence $K = 6$. The contingencies are named according to the disconnected line (e.g., L1, L2, L3, L4, L5, and L6).

7.2.4.2 Detailed Formulation of the OPF Problem

The compact OPF formulation (7.1)–(7.6) is specifically detailed hereafter for the 5-bus system, adopting a rectangular coordinates model for complex bus voltages:

$$\underline{V}_i = e_i + jf_i, \quad i = 1, \dots, 5,$$

where $V_i = \sqrt{e_i^2 + f_i^2}$ is the voltage magnitude at bus i.

The formulation models the most comprehensive operating mode, that is, "also corrective" (see Table 7.3); the simplifying assumptions behind the other modes are explained later on.

The objective (7.1) is economic dispatch (i.e., minimum generation cost):

$$\min[100 + 25P_{G_3}^0 + 0.01(P_{G_3}^0)^2 + 100 + 30P_{G_4}^0 + 0.01(P_{G_4}^0)^2$$
$$+ 100 + 60P_{G_5}^0 + 0.01(P_{G_5}^0)^2] \tag{7.20}$$

The decision variables of this problem are generator active powers and generator voltages, which are limited within some ranges (7.26) and (7.27), respectively.

Equality constraints (7.2) and (7.4) comprise mainly the active and reactive power flow equations at each bus ($i = 1, \dots, 5$) and for each system state ($k = 0, \dots, 6$):

$$P_{G_i}^k - P_{L_i} - [(e_i^k)^2 + (f_i^k)^2] \sum_{j \in N_i^k} G_{ij}$$
$$+ \sum_{j \in N_i^k} [(e_i^k e_j^k + f_i^k f_j^k)G_{ij} + (f_i^k e_j^k - e_i^k f_j^k)B_{ij}] = 0 \tag{7.21}$$

Table 7.3 Values of the objective, decision variables, and critical post-contingency constraints for various operating modes.

Operating modes	Objective cost $/h	Decision variables						Critical constraints			
		P_{G_3} MW	P_{G_4} MW	P_{G_5} MW	V_{G_3} pu	V_{G_4} pu	V_{G_5} pu	V_1^3 pu	V_2^4 pu	I_{L3}^2 pu	I_{L3}^4 pu
unoptimized	65439	700.0	600.0	333.8	1.000	1.000	1.000	0.88	0.90	12.7	11.8
no contingencies	61042	847.0	637.0	150.0	1.050	1.050	1.047	0.94	–	12.5	10.6
preventive*	76709	617.0	319.5	695.5	1.035	1.030	1.050	–	–	11.0	11.5
preventive	76842	693.4	246.4	693.2	1.050	1.029	1.050	–	–	11.0	11.0
also corrective	70550	668.2	442.7	519.1	1.050	1.050	1.050	–	–	10.8	11.0

$$Q_{G_i}^k - Q_{L_i} + [(e_i^k)^2 + (f_i^k)^2] \sum_{j \in N_i^k} (B_{sij} + B_{ij})$$

$$- \sum_{j \in N_i^k} [(e_i^k e_j^k + f_i^k f_j^k)B_{ij} + (e_i^k f_j^k - f_i^k e_j^k)G_{ij}] = 0, \tag{7.22}$$

where N_i^k is the set of adjacent buses to bus i in state k, other notations being self-explanatory.

The reference[3] voltage angle is set at bus 5:

$$f_i^k = 0, \ i = 5, \ k = 0, \dots, K$$

Inequality constraints (7.3) and (7.5) include operational limits on (longitudinal) branch current, voltage magnitude, and generator reactive power:

$$(G_{ij}^2 + B_{ij}^2) \left[(e_i^k - e_j^k)^2 + (f_i^k - f_j^k)^2 \right] \leq (I_{ij}^{max})^2,$$

$$i = 1, \dots, 5, \ j \in N_i^k, \ k = 0, \dots, 6, \tag{7.23}$$

$$(V_i^{min})^2 \leq (e_i^k)^2 + (f_i^k)^2 \leq (V_i^{max})^2, \ i = 1, 2, \ k = 0, \dots, 6, \tag{7.24}$$

$$Q_{G_i}^{min} \leq Q_{G_i}^k \leq Q_{G_i}^{max}, \ i = 3, 4, 5, \ k = 0, \dots, 6 \tag{7.25}$$

and limits on decision variables:

$$P_{G_i}^{min} \leq P_{G_i}^k \leq P_{G_i}^{max}, \ i = 3, 4, 5, \ k = 0, \dots, 6, \tag{7.26}$$

$$(V_i^{min})^2 \leq (e_i^k)^2 + (f_i^k)^2 \leq (V_i^{max})^2, \ i = 3, 4, 5, \ k = 0, \dots, 6 \tag{7.27}$$

Finally, the coupling constraints (7.6) take on the form:

$$|P_{G_i}^k - P_{G_i}^0| \leq \Delta P_{G_i}, \ i = 3, 4, 5, \ k = 1, \dots, 6 \tag{7.28}$$

$$\left| \sqrt{(e_i^k)^2 + (f_i^k)^2} - \sqrt{(e_i^0)^2 + (f_i^0)^2} \right| \leq \Delta V_{G_i}, \ i = 3, 4, 5, \ k = 1, \dots, 6 \tag{7.29}$$

Note that, for the sake of formulation simplicity, the branch current is approximated by the longitudinal current in (7.23).

7.2.4.3 Analysis of Various Operating Modes

Table 7.3 provides, for various operating modes, the values of the objective, the six decision variables, and the critical (i.e., violated, binding, or near-binding) operation constraints in post-contingency states.

Assume first that the system operates in the unoptimized mode (this mode does not consider contingencies). One can observe that by simulating each contingency using a classical power flow tool (G_5 is assumed as slack generator), severe violations of voltage and line current limits occur for some contingencies. Specifically, the lowest voltage corresponds to the third contingency (loss of line L3) and is recorded at bus 1 ($V_1^3 = 0.88$ pu), while line L3 is overloaded by 1.7 pu and 0.8 pu for contingencies two and four, respectively. Note incidentally that the use of a DC model would miss identifying such low voltage problems.

Assume next that the system operates in the "no contingencies" mode where the pre-contingency state is optimized but contingencies are still disregarded. This mode

---

3 Note that, unlike power flow calculations, OPF formulation does not require a slack generator.

corresponds to solving the OPF formulation (7.20)–(7.27) while keeping only the variables and constraints corresponding to the pre-contingency state (i.e., $k = 0$). One can remark that the optimization leads to a good reduction of the generation cost (i.e., from 65 439 \$$/h$ to 61 042 \$$/h$) due to a more economical dispatch of active power in absence of network congestions. Indeed, the most expensive generator G_5 is dispatched at its technical minimum while the cheapest generator G_3 has the largest production increase. This active power re-dispatch has an even stronger negative effect on losses, which slightly increase by 0.2 MW, than the positive impact of reactive power redispatch via an increase in the voltage profile (e.g., almost all voltages at generators reach their upper bound). Also, although not in the optimization objective, constraints violations due to contingencies are alleviated to a certain extent.

7.2.4.4 Iterative OPF Methodology

In order to avoid building unmanageable OPF problems for large real-life applications, practical OPF solution methodologies (see Figure 7.3) include progressively only some contingency constraints in the OPF [2, 10, 20–22]. These constraints should be carefully chosen (e.g., by decreasing order of the overall number of violated constraints). This requires (in its simplest OPF methodology form) iterating between the core optimizer and the security analysis (SA) to check potential constraint violations for contingencies not included in the OPF.

Such a methodology is illustrated in its simplest form, assuming starting "from scratch," for system operation in the (conservative) preventive mode. The latter corresponds to solving the OPF formulation (7.20)–(7.29) while imposing that decision variables have the same values in both pre-contingency and post-contingency states. This can be conceptually expressed[4] with two specific settings in the coupling constraints: (i) $\Delta P_{G_i} = 0$, $i = 3, 4$ in (7.28), which means than only generator G_5 takes care of post-contingency active power imbalance (this ensures coherency with SA, where G_5 is the slack generator), and (ii) $\Delta V_{G_i} = 0$, $i = 3, 4, 5$ in (7.29).

In this methodology one first optimizes the pre-contingency state, which corresponds to the "no contingencies" mode. Then, SA is performed for this state. The results of the SA in terms of post-contingency line currents and bus voltages for the six postulated contingencies are gathered in Table 7.4. Note that only one constraint is violated, specifically line L3 becomes overloaded due to contingency L2. Next, the OPF is solved in preventive mode but including only the harmful contingency L2. The results at this intermediate step of the OPF procedure are presented in Table 7.3 under the label "preventive*." Note that performing SA again on this solution reveals that contingency L4 leads to the overload of line L3. The OPF is run again including both contingencies L2 and L4 and the results are provided in Table 7.3 under the label "preventive." Because no other contingency leads to constraint violation, this is also the optimal solution of the preventive mode.

Obviously the cost to ensure power system security with respect to the contingencies leads to an increase in the generation cost (e.g., from 61 042 \$$/h$ to 76 709 \$$/h$).

Concerning the physical interpretation of the generation dispatch pattern, comparing the solutions in "no contingencies" and "preventive*" modes, one can remark that in

---

4 Obviously, efficient implementation for practical OPF problems will require defining only the set of pre-contingency decision variables and making it intervene in all post-contingency states, dropping the coupling constraints (7.28)–(7.29).

Table 7.4 Results of security analysis performed at the OPF solution in "no contingencies" mode.

Contingency	Longitudinal current (I_{L_j})						Voltage (V_j) at bus				
loss of line	L1	L2	L3	L4	L5	L6	1	2	3	4	5
				pu					pu		
L1	0.0	5.5	6.2	5.4	2.8	3.5	1.009	0.987	1.05	1.05	1.047
L2	0.9	0.0	12.5	4.7	8.1	2.3	0.985	0.987	1.05	1.05	1.047
L3	1.9	10.6	0.0	7.3	1.6	4.5	0.938	0.953	1.05	1.05	1.047
L4	5.6	6.7	10.6	0.0	1.8	1.6	0.985	0.953	1.05	1.05	1.047
L5	2.3	8.2	6.1	3.2	0.0	1.1	1.005	1.001	1.05	1.05	1.047
L6	3.5	6.2	8.7	2.5	2.1	0.0	1.005	1.001	1.05	1.05	1.047

order to diminish the loading of line L3 due to contingency L2, a significant amount of active power is redirected via the L4 (see Figure 7.1). As a consequence, generators G_3 and G_4 reduce their output while G_5 increases its production by 545.5 MW. Further comparing the solutions "preventive*" and "preventive" modes, one can observe that in order to reduce the loading of line L3 due to any of "conflicting" contingencies L2 and L4, the compromise solution consists in further decreasing the production of generator G_4.

Finally, applying the same methodology for the "also corrective" mode, one sees that only contingency L2 is binding at the optimum. Due to the larger number of degrees of freedom[5], this operating mode leads to a reduction in the generation cost compared with the preventive mode (e.g., 70 550 \$/h vs. 76 842 \$/h), but naturally the cost still remains larger than in the "no contingencies" mode, the difference reflecting the cost needed to cover the contingencies. The optimization process in this operating mode takes advantage of the post-contingency generators' active power available and may end up with opposite generation re-dispatch, for example:

- contingency L2: $P_{G_3} = 568.3$ MW, $P_{G_4} = 366.5$ MW, $P_{G_5} = 719.1$ MW;
- contingency L4: $P_{G_3} = 801.8$ MW, $P_{G_4} = 458.5$ MW, $P_{G_5} = 395.7$ MW.

7.3 Risk-Based OPF

7.3.1 Motivation and Principle

In industrial practice, power system security has been addressed so far in a *deterministic* way, mostly adopting the preventive mode (see 7.2.2) [2]. However, although simple and clear, the scope of the deterministic N-1 security criterion has been deemed too narrow by many authors, the following drawbacks being brought forward [3, 23–27]: (i) equiprobable treatment of postulated contingencies (not distinguishing between low likelihood and more likely contingencies), (ii) rigid binary classification (secure vs. insecure) of post-contingency states irrespective of the severity impact (e.g., magnitude of impact and number of violated constraints) and relying on soft operational limits

---

5 Indeed after a contingency, every generator is allowed to reschedule its active power up to $\Delta P_{G_i} = 200$ MW and alter its terminal voltage by $\Delta V_{G_i} = 0.03$ pu.

(e.g., currents and voltages) whose values are arbitrary to some extent, and (iii) disregard of unconventional countermeasures (e.g., post-contingency load shedding) to remove constraint violations.

The direct consequence of these flaws is that the cost of security for low likelihood or low impact contingencies may be significant (e.g., due to effective generators' poor location or high price), while the impact of the contingencies may be counteracted by a small amount of load curtailment [27].

The same authors have advocated the use of a probabilistic, risk-based, security criterion in order to trade-off economy and security in a power system according to either: (i) the optimization of the *expected*[6] cost of system operation [23–25], or (ii) the minimization of only the pre-contingency operational cost under an overall system risk constraint, which accumulates the individual risk[7] related to each contingency [26, 27].

7.3.2 Risk-Based OPF Problem Formulation

In what follows, the presentation adopts the approach in [27]. This approach uses an overall system risk constraint and expresses the contingency severity in terms of the amount of (prompt) load shedding. The latter is a simple and easy to interpret metric which allows replacing the complexity and accuracy of methods which rely on cascading outages simulation (e.g., [24, 25]), but requires field expertise to set the maximum risk allowed risk_{max} in (7.39).

This risk-based OPF (RB-OPF) can be abstractly and generically formulated as follows:

$$\min_{\mathbf{x}_0, \mathbf{u}_0, \mathbf{x}_k, \mathbf{u}_k, \mathbf{c}_k} f_0(\mathbf{x}_0, \mathbf{u}_0) \tag{7.30}$$

$$\text{s.t. } \mathbf{g}_0(\mathbf{x}_0, \mathbf{u}_0) = \mathbf{0} \tag{7.31}$$

$$\mathbf{h}_0(\mathbf{x}_0, \mathbf{u}_0) \le \mathbf{0} \tag{7.32}$$

$$\mathbf{g}_k(\mathbf{x}_k, \mathbf{u}_k, \mathbf{c}_k) = \mathbf{0}, \qquad k = 1, \dots, K \tag{7.33}$$

$$\mathbf{h}_k(\mathbf{x}_k, \mathbf{u}_k, \mathbf{c}_k) \le \mathbf{0}, \qquad k = 1, \dots, K \tag{7.34}$$

$$|\mathbf{u}_k - \mathbf{u}_0| \le \Delta\mathbf{u}_k, \qquad k = 1, \dots, K \tag{7.35}$$

$$\mathbf{0} \le \mathbf{c}_k \le \mathbf{1}, \qquad k = 1, \dots, K \tag{7.36}$$

$$\mathbf{c}_k \, \mathbf{1}^T \, \mathbf{I} \, \mathbf{P}_L \le \Delta\mathbf{P}_L^{\text{max}}, \qquad k = 1, \dots, K \tag{7.37}$$

$$\mathbf{c}_k^T \mathbf{P}_L \le \Delta P^{\text{max}}, \qquad k = 1, \dots, K \tag{7.38}$$

$$\sum_{k=1}^{K} p_k \mathbf{c}_k^T \mathbf{P}_L \le \text{risk}_{\text{max}}, \tag{7.39}$$

where: $\mathbf{c}_k$, the new control variable, is the vector of load curtailment in post-contingency states; $\mathbf{P}_L$ is the vector of base case load active power; $\Delta\mathbf{P}_L^{\text{max}}$ is the vector of maximum allowed amount of bus load curtailment; ΔP^{max} is the maximum allowed overall amount of load shedding if a contingency occurs; $\mathbf{I}$ is the identity matrix of appropriate dimension; p_k is the probability of contingency k; and risk_{max} is the maximum value of overall

---

6 This is defined as the (probability-weighted) sum of pre-contingency cost and the post-contingency cost including the cost of corrective control and load interruption.
7 The risk is the product between the probability of contingency occurrence and contingency severity.

system risk. Constraints (7.36) impose bounds on the percentage of load curtailment; constraints (7.37) impose limits on the allowed amount of load curtailment at any bus; constraints (7.38) model the authorized overall amount of load shedding according to the grid code. The constraint (7.39) enforces that the overall system risk does not exceed a pre-defined value $\text{risk}_{\max}$. Note that if the transmission system operator does not wish to take any operational risk (i.e., imposing $\text{risk}_{\max} = 0$), this leads to recovery of the traditional OPF problem in the "also corrective" mode [9].

7.3.3 Illustrative Example

The main features of the RB-OPF formulation are illustrated using the 5-bus system in Figure 7.1.

7.3.3.1 Detailed Formulation of the RB-OPF Problem

The compact RB-OPF formulation (7.30)–(7.39) is detailed hereafter for the 5-bus system for preventive mode operation.

The objective function (7.30) is the economic dispatch (7.20).

The set of constraints specific to RB-OPF (7.36)–(7.39) take on the form:

$$0 \leq c_i^k \leq 1, \qquad\qquad i = 1, 2, \ k = 1, \dots, 6 \tag{7.40}$$

$$c_i^k P_{L_i} \leq \Delta P_{L_i}^{\max}, \qquad\qquad i = 1, 2, \ k = 1, \dots, 6 \tag{7.41}$$

$$\sum_{i=1}^{2} c_i^k P_{L_i} \leq \Delta P^{\max}, \qquad\qquad k = 1, \dots, 6 \tag{7.42}$$

$$\sum_{k=1}^{6} p_k \sum_{i=1}^{2} c_i^k P_{L_i} \leq \text{risk}_{\max} \tag{7.43}$$

Active and reactive power flow equations (7.21)–(7.22), at each bus ($i = 1, \dots, 5$) and each system state ($k = 0, \dots, 6$), are slightly altered to take into account the new load curtailment control variables c_i^k:

$$P_{G_i}^k - (1 - c_i^k)P_{L_i} - [(e_i^k)^2 + (f_i^k)^2] \sum_{j \in N_i^k} G_{ij}$$
$$+ \sum_{j \in N_i^k} [(e_i^k e_j^k + f_i^k f_j^k)G_{ij} + (f_i^k e_j^k - e_i^k f_j^k)B_{ij}] = 0 \tag{7.44}$$

$$Q_{G_i}^k - (1 - c_i^k)Q_{L_i} + [(e_i^k)^2 + (f_i^k)^2] \sum_{j \in N_i^k} (B_{sij} + B_{ij})$$
$$- \sum_{j \in N_i^k} [(e_i^k e_j^k + f_i^k f_j^k)B_{ij} + (e_i^k f_j^k - f_i^k e_j^k)G_{ij}] = 0 \tag{7.45}$$

In terms of inequality constraints, the RB-OPF includes the set of constraints (7.23)–(7.27) as the deterministic OPF.

Finally, the constraints specific to the preventive mode take on the form:

$$|P_{G_i}^k - P_{G_i}^0| \leq \Delta P_{G_i}, \ i = 5, \ k = 1, \dots, 6 \tag{7.46}$$

$$P_{G_i}^k = P_{G_i}^0, \ i = 3, 4, \ k = 1, \dots, 6 \tag{7.47}$$

$$(e_i^k)^2 + (f_i^k)^2 = (e_i^0)^2 + (f_i^0)^2, \ i = 3, 4, 5, \ k = 1, \dots, 6 \tag{7.48}$$

7.3.3.2 Numerical Results

Two scenarios, called relaxed and constrained, are studied. In both scenarios load shedding is performed under a constant power factor. The relaxed scenario studies the impact of the risk constraint (7.43) only and relaxes constraints (7.41) and (7.42). The constrained scenario considers additionally that load shedding is limited to 10% of the initial power in (7.41) and the maximum allowed overall amount of load shed per contingency is set to $\Delta P^{max} = 200$ MW in (7.42). Unless otherwise specified it is assumed that each contingency has probability $p_k = 0.0001, k = 1, \ldots, 6$.

Figure 7.2 displays (dotted line) the operation cost obtained with the proposed RB-OPF approach, without short-term constraints, for increasing values of system risk between two extremes: (i) *conservative N-1 secure operation* which corresponds to $risk_{max} = 0$ and hence to conventional OPF, as no load shedding is allowed; and (ii) *risky N-0 secure operation*, where the risk constraint is practically relaxed (e.g., for roughly $risk_{max} \geq 4 \times 10^{-2}$). The figure also provides the corresponding overall amount of load shed corresponding to each risk level. One can observe that as the maximum allowed risk increases, the overall load shed increases and the operation cost decreases. This curve could constitute an important basis for a transmission system operator (TSO) to trade-off the cost savings incurred by choosing a more risky operation.

The potential consequences of the operation risk assumed can be evaluated in terms of post-contingency overall and individual load shedding. These are provided in Table 7.5 for contingency L2, the only contingency that is binding in all cases, as it leads to an active current constraint on line L3. Because the curtailment of load in bus 1 has a larger impact on the loading of critical line L3 (and thereby on the cost reduction), as a consequence, it is fully exploited in all cases, as shown in the table.

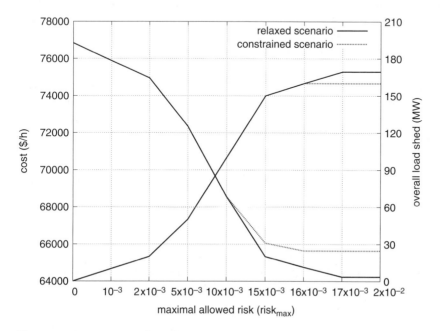

Figure 7.2 Operation cost of RB-SCOPF versus the maximum allowed risk.

Table 7.5 Overall and individual load shedding (MW) for contingency L2.

Risk$_{max}$	Relaxed			Constrained		
	Overall	Bus 1	Bus 2	Overall	Bus 1	Bus 2
0	0.0	0.0	0.0	0.0	0.0	0.0
0.001	10.0	10.0	0.0	10.0	10.0	0.0
0.002	20.0	20.0	0.0	20.0	20.0	0.0
0.005	50.0	50.0	0.0	50.0	50.0	0.0
0.010	100.0	100.0	0.0	100.0	100.0	0.0
0.015	150.0	150.0	0.0	150.0	110.0	40.0
0.016	160.0	160.0	0.0	160.0	110.0	50.0
0.017	169.4	169.4	0.0	160.0	110.0	50.0
0.020	169.4	169.4	0.0	160.0	110.0	50.0

7.4 OPF Under Uncertainty

7.4.1 Motivation and Potential Approaches

This section addresses an extended-scope OPF framework that aims to further take into account *forecast uncertainty* or, in other words, bus power injections uncertainty, stemming mostly from an increasingly significant amount of renewable generation (e.g., wind, solar, etc.). In this increasingly sustainable power systems context, the classical operational planning approach based on the conventional deterministic OPF, which secures only *the most likely forecasted state* (or nominal scenario) of the system for each period of time of the next day, may be insufficient.

The OPF under uncertainty (OPF-UU) problem can be tackled most notably via approaches from robust optimization [28–31], chance-constrained optimization [32, 33], or stochastic optimization [34]. However, whatever the approach taken, compared to conventional OPF, OPF-UU will need to additionally incorporate a certain number of (base case) aggregated scenarios sampling the uncertainty, which significantly increases the already large problem size.

Assuming that probability density functions of uncertainty are not available or trusted the day ahead of operation, this presentation follows a robust optimization approach. Specifically, a special case of the comprehensive framework proposed in [28] is detailed hereafter.

7.4.2 Robust Optimization Framework

Assume first, without loss of generality, a box uncertainty set $\mathcal{W}$ defined by upper and lower bounds on the uncertain power injections, including an infinite number of potential scenarios:

$$\mathcal{W} = \left\{ \mathbf{w} \in \mathbb{R}^n : \underline{\mathbf{w}} \le \mathbf{w} \le \overline{\mathbf{w}} \right\} \tag{7.49}$$

where n is the number of system buses.

The day-ahead robust OPF approach seeks to determine the optimal combination of preventive controls $\mathbf{u}_0$, which are independent of the power injection pattern, and corrective controls $\mathbf{u}_k^w$, which are dependent of both the contingency and the power injection pattern, such that the system operational limits are satisfied whatever postulated contingency $k = 1 \dots, K$ and injection scenario $\mathbf{w} \in \mathcal{W}$ may show up the next day. This robust OPF (R-OPF) can be abstractly formulated as follows:

$$\min_{\mathbf{u}_0, \mathbf{a}_0^w, \mathbf{u}_k^{w}} f(\mathbf{u}_0) \tag{7.50}$$

$$\text{s.t. } \mathbf{g}_0^w(\mathbf{x}_0^w, \mathbf{u}_0, \mathbf{a}_0^w) = \mathbf{0} \qquad\qquad \forall \mathbf{w} \in \mathcal{W} \tag{7.51}$$

$$\mathbf{h}_0^w(\mathbf{x}_0^w, \mathbf{u}_0) \leq \mathbf{0} \qquad\qquad \forall \mathbf{w} \in \mathcal{W} \tag{7.52}$$

$$\mathbf{g}_k^w(\mathbf{x}_k^w, \mathbf{u}_0, \mathbf{a}_0^w, \mathbf{u}_k^w) = \mathbf{0} \qquad k = 1, \dots, K, \forall \mathbf{w} \in \mathcal{W} \tag{7.53}$$

$$\mathbf{h}_k^w(\mathbf{x}_k^w, \mathbf{u}_0, \mathbf{u}_k^w) \leq \mathbf{0} \qquad k = 1, \dots, K, \forall \mathbf{w} \in \mathcal{W} \tag{7.54}$$

$$|\mathbf{u}_k^w - \mathbf{u}_0| \leq \Delta \mathbf{u}_k \qquad k = 1, \dots, K, \forall \mathbf{w} \in \mathcal{W}, \tag{7.55}$$

where constraints (7.52)–(7.55) have the same meaning as constraints (7.2)–(7.6) from the conventional OPF formulation, $\mathbf{w}$ is a vector of uncertain bus active/reactive power injections, and $\mathbf{a}_0^w$ denotes the set of generators that participate in frequency regulation or, in other words, compensate the active power imbalance due to the uncertain power injections. This set of generators is assumed cost-free, according to the objective (7.50), and the nominal scenario corresponds to $\mathbf{a}_0^w = \mathbf{0}$.

By the definition of the set $\mathcal{W}$, the R-OPF problem has an infinite number of constraints. Note that since control variables $\mathbf{u}_0$ are common to all pre-contingency states, the robust OPF solution (if any) immunizes the power system security against any realization of the uncertainty in the set $\mathcal{W}$.

7.4.3 Methodology for Solving the R-OPF Problem

The principle of the proposed methodology for solving the R-OPF formulation (7.50)–(7.55) consists of solving a sequence of problem relaxations, by replacing the infinite set of scenarios $\mathcal{W}$ with a finite subset of *worst-case* (or *worst uncertainty pattern*) scenarios $\mathcal{W}_i$ adjusted to the problem instance at hand. The original infinite dimensional problem is thus reduced to a sequence of finite dimensional subproblems. The proposed algorithm builds up iteratively a growing set of constraining scenarios $\mathcal{W}_0 \subset \dots \subset \mathcal{W}_i \subset \mathcal{W}_{i+1} \subset \dots \subset \mathcal{W}$ as follows:

1) At the first iteration, $\mathcal{W}_0$ comprises the nominal scenario $\tilde{\mathbf{w}}$ (i.e., the most likely forecast for the next day), the computation being reduced to a conventional OPF.
2) Fix $\mathbf{u}_0$ to the OPF optimum $\mathbf{u}_0^{\star}(\mathcal{W}_i)$ derived at the previous step and compute the worst-case scenario with respect to each contingency k as follows:
   (a) Compute the worst-case scenario $\mathbf{w}^{\star}(k)$ with respect to contingency k in the absence of corrective actions by solving the optimization problem:

$$\max_{\mathbf{w}} \mathbf{1}^T \hat{\mathbf{h}}_k(\mathbf{x}_k, \mathbf{u}_0^{\star}, \mathbf{u}_k) \tag{7.56}$$

$$\text{s.t. } \mathbf{g}_0(\mathbf{x}_0, \mathbf{u}_0^{\star}, \mathbf{a}_0, \mathbf{w}) = \mathbf{0} \tag{7.57}$$

$$\mathbf{h}_0(\mathbf{x}_0, \mathbf{u}_0^{\star}) \leq \mathbf{0} \tag{7.58}$$

$$\mathbf{g}_k(\mathbf{x}_k, \mathbf{u}_0^\star, \mathbf{a}_0, \mathbf{w}, \mathbf{u}_k) = \mathbf{0} \tag{7.59}$$

$$\underline{\mathbf{w}} \leq \mathbf{w} \leq \overline{\mathbf{w}}, \tag{7.60}$$

where $\hat{\mathbf{h}}_k$ denotes the subset of violated operation constraints (assumed known beforehand) corresponding to the worst uncertainty pattern. The reader is referred to [30] for an algorithm to identify this subset.

(b) If the worst-case scenario leads to constraint violation, check via a special case of OPF if optimal corrective actions are able to remove the constraint violation:

$$\min_{\mathbf{u}_k, \mathbf{r}} |\mathbf{u}_k - \mathbf{u}_0^\star| + \alpha \mathbf{1}^T \mathbf{r} \tag{7.61}$$

$$\text{s.t. } \mathbf{g}_k(\mathbf{x}_k, \mathbf{u}_0^\star, \mathbf{u}_k, \mathbf{a}_0^\star, \mathbf{w}^\star(k)) = \mathbf{0} \tag{7.62}$$

$$\mathbf{h}_k(\mathbf{x}_k, \mathbf{u}_0^\star, \mathbf{u}_k) \leq \mathbf{r} \tag{7.63}$$

$$|\mathbf{u}_k - \mathbf{u}_0^\star| \leq \Delta\mathbf{u}_k \tag{7.64}$$

$$\mathbf{r} \geq \mathbf{0} \tag{7.65}$$

where $\mathbf{r}$ are positive relaxation variables and α is a large positive penalty term aimed to identify if the original problem is infeasible; otherwise $\mathbf{r} \to \mathbf{0}$.

(c) If this is not the case then preventive controls are required to cover the worst-case scenario; grow the subset of worst-case scenarios: $\mathcal{W}_{i+1} \leftarrow \mathcal{W}_i \cup \mathbf{w}^\star(k)$.

3) Compute a new value $\mathbf{u}_0^\star(\mathcal{W}_{i+1})$ by solving a relaxation of the R-OPF problem (7.50)–(7.55) which includes all scenarios in $\mathcal{W}_{i+1}$ and all contingencies.

4) The process terminates as soon as: (i) a fixed point is reached (i.e., no change in either $\mathbf{u}_0^\star$ or $\mathcal{W}_i$), or (ii) computing budget is exhausted, or (iii) the current relaxation of the R-OPF problem (7.50)–(7.55) becomes infeasible (meaning that the best combination of preventive and corrective controls cannot guarantee the system security over the whole uncertainty set $\mathcal{W}$). Otherwise, go back to step 2.

Note that this iterative algorithm produces a sequence of day-ahead decisions $\mathbf{u}_0^\star$ of growing robustness with respect to uncertainties since, at any intermediate iteration, besides the nominal scenario, a larger set of uncertain patterns than at the previous iteration are covered.

7.4.4 Illustrative Example

A detailed specific formulation of the optimization problems involved in the iterative algorithm for the solution of the R-OPF problem is provided for the 5-bus system.

7.4.4.1 Detailed Formulation of the Worst Uncertainty Pattern Computation With Respect to a Contingency

For simplicity, the worst case scenarios are computed solely with respect to overload and further assume that only one line (known beforehand) is overloaded by the worst uncertainty pattern. The reader is referred to [30] for a detailed comprehensive algorithm to reveal the overall set of overloaded lines whatever the uncertainty pattern and the processing of several overloads in the worst case scenario.

The compact formulation (7.56)–(7.60) is detailed hereafter for contingency c.

The objective function is the maximization of the current in the line linking buses i and j in the state post-contingency c:

$$\max_{P_{ui}, Q_{ui}, P_{G_i}} I_{ij}^c = \sqrt{\left(G_{ij}^2 + B_{ij}^2\right)\left[(e_i^c - e_j^c)^2 + (f_i^c - f_j^c)^2\right]} \tag{7.66}$$

The decision variables of this optimization problem are the uncertain power injections (P_{ui} and Q_{ui}), which are allowed to vary in certain ranges:

$$P_{ui}^{\min} \leq P_{ui} \leq P_{ui}^{\max}, \quad i = 1, 2 \tag{7.67}$$

$$Q_{ui}^{\min} \leq Q_{ui} \leq Q_{ui}^{\max}, \quad i = 1, 2 \tag{7.68}$$

and limits on the active power of the generator responsible for frequency regulation:

$$P_{G_i}^{\min} \leq P_{G_i}^k \leq P_{G_i}^{\max}, \quad i = 5, \ k = 0, c \tag{7.69}$$

The active power and imposed voltage of other generators are fixed, in both states, at the values provided by the conventional OPF ($P_{G_i}^{0\star}$ and $V_{G_i}^{0\star}$):

$$P_{G_i}^c = P_{G_i}^0 = P_{G_i}^{0\star}, \qquad\qquad i = 3, 4 \tag{7.70}$$

$$\sqrt{(e_i^c)^2 + (f_i^c)^2} = \sqrt{(e_i^0)^2 + (f_i^0)^2} = V_{G_i}^{0\star}, \qquad i = 3, 4, 5 \tag{7.71}$$

The power flow equations at each bus ($i = 1, \dots, 5$) including uncertain injections for both the base case and the state post-contingency c (i.e., $k = 0, c$) are:

$$P_{G_i}^k - P_{L_i} + P_{ui} - [(e_i^k)^2 + (f_i^k)^2] \sum_{j \in N_i^k} G_{ij} +$$
$$\sum_{j \in N_i^k} [(e_i^k e_j^k + f_i^k f_j^k)G_{ij} + (f_i^k e_j^k - e_i^k f_j^k)B_{ij}] = 0 \tag{7.72}$$

$$Q_{G_i}^k - Q_{L_i} + Q_{ui} + [(e_i^k)^2 + (f_i^k)^2] \sum_{j \in N_i^k} (B_{sij} + B_{ij}) -$$
$$\sum_{j \in N_i^k} [(e_i^k e_j^k + f_i^k f_j^k)B_{ij} + (e_i^k f_j^k - f_i^k e_j^k)G_{ij}] = 0 \tag{7.73}$$

The reference voltage angle constraint is set as:

$$f_i^k = 0, \ i = 5, \ k = 0, c$$

Operational limits on (longitudinal) branch current, voltage magnitude, and generator reactive power can be written as:

$$(G_{ij}^2 + B_{ij}^2)[(e_i^0 - e_j^0)^2 + (f_i^0 - f_j^0)^2] \leq (I_{ij}^{\max})^2, \ i = 1, \dots, 5, j \in N_i^0, \tag{7.74}$$

$$(V_i^{\min})^2 \leq (e_i^k)^2 + (f_i^k)^2 \leq (V_i^{\max})^2, \ i = 1, 2, 3, 4, 5, \ k = 0, c, \tag{7.75}$$

$$Q_{G_i}^{\min} \leq Q_{G_i}^k \leq Q_{G_i}^{\max}, \ i = 3, 4, 5, \ k = 0, c \tag{7.76}$$

Note that, because the violated line currents in the state post-contingency c are assumed known, other branch current constraints for this state are dropped; hence, constraint (7.74) includes only the set of branch current constraints in the base case.

7.4.4.2 Detailed Formulation of the OPF to Check Feasibility in the Presence of Corrective Actions

The compact formulation (7.61)–(7.65) is detailed hereafter for the 5-bus system.

The objective function is the minimization of the amount of generation redispatch, with respect to the values provided by the conventional OPF, to remove violated constraints in the state post-contingency c:

$$\min_{P^c_{G_i}, e^c_i, f^c_i, r_{ij}} \sum_{i=3}^{5} |P^c_{G_i} - P^{0\star}_{G_i}| + \alpha \sum_{ij} r_{ij} \tag{7.77}$$

The decision variables of this optimization problem are the generators active power and terminal voltage:

$$P^{min}_{G_i} \leq P^c_{G_i} \leq P^{max}_{G_i}, \ i = 3, 4, 5 \tag{7.78}$$

$$(V^{min}_i)^2 \leq (e^c_i)^2 + (f^c_i)^2 \leq (V^{max}_i)^2, \ i = 3, 4, 5 \tag{7.79}$$

The power flow equations at each bus $(i = 1, \ldots, 5)$ for the state post-contingency c are:

$$P^c_{G_i} - P_{L_i} + P^\star_{ui} - [(e^c_i)^2 + (f^c_i)^2] \sum_{j \in N^c_i} G_{ij}$$

$$+ \sum_{j \in N^c_i} [(e^c_i e^c_j + f^c_i f^c_j) G_{ij} + (f^c_i e^c_j - e^c_i f^c_j) B_{ij}] = 0 \tag{7.80}$$

$$Q^c_{G_i} - Q_{L_i} + Q^\star_{ui} + [(e^c_i)^2 + (f^c_i)^2] \sum_{j \in N^c_i} (B_{sij} + B_{ij})$$

$$- \sum_{j \in N^c_i} [(e^c_i e^c_j + f^c_i f^c_j) B_{ij} + (e^c_i f^c_j - f^c_i e^c_j) G_{ij}] = 0 \tag{7.81}$$

The reference voltage angle constraint is set as:

$$f^c_i = 0, \ i = 5.$$

Operational limits on (longitudinal) branch current, voltage magnitude, and generator reactive power can be written as:

$$(G^2_{ij} + B^2_{ij})[(e^c_i - e^c_j)^2 + (f^c_i - f^c_j)^2] \leq (I^{max}_{ij})^2 + r_{ij},$$

$$r_{ij} \geq 0, \ i = 1, \ldots, 5, \ j \in N^c_i, \tag{7.82}$$

$$(V^{min}_i)^2 \leq (e^c_i)^2 + (f^c_i)^2 \leq (V^{max}_i)^2, \ i = 1, 2, 3, 4, 5, \tag{7.83}$$

$$Q^{min}_{G_i} \leq Q^c_{G_i} \leq Q^{max}_{G_i}, \ i = 3, 4, 5. \tag{7.84}$$

Finally, the coupling constraints take on the form:

$$|P^c_{G_i} - P^{0\star}_{G_i}| \leq \Delta P_{G_i}, \ i = 3, 4, 5, \tag{7.85}$$

$$\left| \sqrt{(e^c_i)^2 + (f^c_i)^2} - \sqrt{(e^{0\star}_i)^2 + (f^{0\star}_i)^2} \right| \leq \Delta V_{G_i}, \ i = 3, 4, 5. \tag{7.86}$$

7.4.4.3 Detailed Formulation of the R-OPF Relaxation

A detailed specific formulation of the R-OPF relaxation problem including a finite number of worst case scenarios $(w = 0, \ldots, W)$, where $w = 0$ corresponds to the nominal scenario, is provided hereafter.

The objective function is economic dispatch (i.e., minimum generation cost):

$$\min[100 + 25P_{G_3}^{0,0} + 0.01(P_{G_3}^{0,0})^2 + 100 + 30P_{G_4}^{0,0} + 0.01(P_{G_4}^{0,0})^2$$
$$+ 100 + 60P_{G_5}^{0,0} + 0.01(P_{G_5}^{0,0})^2] \tag{7.87}$$

The decision variables of this problem are the generators active power and terminal voltage, which are limited within some ranges:

$$P_{G_i}^{\min} \leq P_{G_i}^{w,k} \leq P_{G_i}^{\max}, \ i = 3, 4, 5, \ w = 0, \dots, W, \ k = 0, \dots, 6, \tag{7.88}$$

$$(V_i^{\min})^2 \leq (e_i^{w,k})^2 + (f_i^{w,k})^2 \leq (V_i^{\max})^2,$$
$$i = 3, 4, 5, \ w = 0, \dots, W, \ k = 0, \dots, 6 \tag{7.89}$$

The active and reactive power flow equations at each bus $(i = 1, \dots, 5)$ and for each uncertainty scenario $(w = 0, \dots, W)$ and system state $(k = 0, \dots, 6)$ are:

$$P_{G_i}^{w,k} - P_{L_i} - [(e_i^{w,k})^2 + (f_i^{w,k})^2] \sum_{j \in N_i^k} G_{ij}$$

$$+ \sum_{j \in N_i^k} [(e_i^{w,k} e_j^{w,k} + f_i^{w,k} f_j^{w,k}) G_{ij} + (f_i^{w,k} e_j^{w,k} - e_i^{w,k} f_j^{w,k}) B_{ij}] = 0 \tag{7.90}$$

$$Q_{G_i}^{w,k} - Q_{L_i} + [(e_i^{w,k})^2 + (f_i^{w,k})^2] \sum_{j \in N_i^k} (B_{sij} + B_{ij})$$

$$- \sum_{j \in N_i^k} [(e_i^{w,k} e_j^{w,k} + f_i^{w,k} f_j^{w,k}) B_{ij} + (e_i^{w,k} f_j^{w,k} - f_i^{w,k} e_j^{w,k}) G_{ij}] = 0 \tag{7.91}$$

The reference voltage angle constraint is:

$$f_i^{w,k} = 0, \ i = 5, \ w = 0, \dots, W, \ k = 0, \dots, K$$

Operational limits on (longitudinal) branch current, voltage magnitude, and generator reactive power can be written as:

$$(G_{ij}^2 + B_{ij}^2) \left[(e_i^{w,k} - e_j^{w,k})^2 + (f_i^{w,k} - f_j^{w,k})^2 \right] \leq (I_{ij}^{\max})^2,$$
$$i = 1, \dots, 5, \ j \in N_i^k, \ w = 0, \dots, W, \ k = 0, \dots, 6, \tag{7.92}$$

$$(V_i^{\min})^2 \leq (e_i^{w,k})^2 + (f_i^{w,k})^2 \leq (V_i^{\max})^2,$$
$$i = 1, 2, \ w = 0, \dots, W, \ k = 0, \dots, 6, \tag{7.93}$$

$$Q_{G_i}^{\min} \leq Q_{G_i}^{w,k} \leq Q_{G_i}^{\max}, \ i = 3, 4, 5, \ w = 0, \dots, W, \ k = 0, \dots, 6 \tag{7.94}$$

Finally, the coupling constraints (7.55) are:

$$|P_{G_i}^{w,k} - P_{G_i}^{w,0}| \leq \Delta P_{G_i}, \ i = 3, 4, 5, \ w = 0, \dots, W, \ k = 1, \dots, 6 \tag{7.95}$$

$$\left| \sqrt{(e_i^{w,k})^2 + (f_i^{w,k})^2} - \sqrt{(e_i^{w,0})^2 + (f_i^{w,0})^2} \right| \leq \Delta V_{G_i},$$
$$i = 3, 4, 5, \ w = 0, \dots, W, \ k = 1, \dots, 6 \tag{7.96}$$

7.4.4.4 Numerical Results

The main characteristics of the abstract OPF-UU formulation are best shown for educational purposes using the 5-bus system in Figure 7.1.

Uncertainty can be assumed to stem from renewable generation present in lower voltage grids, which have not been explicitly modeled, the renewable generation being simply aggregated to load buses 1 and 2. Uncertainty is modeled as variable active and reactive power injections at the two load buses in the range of −5% to +5% of the nominal active/reactive load.

Following the proposed methodology for solving the R-OPF problem, one first computes a reference schedule for the nominal scenario via classical OPF formulation in "also corrective" mode including the six postulated contingencies (see Table 7.6).

The solution provided by this classical OPF computes, via the formulation (7.66)–(7.76), the worst uncertainty pattern for each contingency with respect to thermal overload. Note that, due to the network simplicity, only one worst uncertainty pattern, common for all contingencies, is revealed. The worst uncertainty pattern corresponds to a case where the output of renewable generation at buses 1 and 2 is zero. Accordingly, the pre-contingency state corresponding to the worst uncertainty pattern differs, with respect to the classical OPF solution, by additional active/reactive power drawn from the network (i.e., 55 MW/20 MVAr at bus 1 and 25 MW/10 MVAr at bus 2, respectively), the active power balance being ensured by an increase of 83.3 MW at G5, while the generators' reactive power production change so as to maintain their imposed voltages in the prevailing operation conditions.

Further, only two out of six contingencies lead to overloads for their worst uncertainty pattern. Specifically, contingency L2 overloads line L3 with 0.95 pu and contingency L4 also overloads line L3 with 0.9 pu. In both cases, upon solving the formulation (7.77)–(7.86), one observes that the overload cannot be fully removed despite the best use of corrective actions. Specifically, for contingency L2, the best redispatch scheme consists in reducing the active power of G4 and increasing the active power of G5, which creates a counterflow in the overloaded line L3. The latter line current cannot be reduced to more than 11.53 pu. For contingency L4, despite the best redispatch between generators G3 and G4 (where the former increases its production), which creates a counterflow in the overloaded line L3, its latter current is reduced only to 11.12 pu.

Next, in order to compute the intermediate robust optimal solution, one solves the R-OPF formulation (7.87)–(7.96) which includes the reference scenario and the sole worst case scenario found (common to both contingencies).

Table 7.6 Values of the objective, decision variables, and critical post-contingency constraints for various operating modes.

Operating modes	Objective cost $/h	Decision variables						Critical constraints	
		P_{G_3} MW	P_{G_4} MW	P_{G_5} MW	V_{G_3} pu	V_{G_4} pu	V_{G_5} pu	I^2_{L3} pu	I^4_{L3} pu
also corrective	70550	668.2	442.7	519.1	1.050	1.050	1.050	10.8	11.0
robust solution	77235	700.5	228.7	703.2	1.050	1.047	1.050	10.5	10.8

Table 7.6 provides, for various operating modes, the values of the objective, the six decision variables, and the binding operation constraints. Note that the robust optimal solution encompasses both the active power and the terminal voltage of each generator.

In the second loop of the algorithm, one notices that the worst uncertainty pattern does not change. Because the latter scenario is already covered by the R-OPF solution, this means that the algorithm reaches a fixed point and hence this schedule is the robust optimal solution. The operation cost of robustness exceeds roughly 10% the cost of the most likely scenario (see Table 7.6).

7.5 Advanced Issues and Outlook

7.5.1 Conventional OPF

7.5.1.1 Overall OPF Solution Methodology
In order to facilitate the understanding of the major basic concepts of an OPF problem, the latter has been assumed as a clean mathematical programming problem (e.g., NLP) which can be solved directly by an appropriate optimization method (e.g., IPM). However, real-world OPF problems are far more complex. For instance, the major challenges to today's deterministic OPF problems are [2–6]: (i) the huge size, (ii) the management of discrete control variables, and (iii) the incorporation of several problem peculiarities that reflect TSO practical needs (e.g., usable solution in terms of number and sequence of corrective actions, etc.) and operating rules (e.g., conditional corrective actions, constraints and control variables priorities, etc.). Regarding item (iii), it must be emphasized that some practical aspects of the OPF problem may not be formulated analytically in an advantageous form exploitable by an optimizer, requiring one to resort to heuristic techniques during the OPF solution methodologies [2]. However, since benchmarks of practical real-world OPF problems are not available,[8] the academic realm typically only addresses items (i) and (ii), interpreting, in the most ambitious vision, a real-life OPF problem as a *large scale* non-convex mixed-integer nonlinear programming (MINLP) problem [22].

The huge size of real-life OPF problems stems from the large number of contingencies that have to be secured (e.g., at least all N-1 events) and the large size of the power system. Because taking a direct approach to such a problem is computationally inefficient if even feasible, methodologies for solving practical OPF problems require decomposition into (and *iterations* over) a certain number of modules (see Figure 7.3) including in the most generic also corrective mode:

- Core NLP optimizers for the *master problem* and *slave (i.e., contingency)* problems: The master problem includes constraints for the base case and some potentially binding contingencies while each slave problem relates to a post-contingency state and checks whether it is feasible thanks to corrective actions.
- Manager of discrete variables module: sets discrete control variables according to some approach (e.g., round-off [36], progressive round-off [22, 37], sensitivity-based [38]) or dedicated techniques for topology control (transmission switching) [39]. Note that this module may interact several times during the solution process with optimizers for both master and slave problems.

---

8 Some large scale system data became available very recently on the MATPOWER website [35].

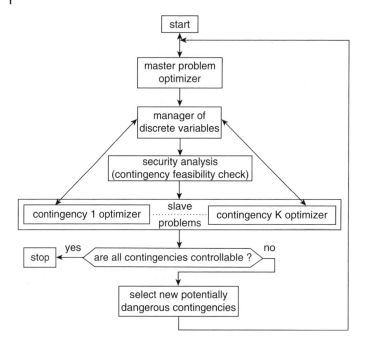

Figure 7.3 High level flowchart of the OPF solution methodology in the also corrective mode.

- Security analysis module: performs classical post-contingency power flow calculations to check whether constraints violations occur.
- Contingency selection (or filtering) module: selects from the set of contingencies that violate constraints some promising potentially binding contingencies to be added to the master optimizer [20, 21, 40, 41].

An acceptable solution of the OPF problem is found (and iterations stop) when no contingency leads to constraint violation in the presence of corrective actions.

Significant progress in the OPF state-of-the-art has been achieved recently in the following methodologies for problem decomposition: all potentially binding contingencies together (possibly with post-contingency network compression) [20–22]; adaptive Benders decomposition [42] (which improves solution quality but requires larger computational effort compared to the classical Benders decomposition-based OPF methodology [9, 21]); and alternating direction method of multipliers [42]. The core optimizer in these methodologies utilizes powerful general-purpose local solvers (mostly based on the IPM) for NLP (e.g KNITRO and IPOPT). The use of high performance computations renders these computationally demanding OPF methodologies practical for solving realistic large-scale OPF problems (e.g., for large systems up to roughly 10 000 buses and 12 000 contingencies [22]), even starting "from scratch." Because the core optimizer in these methodologies is the IPM, further improvement can be obtained by decomposing the structure of the linear system of equations to be solved in the interior point method [40]. Anyway, to cope with stringent OPF running time industry needs, powerful computing infrastructure is deployed enabling distributed (parallel) computations and hence fostering OPF decomposition methods.

7.5.1.2   Core Optimizers: Classical Methods Versus Convex Relaxations

In most today OPF commercial software used by the industry, the core optimizer relies on SLP, SQP, Newton's method, and IPM. Although solvers for LP and QP are reliable and fast, SLP and SQP methodologies may provide approximate solutions of the NLP core problem, especially for highly nonlinear OPF problems (e.g., reactive power dispatch). Newton's method and IPM can provide exact (at least) local optimum for the underlying NLP core problem, but while Newton's method may have difficulties in handling inequality constraints, IPM (both general-purpose and produced by academia, tailored specifically for the OPF problem) is not fully reliable [17] and does not warm/hot start well. The warm/hot start feature of an algorithm (via the latest solution at hand or an intermediate approximated solution at a given iteration) is supposed to generally accelerate OPF solution methodologies, especially if successive master problem solutions are close to each other, or their active set of constraints do not change significantly.

A research framework that has attracted much interest recently is devoted to convex relaxations[9] of the non-convex NLP problems intervening in the core optimizers [44–46]. The major goal of these approaches is to provide, in certain conditions, the global optimum of the non-convex NLP continuous relaxation of the OPF problem. Other notable advantages of these approaches are: provision of a lower bound of the NLP problem (which allows evaluation, provided that the relaxation is tight, of the quality of the local optimizer solution), problem infeasibility guarantee, reliability, and polynomial solution time. Most of these approaches rely on either semi-definite programming (SDP) relaxation [44] or the more general high-order moment-based relaxation [45] (in fact SDP relaxation corresponds to the first order moment).

Various research with convex relaxations report very useful empirical evidence that, generally, the solution provided by local optimizers is the global optimum (or a solution of very high quality) [42, 45, 46]. Next, extensive experiments conducted using a diverse data set point out that [46]: (i) in most cases SDP relaxation does not recover the global solution (since the duality gap is greater than 0%) and (ii) convex relaxations methods are generally slower and scale less well than local optimizers on large OPF problems. Further research with convex relaxations is therefore needed in order to reduce the computational burden, improve the convergence precision of SDP [45], solve real-life OPF problems (which, inter alia, may present negative Lagrange multipliers), and so on.

"It is important to emphasize that, due to today's limitations of the MINLP state-of-the-art, when solving real-life large-scale AC OPF problems under stringent time requirements, the aspect of whether the solution of the OPF continuous relaxation dealt with by the core optimizer is *the global optimum or not becomes definitely of secondary importance* compared to the primary aim which is to provide, within the allowed computational budget, a (possibly sub-optimal) solution which satisfies the constraints." [6]. Furthermore, it is worth mentioning that the core algorithm classification (approximate, at least local, or global) relates to the ideal case of a perfect system state forecast, which is less and less the case in today's increasingly uncertain operational environment (especially in day-ahead, which has been the main OPF application timeframe).

Although there is definitely room to improve convex relaxations in the coming years and even make them usable by industry as a complement for classical well-established

---

9 A comprehensive survey of existing approaches up to 2014 can be found in [43].

non-convex local optimizers (in terms of solution quality and certificate of problem infeasibility), it is unlikely that these relaxations will eventually supersede local optimizers.

Finally, in order to efficiently solve increasingly complex practical OPF problems, there will be no single winner in terms of method adopted by the core optimizer, but rather modern optimizers will embed different solution methods (like the powerful solver KNITRO), being able to seamlessly switch between them in order to adapt on-the-fly to the features of the optimization problem at hand.

7.5.2 Beyond the Scope of Conventional OPF: Risk, Uncertainty, Smarter Sustainable Grid

Driven by stringent sustainability commitments, the fast-paced integration of a significant amount of intermittent renewable generation at all voltage levels requires smarter operation of the grid. The latter can be achieved first of all through the provision and smart management of operation flexibility, which enables the system to maintain security while adapting to unforeseen prevailing operational conditions. This increases the reliance upon effective controllable devices (e.g., phase shifters, storage, FACTS devices, HVDC links, topology control, etc.) as *fast* corrective actions for controlling both post-contingency and poorly predicted operation conditions. As a consequence, the decision making process close to and in real-time will gain more attention and utilization. This calls for faster OPF solution methodologies, possibly different to those developed for the main (day-ahead) OPF purpose, tailored for closer to real-time application.

Furthermore, such smarter operation of increasingly sustainable power systems will require in turn a *flexible* decision making process balancing *risk*-based security criteria and renewable generation forecast *uncertainty*. This very ambitious goal will lead in the future to the development of the first generation of *risk-based* AC OPF *under uncertainty* tools. This highly complex and computationally challenging problem is today out of reach using the AC grid model. For instance, [34] provides the most computationally advanced risk-based OPF under uncertainty tool in terms of optimization problem features (e.g., a few hundred thousand variables and constraints are considered). However, this work adopts the DC approximation, highlighting the immense computational challenge (and hence tool capability limitations) of handling simultaneously multiple periods, contingencies, uncertain power injections scenarios, many binary variables, and so on. There is therefore an enormous research potential in this area for developing not only frameworks [33, 34, 47] but, more importantly, scalable algorithms. Specifically, in lieu of tackling this highly complex paradigm directly, there is a need to properly address both its sub-fields, that is, risk-based OPF and OPF under uncertainty, while improving the algorithms of the deterministic AC OPF framework, which can yield efficient common building blocks. In terms of large scale application, both sub-fields are still insufficiently mature, unless one adopts a DC-based approach that has proven scalability to large systems [26, 29]. "However, this paper argues that scalable methodologies for deterministic AC OPF can be adapted to risk-based AC OPF problems, providing that the latter are formulated in a decomposable way, fostering the leverage of RB-OPF to large scale systems. As regards AC OPF under uncertainty, although some promising results have been obtained within the robust optimization framework for a medium size system [28], such an approach remains very computationally heavy" [6]. In order to provide more tractable algorithms, the major focus should be devoted to the *reduction of the number of critical scenarios modeling the uncertainty* (e.g., through efficient scenario aggregation), which is responsible for the most additional computational burden compared to deterministic OPF.

The time line of decision making should be revisited depending on the degree of trust in available uncertainty models. Thus, one can envision favoring robust optimization in day-ahead applications, chance constrained or stochastic optimization in intraday, and deterministic optimization close to real-time.

Another research line beyond the scope of deterministic OPF consists in seeking for optimal solutions while additionally taking into account (voltage, transient, and small-disturbance) stability constraints via either deriving approximations of the stability boundaries and including them directly into the OPF or coupling OPF and time-domain simulation [48–50]. This is a very important extension of the OPF scope but remains with a high computational burden for which scalable approaches of reasonable accuracy need to be found.

Another essential group of additional control options to be investigated are demand response management techniques, which offer additional degrees of freedom in terms of load shifting and large scale deployment of storage devices, for which fast technological progress is envisioned. These new control means will require optimization to be performed on a longer time horizon leading to consideration of multi-period OPF applications.

Finally, attention should also be paid to the interface with distribution grids, which are increasingly active in terms of distributed generation and may offer useful ancillary services in the future.

References

1 Jacques, C. (1962) Contribution à l'étude du dispatching économique, in *Bulletin de la Societé Française d'Electricité*, vol. **3**, pp. 431–447.

2 Stott, B. and Alsac, O. (2012) Optimal power flow - basic requirements for real-life problems and their solutions (white paper), in *SEPOPE XII Symposium, Rio de Janeiro, Brazil*.

3 Capitanescu, F., Martinez Ramos, J.L., Panciatici, P., Kirschen, D., Marano Marcolini, A., Platbrood, L., and Wehenkel, L. (2011) State-of-the-art, challenges, and future trends in security constrained optimal power flow. *Elec. Pow. Syst. Research*, **81** (8), 1731–1741.

4 Momoh, J., Koessler, R., Bond, M., Stott, B., Sun, D., Papalexopoulos, A., and Ristanovic, P. (1997) Challenges to optimal power flow. *Power Systems, IEEE Transactions on*, **12** (1), 444–455.

5 Stott, B., Alsac, O., and Monticelli, A.J. (1987) Security analysis and optimization. *Proceedings of the IEEE*, **75** (12), 1623–1644.

6 Capitanescu, F. (2016) Critical review of recent advances and further developments needed in AC optimal power flow. *Elec. Pow. Syst. Research*, **136**, 57–68.

7 Overbye, T.J., Cheng, X., and Sun, Y. (2004) A comparison of the AC and DC power flow models for LMP calculations, in *System Sciences, 2004. Proceedings of the 37th Annual Hawaii International Conference on*, IEEE.

8 Alsac, O. and Stott, B. (1974) Optimal load flow with steady-state security. *Power Apparatus and Systems, IEEE Transactions on*, **93** (3), 745–751.

9 Monticelli, A.J., Pereira, M.V.P., and Granville, S. (1987) Security-constrained optimal power flow with post-contingency corrective rescheduling. *Power Systems, IEEE Transactions on*, **2** (1), 175–182.

10 Alsac, O., Bright, J., Prais, M., and Stott, B. (1990) Further developments in LP-based optimal power flow. *Power Systems, IEEE Transactions on*, **5** (3), 697–711.

11 Burchett, R., Happ, H., and Vierath, D. (1984) Quadratically convergent optimal power flow. *Power Apparatus and Systems, IEEE Transactions on*, **103** (11), 3267–3275.

12 Sun, D.I., Ashley, B., Brewer, B., Hughes, A., and Tinney, W.F. (1984) Optimal power flow by Newton approach. *Power Apparatus and Systems, IEEE Transactions on*, **103** (10), 2864–2880.

13 Wu, Y.C., Debs, A.S., and Marsten, R.E. (1994) A direct nonlinear predictor-corrector primal-dual interior point algorithm for optimal power flows. *IEEE Trans. Power Syst.*, **9** (2), 876–883.

14 Granville, S. (1994) Optimal reactive dispatch through interior point methods. *IEEE Trans. Power Syst.*, **9** (1), 136–146.

15 Torres, G.L. and Quintana, V.H. (1998) An interior-point method for nonlinear optimal power flow using rectangular coordinates. *IEEE Trans. Power Syst.*, **13** (4), 1211–1218.

16 Torres, G.L. and Quintana, V.H. (2001) On a nonlinear multiple-centrality-corrections interior-point method for optimal power flow. *IEEE Trans. Power Syst.*, **16** (2), 222–228.

17 Capitanescu, F. and Wehenkel, L. (2013) Experiments with the interior-point method for solving large scale optimal power flow problems. *Elec. Pow. Syst. Research*, **95** (2), 276–283.

18 Fiacco, A.V. and McCormick, G.P. (1968) *Nonlinear programming: Sequential unconstrained minimization techniques*, in *John Willey & Sons*.

19 Martinez Ramos, J.L. and Quintana, V.H. (2009) Optimal and secure operation of transmission systems. *Electric Energy Systems: Analysis and Operation, CRC Press, Boca Raton*, pp. 211–264.

20 Capitanescu, F., Glavic, M., Ernst, D., and Wehenkel, L. (2007) Contingency filtering techniques for preventive security-constrained optimal power flow. *Power Systems, IEEE Transactions on*, **22** (4), 1690–1697.

21 Capitanescu, F. and Wehenkel, L. (2008) A new iterative approach to the corrective security-constrained optimal power flow problem. *Power Systems, IEEE Transactions on*, **23** (4), 1533–1541.

22 Platbrood, L., Capitanescu, F., Merckx, C., Crisciu, H., and Wehenkel, L. (2014) A generic approach for solving nonlinear-discrete security-constrained optimal power flow problems in large-scale systems. *Power Systems, IEEE Transactions on*, **29** (3), 1194–1203.

23 Condren, J., Gedra, T.W., and Damrongkulkamjorn, P. (2006) Optimal power flow with expected security costs. *Power Systems, IEEE Transactions on*, **21** (2), 541–547.

24 He, J., Cheng, L., Kirschen, D.S., and Sun, Y. (2010) Optimising the balance between security and economy on a probabilistic basis. *IET Gener. Transm. Distrib.*, **4** (12), 1275–1287.

25 Karangelos, E., Panciatici, P., and Wehenkel, L. (2013) Whither probabilistic security management for real-time operation of power systems ?, in *IREP Symposium - Bulk Power Systems Dynamics and Control-IX (IREP), Rethymnon, Greece.*

26 Wang, Q., McCalley, J.D., Zheng, T., and Litvinov, E. (2016) Solving corrective risk-based security-constrained optimal power flow with lagrangian relaxation and benders decomposition. *International Journal of Electrical Power & Energy Systems*, **75** (2), 255–264.

27 Capitanescu, F. (2015) Enhanced risk-based scopf formulation balancing operation cost and expected voluntary load shedding. *Elec. Pow. Syst. Research*, **128** (1), 151–155.

28 Capitanescu, F., Fliscounakis, S., Panciatici, P., and Wehenkel, L. (2012) Cautious operation planning under uncertainties. *Power Systems, IEEE Transactions on*, **27** (4), 1859–1869.

29 Fliscounakis, S., Panciatici, P., Capitanescu, F., and Wehenkel, L. (2013) Contingency ranking with respect to overloads in very large power systems taking into account uncertainty, preventive, and corrective actions. *Power Systems, IEEE Transactions on*, **28** (4), 4909–4917.

30 Capitanescu, F. and Wehenkel, L. (2013) Computation of worst operation scenarios under uncertainty for static security management. *Power Systems, IEEE Transactions on*, **28** (2), 1697–1705.

31 Korad, A. and Hedman, K. (2013) Robust corrective topology control for system reliability. *Power Systems, IEEE Transactions on*, **28** (4), 4042–4051.

32 Roald, L., Vrakopoulou, M., Oldewurtel, F., and Andersson, G. (2015) Risk-based optimal power flow with probabilistic guarantees. *International Journal of Electrical Power & Energy Systems*, **72**, 66–74.

33 Hamon, C., Perninge, M., and Söder, L. (2015) A computational framework for risk-based power systems operations under uncertainty. part I: Theory. *Electric Power Systems Research*, **119**, 45–53.

34 Murillo-Sanchez, C., Zimmerman, R., Lindsay Anderson, C., and Thomas, R. (2013) Secure planning and operations of systems with stochastic sources, energy storage and active demand. *Smart Grid, IEEE Transactions on*, **4** (4), 2220–2229.

35 Zimmerman, R., Murillo-Sanchez, C., and Thomas, R. (2011) MATPOWER: Steady-state operations, planning and analysis tools for power systems research and education. *Power Systems, IEEE Transactions on*, **26** (1), 12–19.

36 Papalexopoulos, A., Imparato, C., and Wu, F. (1989) Large-scale optimal power flow: Effects of initialization, decoupling & discretization. *Power Systems, IEEE Transactions on*, **4**, 748–759.

37 Macfie, P., Taylor, G., Irving, M., Hurlock, P., and Wan, H. (2010) Proposed shunt rounding technique for large-scale security constrained loss minimization. *Power Systems, IEEE Transactions on*, **25** (3), 1478–1485.

38 Capitanescu, F. and Wehenkel, L. (2010) Sensitivity-based approaches for handling discrete variables in optimal power flow computations. *Power Systems, IEEE Transactions on*, **25** (4), 1780–1789.

39 Rolim, J. and Machado, L. (1999) A study of the use of corrective switching in transmission systems. *Power Systems, IEEE Transactions on*, **14** (1), 336–341.

40 Jiang, Q. and Xu, K. (2014) A novel iterative contingency filtering approach to corrective security-constrained optimal power flow. *Power Systems, IEEE Transactions on*, **29** (3), 1099–1109.

41 Ardakani, A.J. and Bouffard, F. (2013) Identification of umbrella constraints in dc-based security-constrained optimal power flow. *Power Systems, IEEE Transactions on*, **28** (4), 3924–3934.

42 Phan, D. and Kalagnanam, J. (2014) Some efficient methods for solving the security-constrained optimal power flow problem. *Power Systems, IEEE Transactions on*, **29** (2), 863–872.

43 Panciatici, P., Campi, M., Garatti, S., Low, S., Molzahn, D., Sun, A., and Wehenkel, L. (2014) Advanced optimization methods for power systems, in *Power Systems Computation Conference (PSCC), 2014*, IEEE, pp. 1–18.

44 Lavaei, J. and Low, S. (2012) Zero duality gap in optimal power flow problem. *Power Systems, IEEE Transactions on*, **27** (1), 92–107.

45 Molzahn, D.K. and Hiskens, I.A. (2015) Sparsity-exploiting moment-based relaxations of the optimal power flow problem. *Power Systems, IEEE Transactions on*, **30** (6), 3168–3180.

46 Coffrin, C., Hijazi, H., and Van Hentenryck, P. (2016) The QC relaxation: A theoretical and computational study on optimal power flow. *Power Systems, IEEE Transactions on*.

47 Varaiya, P.P., Wu, F.F., and Bialek, J.W. (2011) Smart operation of smart grid: risk-limiting dispatch. *Proceedings of the IEEE*, **99** (1), 40–57.

48 Bruno, S., De Tuglie, E., and La Scala, M. (2002) Transient security dispatch for the concurrent optimization of plural postulated contingencies. *Power Systems, IEEE Transactions on*, **17** (3), 707–714.

49 Capitanescu, F., Van Cutsem, T., and Wehenkel, L. (2009) Coupling optimization and dynamic simulation for preventive-corrective control of voltage instability. *Power Systems, IEEE Transactions on*, **24** (2), 796–805.

50 Hamon, C., Perninge, M., and Soder, L. (2013) A stochastic optimal power flow problem with stability constraints - part I: Approximating the stability boundary. *Power Systems, IEEE Transactions on*, **28** (2), 1839–1848.

8

Modeling Preventive and Corrective Actions Using Linear Formulation

Tom Van Acker and Dirk Van Hertem

Research Group Electa, Department of Electrical Engineering, University of Leuven, Belgium

8.1 Introduction

Power system reliability management can be defined as a sequence of decisions which are taken under uncertainty in order to meet a certain reliability criterion (see also Chapter 2). A reliability criterion is a principle imposing a basis to determine whether or not the reliability level of a given operating state of the power system is acceptable [1]. This can be expressed as a set of constraints that must be satisfied by the decisions taken by the system operator (SO). An example of such a reliability criterion is the N-1 criterion, which states that the considered power system must be able to withstand any credible single contingency, for example, the loss of a generator, in such a way that a new operational set point can be reached without violating the security constraints of that power system. The SO makes use of flexibility options that are inherent to a power system to reach a new, secure operating point after a contingency. These flexibility options are also referred to as actions and the distinction can be made between preventive and corrective or curative actions.

Preventive actions are those flexibility options preventively activated to adapt the system state. Through preventive actions, a set-point is chosen which is inherently safe and anticipates contingencies by building in reserves. As such, there is no need for corrective operator intervention after a certain contingency occurs.

Corrective actions are those flexibility options available to adapt the system state immediately after a contingency has occurred, sufficiently quickly to avoid cascading of the event. Typically, corrective actions are (partly) automatic actions.

It is the responsibility of the SO to select the actions, both preventive and corrective, that result in optimal reliability, which ideally leads to a minimal global socio-economic cost. In order to obtain such optimality, the system operator needs to implement all considered actions in an Optimal Power Flow (OPF) [2], which is adapted to take the possibility of contingencies into account. This type of OPF is generally referred to as a Security Constrained OPF (SCOPF) [3].

This chapter presents the various reliability management actions and a possible formulation to integrate these actions into a security constrained OPF environment. The different actions are represented using a linearized formulation. Linearization allows

Dynamic Vulnerability Assessment and Intelligent Control for Sustainable Power Systems, First Edition.
Edited by José Luis Rueda-Torres and Francisco González-Longatt.

to keep the computational costs low while retainting a sufficiently accurate approximation of the behavior of the system [4]. Many other formulations can be found in the literature. The main objective is to show how each of the different control actions can be implemented in a consistent manner and to illustrate the impact of such control actions.

8.2 Security Constrained OPF

The general formulation of a SCOPF is given by (8.1)–(8.5):

$$\min_{u_0, u_c} \quad f(u_0, u_c) \tag{8.1}$$

$$\text{subject to:} \quad G_0(x_0, u_0, y_0) = 0 \tag{8.2}$$

$$H_0(x_0, u_0, y_0) \geq 0 \tag{8.3}$$

$$G_c(x_c, u_c, y_c) = 0, \qquad \forall c \tag{8.4}$$

$$H_c(x_c, u_c, y_c) \geq 0, \qquad \forall c \tag{8.5}$$

In this formulation, u, x and y represent the control variables, state variables and parameters respectively. The index 0 and c indicate the preventive base case and the corrective contingency cases. Equations (8.2)–(8.3) respectively ensure that the power flow equations and the operating limits are respected for the preventive base case while (8.4)–(8.5) does this for the considered corrective contingency cases. Equations (8.2) and (8.4) contain the power flow equations and other equality constraints of the OPF. The inequality constraints, (8.3) and (8.5), are for instance used to represent the limits of the power system components. For the purpose of this chapter, a DC SCOPF [5] is considered and all actions are formulated towards their implementation in a DC SCOPF. In a DC SCOPF, the power flow equations are written in their approximated linear version, the DC power flow equations. However, the techniques presented in this chapter can be extended toward other SCOPF formulations, taking into account the assumptions related to the alternative SCOPF formulation.

8.3 Available Control Actions in AC Power Systems

In this section, the most common available actions in an AC power system are listed and described. There are four main type of actions:

1. Actions that affect the active power injections in the system: generator redispatch, demand side management and load curtailment [6].
2. Topological actions influencing power flows through the system: phase shifting transformer control [7], switching actions [8].
3. Reactive power and voltage management: generator voltage set point changes, on-load tap changer transformer control.
4. Special protection schemes: Specific actions triggered by specific grid related events.

This chapter focuses on the first two types of actions.

8.3.1 Generator Redispatch

Generator redispatch uses the available upward and downward power reserves in the power system to change the power injection and consequently change the power flows within the grid [9]. Generator redispatch can be used to alleviate congestion or to compensate for a generator outage. The implementation of generator redispatch is very case specific and depends on the market structure as well as the SO. These power reserves are stipulated in contracts or purchased on the market in a liberalized system. The SO can ask to shut down, start up or change the power output of a generator. These actions need to be financially compensated by either the SO or the generator, depending on the type of redispatch. In a vertically integrated system, these reserves are made available to the SO.

8.3.2 Load Shedding and Demand Side Management

As an alternative to generator redispatch, the demand can also be altered in order to keep the power balance within the system. Two methods exist that facilitate a change of the power demand: load shedding and demand side management [10].

The unannounced interruption of supply is called load shedding. The cost of load shedding corresponds to the value of lost load, which is the estimated amount a customer is willing to pay to avoid a disruption in their electricity service. The value of the loss of electricity can be expressed as a customer damage function (CDF), which is a non-linear function of load type, time and duration of the interruption. Demand side management is the modification of consumer demand in exchange for financial compensation, based on a prearranged agreement. Demand side management allows shifting a part of a consumer's energy requirements to another point in time. The cost of demand side management is directly linked to either the electricity price at a specific point in time or a predefined price.

8.3.3 Phase Shifting Transformer

A phase shifting transformer (PST) is able to control the active power flow through a certain transmission line, thus influencing the power flows throughout the entire grid [11].

A PST can be represented as a reactance X_{PST} in series with a phase shift as depicted in Figure 8.1 [11]. The power flow through the line is altered by inserting an angle α, which changes the phase angle over the line from θ to $\theta + \alpha$, and consequently the voltage drop over the line impedance $X_L + X_{PST}$ and the power flow through it. As the flow through the line with the phase shifter is changed, the flows throughout the meshed system will change following Kirchhoff's laws. The angle α of the phase shifting transformer is controllable within certain limits, α^{max} and α^{min}. A PST operated by the SO can significantly influence the power flows in the system while the cost associated with its control is very low.

Figure 8.1 Representation of transmission line with a PST.

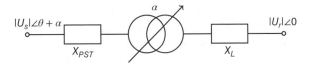

8.3.4 Switching Actions

Reconfiguration of the grid topology using switching actions allows altering the power flow in the system, alleviating possible congestion. Switching as a means of control in a power system is described in [12]. This can enable cheaper redispatch or avoid the need for load shedding when balancing the system. Two main methods exist to alter the topology of a transmission system. The first method, in literature often referred to as transmission switching, allows for the connection or disconnection of a branch within a transmission grid. A second method, named bus bar switching, allows adapting the topology of a transmission grid at breaker level, allowing the substation topology to be altered. Bus bar switching also includes transmission switching. When including switching actions in calculations, each switching action adds a possible topology to the solution set, increasing the complexity of the problem.

8.3.5 Reactive Power Management

Reactive power management is possible through five main control actions: generator excitation controls, switching capacitors or inductors, the use of FACTS controllers, the use of under load tap-changing transformers or the adjustment of the set point of inverter connected generation.

In transmission systems, which are largely inductive in nature, the effect of reactive power compensation has a direct effect on the voltage near the compensation. From a reliability point of view, reactive power compensation is crucial to maintain the voltage within acceptable limits and to avoid cascading of events or voltage instability.

8.3.6 Special Protection Schemes

Special protection schemes form a special type of power system protection components. They are designed to detect a particular system condition that is known to cause unusual stress to the power system and to take a specific and predefined action to counteract the observed condition in a controlled manner [13]. Special protection schemes are linked to a particular (set of) event(s) and control actions. These control actions can include line switching, generator tripping, load shedding and fast HVDC redispatch.

8.4 Linear Implementation of Control Actions in a SCOPF Environment

In this section, the linear implementation of different control actions in an SCOPF environment is discussed. As voltage or reactive power cannot be represented by a DC power flow, reactive power management is not covered. Also, special protection schemes are not shown as they form particular solutions for which no general formulation is possible. The other control actions of 8.3 are treated. The presented formulation generates a finite set of linear constraints that produces a search space in the form of a convex polytope. Such a formulation results in a minimal computational burden. Implementation of the preventive and corrective formulation for the same action are fairly similar in notation. In each subsection, the formulation of the preventive implementation of an action is discussed, after which the changes for the corrective implementation are highlighted. The index *0* for the preventive variables introduced in Section 8.2 in (8.1)–(8.3) is omitted as it is redundant.

8.4.1 Generator Redispatch

The implementation of preventive generation redispatch within an OPF requires the addition of a cost term (8.6) to the objective function (8.1) and the addition of constraints (8.7) to (8.9).

$$\Delta O^{\text{prev,g}} = \sum_g C_g^{\text{prev}} \Delta P_g^{\text{prev}} \qquad (8.6)$$

The generator redispatch cost for a generator g is equal to the marginal price of preventive generator redispatch C_g^{prev} multiplied by the redispatched generation ΔP_g^{prev}. The total preventive redispatch cost is the summation of the cost of preventive generation redispatch for all generators g. The marginal price parameter C_g^{prev} changes depending on the type of generator and on the nature of the redispatch. Piece-wise linear formulation [14] of the objective function allows integration of such a non-constant parameter C_g^{prev}. This formulation is convex as long as the marginal price parameter $C_g^{\text{prev,up}}$ for upward preventive redispatch of a generator g exceeds that of downward preventive redispatch $C_g^{\text{prev,down}}$. An example of such a piece-wise linear representation of the preventive redispatch cost is depicted Figure 8.2.

$$P_g^{\text{min}} \le P_g^{\text{prev}} \le P_g^{\text{max}}, \qquad \forall g \qquad (8.7)$$

$$\Delta P_g^{\text{prev}} = P_g^{\text{prev}} - P_g^{\text{ini}}, \qquad \forall g \qquad (8.8)$$

$$R_g^{\text{prev,down}} \le \Delta P_g^{\text{prev}} \le R_g^{\text{prev,up}}, \qquad \forall g \qquad (8.9)$$

Equation (8.7) ensures that the output limits of each generator g are respected. Equation (8.8) determines the preventive generator redispatch ΔP_g^{prev}, which is equal to the difference between the generator set point after preventive generator redispatch

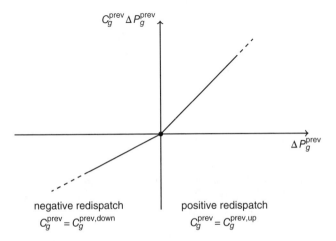

Figure 8.2 Piece-wise linear representation of the preventive generator redispatch cost.

P_g^{prev} and the set point provided by the market after the closure of the day-ahead market P_g^{ini} in the case of a liberalized market structure. In the case of a vertically integrated system, the P_g^{ini} set point reflects the generator outputs made available to the SO. Equation (8.9) ensures that the preventive generator redispatch ΔP_g^{prev} is within the downward $R_g^{\text{prev,down}}$ and the upwards reserves $R_g^{\text{prev,up}}$.

The implementation of corrective generation redispatch within an OPF requires the addition of a cost (8.10) to the objective function (8.1) and the addition of constraints (8.11) to (8.13).

$$\Delta O^{cor,g} = \sum_g \sum_c \lambda_c \zeta_{g,c} C_g^{\text{cor}} \Delta P_{g,c}^{\text{cor}} \tag{8.10}$$

The corrective redispatch cost for a generator g is equal to the marginal price of corrective generator redispatch C_g^{cor} multiplied by the corrective redispatched generation $\Delta P_{g,c}^{\text{cor}}$, taking into account the probability of the occurrence of a certain contingency λ_c. If a contingency c occurs on a certain generator g, that generator can be excluded from the total cost. This can be done by multiplying the cost by a parameter $\zeta_{g,c}$. The constant $\zeta_{g,c}$ is equal to zero when the generator g is unavailable during a contingency c, and one in all other instances of $\zeta_{g,c}$. The integration of different marginal price parameters for upward and downward redispatch can be done using piece-wise linear constraints, similar to preventive redispatch. The total cost is the summation of the cost of redispatch for all generators g and contingencies c.

$$\zeta_{g,c} P_g^{\text{min}} \leq P_{g,c}^{\text{cor}} \leq \zeta_{g,c} P_g^{\text{max}}, \qquad \forall g, c \tag{8.11}$$

$$\Delta P_{g,c}^{\text{cor}} = P_{g,c}^{\text{cor}} - P_g^{\text{prev}}, \qquad \forall g, c \tag{8.12}$$

$$R_g^{\text{cor,down}} - M(1 - \zeta_{g,c}) \leq \Delta P_{g,c}^{\text{cor}} \leq R_g^{\text{cor,up}}, \qquad \forall g, c \tag{8.13}$$

Equation (8.11) ensures that the output limitations of the generators are respected. The addition of $\zeta_{g,c}$ on both the left and the right hand side of the equation ensures that during a contingency involving the generator g, the power output $P_{g,c}^{\text{cor}}$ of the generator is zero. Equation (8.12) determines the corrective redispatch ΔP_g^{prev}, which is equal to the differences between the corrective and preventive generator set points. Equation (8.13) ensures that the corrective redispatch ΔP_g^{prev} is within the contracted downwards $R_g^{\text{cor,down}}$ and the upwards reserves $R_g^{\text{cor,down}}$, which takes the ramp rates of the generators into account. The downward redispatch limit is disabled for a contingent generator using the term $M(1 - \zeta_{g,c})$, where M is a large number.

8.4.2 Load Shedding and Demand Side Management

In order to implement load shedding as linear constraints, each load can be subdivided into smaller loads, each with a constant load shedding cost. The implementation of preventive load shedding within an OPF requires the addition of a cost term (8.14) to the objective function (8.1) and the addition of constraints (8.15) to (8.16).

$$\Delta O^{\text{prev,l}} = \sum_l C_l^{\text{prev}} \Delta P_l^{\text{prev}} \tag{8.14}$$

The load shedding cost for a load l is equal the marginal price of load shedding C_l^{prev} multiplied by the amount of shed load ΔP_l^{prev}, and the total cost is the summation of the cost of load shedding for all loads l.

$$P_l^{\min} \leq P_l^{\text{prev}} \leq P_l^{\max}, \qquad\qquad \forall l \tag{8.15}$$

$$\Delta P_l^{\text{prev}} = P_l^{\text{ini}} - P_l^{\text{prev}}, \qquad\qquad \forall l \tag{8.16}$$

Equation (8.15) ensures that the active power consumed by the load l is limited between $P_l^{\min}$ and $P_l^{\max}$. Equation (8.16) ensures that the shed load ΔP_l^{prev} is equal to the difference between the initial load set point P_l^{ini} and the actual preventive load set point P_l^{prev}. The implementation of corrective load shedding uses a similar formulation. The cost of corrective load shedding (8.17) needs to be calculated for each contingency c taking into account the probability of each contingency λ_c.

$$\Delta O^{\text{cor,l}} = \sum_l \sum_c \lambda_c C_l^{\text{cor}} \Delta P_{l,c}^{\text{cor}} \tag{8.17}$$

The constraints are again similar to the preventive actions, but need to be generated for all contingencies c. Constraint (8.16) needs to be adapted into (8.18) for corrective load shedding in order to reflect the difference between the preventive set point P_l^{prev} and the corrective set point $P_{l,c}^{\text{cor}}$ instead of the difference between the initial set point and the preventive set point.

$$\Delta P_{l,c}^{\text{cor}} = P_l^{\text{prev}} - P_{l,c}^{\text{cor}}, \qquad\qquad \forall l, c \tag{8.18}$$

Demand side management can be implemented in the same way as load shedding, taking into account that both an increase as a decrease of the load power is possible.

8.4.3 Phase Shifting Transformer

A transmission line equipped with a PST can be modeled as depicted in Figure 8.3. The effect of the phase shifting angle α can be modeled as two additional active power injections with opposite signs on each side of the PST, represented by the impedance X_{PST}. The power injections $P^+(\alpha)$ and $P^-(\alpha)$ alter the power flow through the transmission line. The sum of the injections of both fictive generators needs to be zero, as the PST only shifts power, but does not generate or consume power. Their injections are also be limited, taking into account the technical limits $\alpha^{\min}$ and $\alpha^{\max}$ of the PST.

The implementation of a PST within an OPF requires the addition of a cost term (8.19) to the objective function (8.1) and the addition of constraints (8.20) to (8.22).

$$\Delta O^{\text{prev,p}} = \sum_p C_p^{\text{prev}} \Delta P_p^{\text{prev}} \tag{8.19}$$

The cost for a PST p is equal to the product of the marginal price of preventive usage of the phase shifting transformer C_p^{prev}, accounting for the wear of the transformer, and

Figure 8.3 Model of a PST.

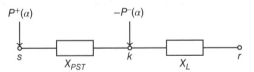

the difference of the phase shifting transformer setting ΔP_p^{prev}. The total cost is the summation of the cost of phase shifting for all phase shifting transformers p.

$$P_p^{\text{prev},+} - P_p^{\text{prev},-} = 0, \qquad \forall p \tag{8.20}$$

$$\Delta P_p^{\text{prev},+,\min} \leq \Delta P_p^{\text{prev},+} \leq \Delta P_p^{\text{prev},+,\max}, \qquad \forall p \tag{8.21}$$

$$|P_p^{\text{ini},+} - P_p^{\text{prev},+}| = \Delta P_p^{\text{prev}}, \qquad \forall p \tag{8.22}$$

Equation (8.20) ensures that the fictive power $P_p^{\text{prev},+}$ injected at node s is equal to the power $P_p^{\text{prev},-}$ extracted at node k (Fig. 8.3). Equation (8.21) limits the fictive power injection $P_p^{\text{prev},+}$, and consequently also $P_p^{\text{prev},-}$, taking into account the limits on α. Equation (8.22) determines ΔP_p^{prev}, which is equal to the absolute difference between the initial set point $P_p^{\text{ini},+}$ and the set point $P_p^{\text{prev},+}$ of the phase shifting transformer in the preventive grid state. As it is the objective of the optimization to minimize the cost, the absolute value $|P_p^{\text{ini},+} - P_p^{\text{prev},+}|$ can be implemented in a linear way by replacing equations (8.21) and (8.22) with equations (8.23), (8.24) and (8.25). Either $\Delta P_p^{\text{prev,up}}$ or $\Delta P_p^{\text{prev,down}}$ deviates from zero if the set point of the PST is changed, representing the absolute value $|P_p^{\text{ini},+} - P_p^{\text{prev},+}|$. The other variable is then equal to zero.

$$\Delta P_p^{\text{prev}} = \Delta P_p^{\text{prev,up}} + \Delta P_p^{\text{prev,down}}, \qquad \forall p \tag{8.23}$$

$$\left.\begin{matrix} 0 \\ P_p^{\text{prev},+} - P_p^{\text{ini},+} \end{matrix}\right\} \leq \Delta P_p^{\text{prev,up}} \leq \Delta P_p^{\text{prev},+,\max}, \qquad \forall p \tag{8.24}$$

$$\left.\begin{matrix} 0 \\ P_p^{\text{ini},+} - P_p^{\text{prev},+} \end{matrix}\right\} \leq \Delta P_p^{\text{prev,down}} \leq \left|\Delta P_p^{\text{prev},+,\min}\right|, \qquad \forall p \tag{8.25}$$

The implementation of corrective actions with a PST is similar to the preventive formulation. The cost for corrective PST usage (8.26) needs to be calculated for each contingency c, taking into account the probability of each contingency λ_c.

$$\Delta O^{\text{cor,P}} = \sum_p \sum_c \lambda_c C_p^{\text{cor}} \Delta P_{p,c}^{\text{cor}} \tag{8.26}$$

The constraints are similar to the preventive formulation but need to be generated for all contingencies c individually. Constraint (8.22) needs to be adapted into (8.27) for the corrective usage of the PST in order to reflect the difference between the preventive set point $P_p^{\text{prev},+}$ and the corrective set point $P_{p,c}^{\text{cor},+}$ instead of the difference between the initial set point and the preventive set point.

$$\left|P_p^{\text{prev},+} - P_{p,c}^{\text{cor},+}\right| = \Delta P_{p,c}^{\text{cor}}, \qquad \forall p, c \tag{8.27}$$

8.4.4 Switching

Breakers can be modeled in an OPF as lossless elements using on/off constraints [15]. An on/off constraint is a constraint that is activated when the corresponding binary variable δ_b of the considered breaker b is equal to one. The implementation of preventive switching of a breaker within an OPF requires the addition of a cost (8.28) to the objective function (8.1) and the addition of constraints (8.29) to (8.31).

$$\Delta O^{\text{prev,b}} = \sum_b C_b^{\text{prev}} \Delta \delta_b^{\text{prev}} \tag{8.28}$$

The cost for operating a breaker b is equal to the product of the marginal price of preventive usage of the breaker C_b^{prev}, accounting for the wear of the breaker, and the change of the state of the breaker $\Delta\delta_b$. The total cost is the summation of the cost of breaker usage for all breakers b.

$$\delta_b^{\text{prev}} P_b^{\text{min}} \leq P_b^{\text{prev}} \leq \delta_b^{\text{prev}} P_b^{\text{max}}, \qquad \forall b \qquad (8.29)$$

$$-M(1 - \delta_b^{\text{prev}}) \leq \theta_i^{\text{prev}} - \theta_j^{\text{prev}} \leq M(1 - \delta_b^{\text{prev}}), \qquad \forall b \qquad (8.30)$$

$$\left| \delta_b^{\text{ini}} - \delta_b^{\text{prev}} \right| = \Delta\delta_b^{\text{prev}}, \qquad \forall b \qquad (8.31)$$

Equation (8.29) ensures that the power flow P_b^{prev} through a breaker b is equal to zero if the breaker is open ($\delta_b^{\text{prev}} = 0$). If the breaker is closed ($\delta_b^{\text{prev}} = 1$), (8.29) ensures that the active power limits of the breaker are respected. The actual power flow P_b^{prev} through the breaker is determined by the power balance equations included in the standard OPF formulation. Equation (8.30) sets the voltage angles θ of both end nodes of the breaker equal to each other when the breaker b is closed. If the breaker is open, (8.30) ensures that the voltage angles are independent from each other. Equation (8.31) determines $\Delta\delta_b^{\text{prev}}$ which is the absolute difference between the status of the breakers for their initial state δ_b^{ini} and their preventive state δ_b^{prev}. The absolute value in (8.31) can be linearized using the same technique as described for the phase shifting transformers.

The implementation of corrective switching of a breaker is again similar to the preventive formulation. The cost for breaker usage (8.32) needs to be calculated for each contingency c taking into account the probability of each contingency λ_c.

$$\Delta O^{\text{cor,b}} = \sum_b \sum_c \lambda_c C_b^{\text{cor}} \Delta\delta_{b,c}^{\text{cor}} \qquad (8.32)$$

Also the constraints are similar, but need to be generated for all contingencies c. Constraint (8.31) becomes (8.33) for the corrective usage of the breakers in order to reflect the difference between the preventive set point δ_b^{prev} and the corrective set point $\delta_{b,c}^{\text{cor}}$ instead of the difference between the initial set point and the preventive set point.

$$\left| \delta_b^{\text{prev}} - \delta_{b,c}^{\text{cor}} \right| = \Delta\delta_{b,c}^{\text{cor}}, \qquad \forall b, c \qquad (8.33)$$

8.5 Case Study of Preventive and Corrective Actions

To show the effect of different preventive and corrective actions, they are implemented in a DC SCOPF and tested on the Roy Billinton Test System (RBTS) (Figure 8.4). The RBTS consists of five substations. A load (Table 8.2) is connected to each substation, except for substation S1. The nodes of the RBTS are interconnected by seven transmission lines (Table 8.3). Eleven generators (Table 8.4) are located in the grid, of which four are connected to substation S1 and seven to substation S2.

In this section, three case studies are conducted. Each case study is a one-hour Day Ahead Congestion Forecast (DACF) where all possible N-1 contingencies are considered. The Day Ahead Congestion Forecast (DACF) corresponds to the instance in the grid scheduling where the system operator receives all planned market positions from the different stakeholders and from the neighboring systems. The DACF is used to assess whether the expected operating points are within the security boundaries. If not, the

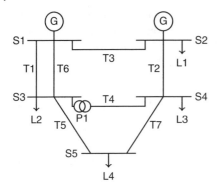

Figure 8.4 Roy Billinton Test System [16].

system operator will take preventive actions, or prepare corrective actions. In order to keep the results clear, it is assumed that load and renewable generation remain constant during the considered hour. During each case study, the SO has different preventive and corrective actions at his disposal. The available actions during each case study are depicted in Table 8.1.

The preliminary set points of all generators (P_g^{ini}) are submitted to the SO. Based on that data, a DACF analysis is conducted by the SO in order to determine the preventive and corrective actions it has to take in order to have a stable system at all times. During this analysis, the considered contingencies are failure of one generator or one transmission line, with the probability of failure λ_c denoted in their respective tables.

8.5.1 Case Study 1: Generator Redispatch and Load Shedding (CS1)

In the first case study, preventive generation redispatch, corrective generation redispatch and corrective load shedding are the only available actions. The preventive redispatch allows adapting the generation set point compared to those set by the market P_g^{ini}.

Table 8.1 Available actions for each case study.

	Preventive Actions			Corrective Actions			
	Redisp.	PST	Switch	Redisp.	Shed	PST	Switch
1	✓	–	–	✓	✓	–	–
2	✓	✓	–	✓	✓	✓	–
3	✓	–	✓	✓	✓	–	✓

Table 8.2 Load data.

Load	Sub.	$P_l^{\min}$ [MW]	$P_l^{\max}$ [MW]	C_l^{cor} [€/MWh]
L1	S2	10	20	10206
L2	S3	75	85	10206
L3	S4	30	40	10206
L4	S5	10	20	10206

Table 8.3 Transmission line data.

Line	Length [km]	Admittance [p.u.]	P_t^{max} [MW]	λ_c [-]
T1, T6	48	5.555	85	0.001713
T2	160	1.666	50	0.005710
T3	128	2.083	50	0.004568
T4, T5, T7	32	8.333	71	0.001142

As preventive generation redispatch deviates from the ideal market scheduling, it comes at a cost. It is only activated if the market set points cause congestion in the system or if preventive redispatch is cheaper then corrective redispatch taking into account the probability of the contingencies. The market set points P_g^{ini} provided in Table 8.4 cause congestion on the transmission lines T2 and T3 as a result of an excessive generation output on substation S2. Preventive generation redispatch alleviates this congestion by reducing the generation of generator G5 and G10 of substation S2 and increasing the generation of generators G1 and G2 of substation S1 (Fig. 8.5).

Unbalances in generation and load can be caused by a contingency, either by the failure of a generator or because of congestion caused by an outage of a transmission line. In this first case study, these unbalances can only be corrected by corrective generation redispatch or by shedding load. The corrective generation redispatch and shed load are depicted in Figure 8.6 for contingency C1 to C18.

8.5.2 Case Study 2: Generator Redispatch, Load Shedding and PST (CS2)

The implementation of a PST can help to alleviate congestion during certain contingencies and consequently enable cheaper generation redispatch. Case study 2 introduces a PST P1 to transmission line T4 (Figure 8.4). For the purpose of this case study, the angle limits are set to ±0.4 (rad). The usage of the PST causes wear of the transformer tap changer and consequently comes at a cost. The cost of using the PST is set equal to an arbitrary value of 400 €/(rad). The implementation of the PST P1 reduces congestion and realizes less or cheaper generator redispatch during contingencies C12 and C17 (Figure 8.7).

Contingencies C12 and C17 exclude either transmission line T1 or T6, which are equivalent, resulting in the same grid state. This causes congestion on both T6/T1 and T2. The total power that can be transported by T6/T1 (85 MW) and T2 (50 MW) is 135 MW, which is insufficient to supply the loads connected to substations S3, S4 and S5 with a total demand of 145 MW. In order to correct this unbalance, 10 MW of load L2 is shed. To supply the remaining load, the flow through the grid needs to be adapted. In case study 1, this was accomplished by downwards redispatch of the generators G2, G3 and G4 of substation S1 and upwards generator redispatch of G5 and G11 of substation S2. The downwards redispatch of G3 and G4 comes at a high cost as these are renewable generators and their downwards redispatch cost significantly exceeds that of a conventional generator.

Table 8.4 Generator data.

Gen.	Sub.	P_g^{max} [MW]	P_g^{ini} [MW]	λ_c [-]	Preventive Generator Data				Corrective Generator Data			
					$C_g^{pre,down}$ [€/MWh]	$C_g^{pre,up}$ [€/MWh]	$R_g^{pre,down}$ [MW]	$R_g^{pre,up}$ [MW]	$C_g^{cor,down}$ [€/MWh]	$C_g^{cor,up}$ [€/MWh]	$R_g^{cor,down}$ [MW]	$R_g^{cor,up}$ [MW]
G1	S1	40	4.2	0.015	4.45	12.05	−40	40	44.5	120.5	0	10
G2	S1	40	15.8	0.015	4.35	12.05	−40	40	43.5	120.5	−10	10
G3	S1	10	8.7	0.015	–	–	0	0	−3000.0	–	−10	0
G4	S1	20	16.3	0.015	–	–	0	0	−3000.0	–	−20	0
G5	S2	40	38.4	0.005	5.00	14.00	−40	40	50.0	140.0	−10	10
G6	S2	20	19.7	0.005	4.30	11.90	−20	20	43.0	119.0	−20	0
G7	S2	20	19.2	0.005	–	–	0	0	−3000.0	–	−20	0
G8	S2	20	19.8	0.005	–	–	0	0	−3000.0	–	−20	0
G9	S2	20	19.3	0.005	–	–	0	0	−3000.0	–	−20	0
G10	S2	5	3.6	0.005	5.45	15.35	−5	5	54.5	153.5	0	5
G11	S2	5	0	0.005	4.40	12.20	−5	5	44.0	122.0	0	5

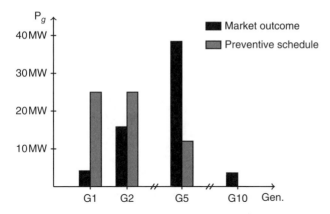

Figure 8.5 Case Study 1: Preventive generator redispatch.

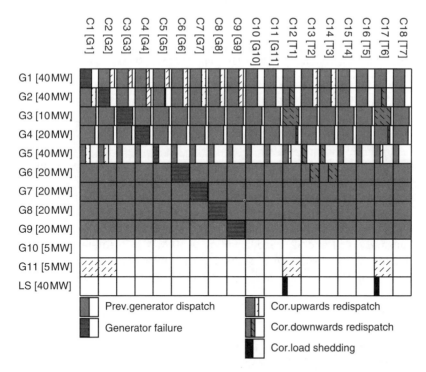

Figure 8.6 Case Study 1: Corrective generator redispatch. Each square of the first eleven rows depicts the corrective generator set point $P^{cor}_{g,c}$ relative to the maximum set point P^{max}_g of that generator for a contingency. The different contingencies are shown in columns. The corrective generator set point is a combination of the preventive generator set point P^{prev}_g, depicted in gray, adjusted with possible upwards or downwards corrective generation redispatch, respectively depicted in green and orange. Each square in the last row depicts the part of the total load that is shed during a contingency.

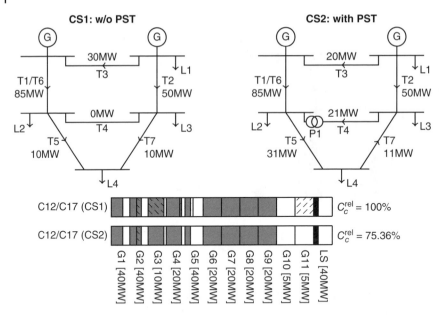

Figure 8.7 Case study 2: The difference in corrective generation redispatch and load shedding between CS1 without PST in the grid and CS2 with a PST connected to transmission line T4 for contingencies C12 (T1) and C17 (T7).

$$\Delta P^{\text{cor}}_{L2,C12} + \Delta P^{\text{cor,up}}_{G5,C12} + \Delta P^{\text{cor,up}}_{G11,C12}$$

$$=$$

$$\Delta P^{\text{cor,down}}_{G2,C12} + \Delta P^{\text{cor,down}}_{G3,C12} + \Delta P^{\text{cor,down}}_{G4,C12}$$

The redispatch can be avoided by adapting the power flow in the power system using a PST. In case study 2, the set point of the PST for contingencies C12 and C17 is adapted to change the power flow through T4 from 0 to 21 MW. This influences the power flow in the system in such a way that the power flow through T3 is reduced to 20 MW compared to 30 MW in the case without a PST, negating the need for redispatch between generators of substations S1 and S2. Load shedding of 10 MW (L2) still takes place as the lines T6/T1 and T2 are incapable of supplying the entire load of substation S3, S4 and S5. The corresponding downward redispatch is provided by generator G2 of substation S1.

$$\Delta P^{\text{cor}}_{L2,C12} = \Delta P^{\text{cor,down}}_{G2,C12}$$

8.5.3 Case Study 3: Generator Redispatch, Load Shedding and Switching (CS3)

An alternative approach to adapting the power flow in a transmission system is changing its topology. Case study 3 introduces transmission line switching, which allows eliminating transmission lines by opening their breakers. This causes wear of the breakers and consequently comes at a cost. For the purpose of this case study, the cost of breaker operation is set at an arbitrary value of 400 €. Elimination of transmission lines T4 and T5 influences the power flow in the system is such a way that the power flow through T3 is reduced to 10 MW (Fig. 8.8). This allows the negative redispatch needed to compensate for the load shedding of 10 MW (L4) to be done by generator G5.

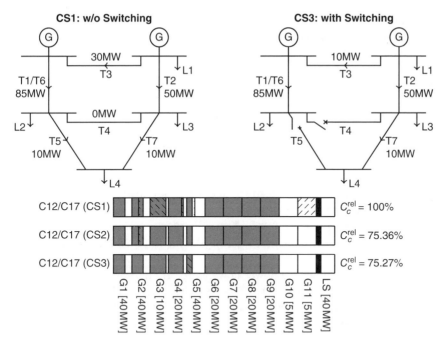

Figure 8.8 Case study 3: The difference in corrective generation redispatch and load shedding between CS1 without transmission line switching in the grid and CS3 with transmission line switching for contingencies C12 (T1) and C17 (T7).

$$\Delta P^{cor}_{L4,C12} = \Delta P^{cor,down}_{G5,C12}$$

A note must be made of the fact that these switching actions reduce the RBTS system to a radial system, which in reality may not be an acceptable strategy.

8.6 Conclusions

This chapter presents the main reliability control actions. These actions allow the system operator to influence the operating point. They allow the operator to either work with a sufficient margin through preventive control, or to quickly return to the safe operating area using corrective control actions. In order to calculate the optimal use of these control actions, an optimal power flow program can be used. Each of the control actions add a number of constraints to the optimization program, and alter the objective function. In this chapter, the linear implementation of control actions in an OPF is given. An extensive case study is used to show how different actions influence both system reliability and system cost on a small test system.

References

1 Kundur, P., Paserba, J., Ajjarapu, V., Andersson, G., Bose, A., Van Cutsem, T., Canizares, C., Hatziargyriou, N., Hill, D., Vittal, V., Stankovic, A., and Taylor, C. (2004) Definition and classification of power system stability IEEE/CIGRE joint task

force on stability terms and definitions. *IEEE Transactions on Power Systems*, **19** (3), 1387–1401, doi: 10.1109/TPWRS.2004.825981.

2 Wood, A.J. and Wollenberg, B.F. (1996) *Power generation, operation and control, 2nd Edition*, John Wiley & Sons, Ltd.

3 Capitanescu, F., Martinez Ramos, J.L., Panciatici, P., Kirschen, D., Marano Marcolini, A., Platbrood, L., and Wehenkel, L. (2011) State-of-the-art, challenges, and future trends in security constrained optimal power flow. *Electric Power Systems Research*, **81** (8), 1731–1741, doi: 10.1016/j.epsr.2011.04.003.

4 Conejo, A.J., Castillo, E., Minguez, R., and Garcia-Bertrand, R. (2006) *Decomposition techniques in mathematical programming: engineering and science applications*, Springer Science & Business Media.

5 Stott, B., Jardim, J., and Alsaç, O. (2009) Dc power flow revisited. *IEEE Transactions on Power Systems*, **24** (3),1290–1300.

6 Christie, R.D., Wollenberg, B.F., and Wangensteen, I. (2000) Transmission management in the deregulated environment. *Proceedings of the IEEE*, **88** (2), 170–195, doi: 10.1109/5.823997.

7 Han, Z.X. (1982) Phase shifter and power flow control. *IEEE Transactions on Power Apparatus and Systems*, **PAS-101** (10), 3790–3795, doi: 10.1109/TPAS.1982.317064.

8 Hedman, K.W., Neill, R.P.O., Fisher, E.B., and Oren, S.S. (2009) Optimal transmission switching with contingency analysis. *IEEE Transactions on power systems*, **24** (3), 1577–1586.

9 Van den Bergh, K., Couckuyt, D., Delarue, E., and D'haeseleer, W. (2015) Redispatching in an interconnected electricity system with high renewables penetration. *Electric Power Systems Research*, **127** (January), 64–72, doi: 10.1016/j.epsr.2015.05.022.

10 Capozza, A., D'Adamo, C., Mauri, G., and Pievatolo, A. (2005) Load shedding and demand side management enhancements to improve the security of a national electrical system. *2005 IEEE Russia Power Tech, PowerTech*, pp. 1–7, doi: 10.1109/PTC.2005.4524568.

11 Verboomen, J., Van Hertem, D., Schavemaker, P., Kling, W., and Belmans, R. (2005) Phase shifting transformers: principles and applications. *2005 International Conference on Future Power Systems*, (February 2016), p. 6, doi: 10.1109/FPS.2005.204302.

12 Glavitsch, H. (1985) Switching as means of control in the power system. *International Journal of Electrical Power & Energy Systems*, **7** (2), 92–100, doi: 10.1016/0142-0615(85)90014-6.

13 Anderson, P.M. and LeReverend, B.K. (1996) Industry experience with special protection schemes. *IEEE Transactions on Power Systems*, **11** (3), 1166–1179, doi: 10.1109/59.535588.

14 Somuah, C.B. and Khunaizi, N. (1990) Application of linear programming redispatch technique to dynamic generation allocation. *IEEE Transactions on Power Systems*, **5** (1), 20–26, doi: 10.1109/59.49081.

15 Henneaux, P. and Kirschen, D.S. (2016) Probabilistic security analysis of optimal transmission switching. *IEEE Transactions on Power Systems*, **31** (1), 508–517.

16 Billinton, R. and Allan, R.N. (1996) *Reliability evaluation of power systems*, Plenum Press.

9

Model-based Predictive Control for Damping Electromechanical Oscillations in Power Systems

Da Wang

Researcher of Intelligent Electrical Power Systems, Intelligent Electrical Power Grids (IEPG), TU Delft, Netherlands

9.1 Introduction

Electromechanical oscillations mean relative motions of generator shafts and resultant fluctuations of electrical variables (current, voltage, etc.) In modern large-scale electrical power systems, long transmission distances over weak grids, highly variable generation patterns, heavy loading, and so on, tend to increase the probability of appearance of sustained wide area electromechanical oscillations [1, 2]. Moreover, higher and higher penetration of renewable energy also changes current damping characteristics and increases the risk of oscillation instability [3]. Such oscillations threaten the security of power systems, and if they cause cascading faults, can lead to large-scale blackouts [4].

It is usually considered that electromechanical oscillations are caused by insufficient damping of power systems. Consequently, most researches on damping oscillations focus on developing different controllers, like Power System Stabilizers (PSSs), and Thyristor Controlled Series Compensators (TCSCs), in order to improve system damping [5]. Conventionally, these controllers use local measurements at their inputs. The control rules are determined in off-line studies using time domain simulations such as Prony or Eigen analysis, and remain fixed in practice. With recent developments in Wide Area Measurement Systems (WAMSs), further research is being devoted to improving the ability to damp inter-area oscillations by introducing global signals [6, 7].

Indeed, the damping depends on the structure, controller setting, and operation mode of power systems, but the latter continuously varies in practice. Therefore, an expected damping controller should first be able to adaptively adjust its control behaviors following changes of operation conditions. Moreover, since global damping effects are decided by all damping controllers scattered into different areas and installed at different times, a good damping controller is also expected to coordinate with other controllers in order to obtain satisfactory global effects.

Based on advances in power system modeling and large-scale optimization, a new idea is to apply Model Predictive Control (MPC) to design such an adaptive and coordinated damping controller. MPC is a proven technique with numerous real-life applications, whose success comes from three factors: an explicit process model, a future temporal horizon, and an ability to deal with different constraints [8, 9]. These factors allow MPC to deal with all significant dynamics over a future temporal horizon and to drive the

Dynamic Vulnerability Assessment and Intelligent Control for Sustainable Power Systems, First Edition.
Edited by José Luis Rueda-Torres and Francisco González-Longatt.
© 2018 John Wiley & Sons Ltd. Published 2018 by John Wiley & Sons Ltd.

outputs of the controlled plant to evolve more closely along a desired trajectory, under the condition of satisfying different constraints on inputs, states, and outputs.

MPC has been applied to optimal power flow, voltage regulation, and dynamic stability of power systems [10–12]. As far as damping control of electromechanical oscillations is concerned, one of the earliest MPC applications is presented in [13] where generalized predictive control is used to switch capacitors. The control strategy is computed by minimizing a quadratic cost function that combines local system outputs and rates-of-change of controls over a prediction horizon. An MPC for step-wise series reactance modulation of a TCSC is presented in [14] to stabilize electromechanical oscillations, where a reduced two-machine model of the power system is used and updated using local measurements. Penalizing deviations of predicted outputs and control increments through an objective function, reference [15] proposes a model predictive adaptive controller to damp inter-area oscillations in a four-generator system. In references [16, 17], a bank of linearized system models is developed to represent system response under oscillations. For each model in the bank, an observer-based state feedback controller is designed a priori, and MPC optimizes the weights for individual controllers.

In this chapter, MPC is applied to calculate supplementary signals for existing damping controllers (exciters and TCSC). These signals are superimposed on the inputs of controllers and continuously updated in order to improve and coordinate the damping effects of the controllers. Defining angular speeds as outputs, MPC treats damping control as a multi-step optimization problem with discrete dynamics and costs, which penalizes deviations of angular speeds from the reference speed over a future horizon. At each control time, MPC collects current system states, solves this optimization problem, and calculates supplementary signals. Utilizing these signals, it forces angular speeds to return to the reference speed and remain as close to the latter as possible. When all generators are running at the reference speed, oscillations are damped.

In order to implement this idea, a centralized MPC controller is first designed based on a linearized, discrete state space model. Its performance is evaluated both in ideal conditions and considering realistic State Estimation (SE) errors and computation and communication delays. Different damping controllers are also studied in order to assess their versatility. Next, the centralized MPC is further extended into a distributed one with the aim of making it more viable for large-scale or multi-area systems. The decoupling and coordination between subsystems is analyzed. Finally, a robust hierarchical MPC is proposed that introduces a second layer of MPC controllers on local generators. The performances of three MPC schemes are demonstrated and compared on a 16-generator, 70-bus system.

9.2 MPC Basic Theory & Damping Controller Models

9.2.1 What is MPC?

The basic idea of MPC is illustrated in Figure 9.1, which describes dynamic control of a single-input single-output system [18]. System dynamics are discretized by a small simulation step of δ seconds. Every Δt seconds, MPC is carried out one time. It is assumed

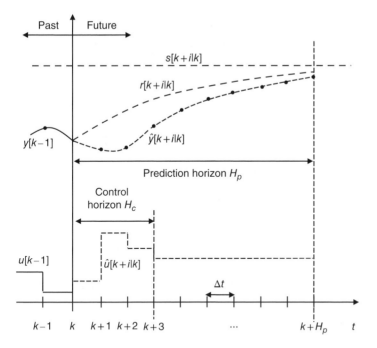

Figure 9.1 MPC concept.

that Δt is an integral multiple of δ. y and u denote system out and control input, respectively. Assuming the current control time is labeled as k (namely $k\Delta t$ seconds), MPC works as follows:

– It collects current state $x[k|k]$, and corresponding to a series of $\hat{u}[k + i|k]$, predicts $\hat{y}[k + i|k]$ over a future horizon $k + 1$, $k + 2 \ldots k + H_p$, by iterating a state space model of controlled plant given in (9.1):

$$\hat{x}[k + 1|k] = \mathbf{A}x[k|k] + \mathbf{B}\hat{u}[k|k]$$
$$\hat{y}[k + 1|k] = \mathbf{C}\hat{x}[k + 1|k] \tag{9.1}$$

$\hat{u}[k + i|k]$ varies over the first $H_c\Delta t$ seconds but remains constant thereafter. H_c is called the control horizon. H_p is the prediction horizon. The sign "^" means the predicted value of a variable.

– A cost function is defined as

$$V[k] = \sum_{i=N_w}^{N_p} \|\hat{y}[k + i|k] - r[k + i|k]\|_{Q_i}^2 + \sum_{i=0}^{N_c} \|\Delta\hat{u}[k + i|k]\|_{R_i}^2 \tag{9.2}$$

It penalizes deviations of $\hat{y}[k + i|k]$ from a reference trajectory $r[k + i|k]$ and increments of $\hat{u}[k + i|k]$. $r[k + i|k]$ defines an ideal trajectory along which the plant returns to a set-point trajectory $s[k + i|k]$, and it can be replaced by $s[k + i|k]$ in (9.2). Q_i and R_i are weight matrices. N_p is the number of simulations steps covered by H_p, that is to say $N_p = H_p\Delta t / \delta$. Similarly, $N_c = (H_c\text{-}1)\Delta t / \delta$. N_w means that the deviations of $\hat{y}[k + i|k]$ are penalized from the N_wth simulation step. During one Δt, it is assumed that $\hat{u}[k + i|k]$ is constant at each simulation step.

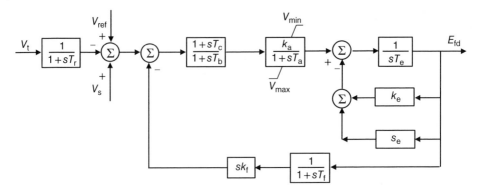

Figure 9.2 Exciter block diagram.

- The optimal sequence of $\hat{u}[k + i|k]$ is the one that minimizes $V[k]$:

$$\min_{\hat{u}[\cdot]} V[k] = \min_{\hat{u}[\cdot]} \sum_{i=N_w}^{N_p} \|\hat{y}[k + i|k] - r[k + i|k]\|_{Q_i}^2 + \sum_{i=0}^{N_c} \|\Delta\hat{u}[k + i|k]\|_{R_i}^2 \qquad (9.3)$$

This is a Quadratic Programming (QP) problem, which can be solved by interior point methods, active set methods, and so on.

- Once an optimal sequence of $\hat{u}[k + i|k]$ is chosen, only the first element $\hat{u}[k|k]$ is applied to the plant.
- The above cycle of prediction, optimization, and application is repeated at subsequent control times, in order to make $\hat{y}[k + i|k]$ return back to $s[k + i|k]$.

9.2.2 Damping Controller Models

In this section, the models of damping controllers considered in this chapter are described to facilitate readers' understanding of the developed MPC controllers. First, an IEEE Type DC1 model is adopted to represent exciter dynamics, as shown in Figure 9.2.

V_t is terminal voltage of the generator after load compensator; E_{fd} is excitation voltage; V_s are supplementary inputs for the exciter, which could be PSS/MPC signals. Defining $V_{tr}, V_{as}, V_r, E_{fd}, R_f$ as state variables, the dynamics are described by

$$
\begin{bmatrix} \dfrac{dV_{tr}}{dt} \\[2mm] \dfrac{dV_{as}}{dt} \\[2mm] \dfrac{dV_r}{dt} \\[2mm] \dfrac{dE_{fd}}{dt} \\[2mm] \dfrac{dR_f}{dt} \end{bmatrix} =
\begin{bmatrix}
-\dfrac{1}{T_r} & & & & \\[2mm]
-\dfrac{1}{T_b} & -\dfrac{1}{T_b} & & -\dfrac{k_f}{T_b T_f} & \dfrac{k_f}{T_b T_f} \\[2mm]
-\dfrac{K_a T_c}{T_a T_b} & \dfrac{K_a(T_b - T_c)}{T_a T_b} & -\dfrac{1}{T_a} & -\dfrac{k_a T_c k_f}{T_a T_b T_f} & \dfrac{k_a T_c k_f}{T_a T_b T_f} \\[2mm]
& & \dfrac{1}{T_e} & -\dfrac{k_e + s_e}{T_e} & \\[2mm]
& & & \dfrac{1}{T_f} & -\dfrac{1}{T_f}
\end{bmatrix}
\begin{bmatrix} V_{tr} \\[2mm] V_{as} \\[2mm] V_r \\[2mm] E_{fd} \\[2mm] R_f \end{bmatrix}
$$

$$+ \begin{bmatrix} \dfrac{1}{T_r} & & \\ & \dfrac{1}{T_b} & \dfrac{1}{T_b} \\ & \dfrac{K_a T_c}{T_a T_b} & \dfrac{K_a T_c}{T_a T_b} \end{bmatrix} \begin{bmatrix} V_t \\ V_s \\ V_{ref} \end{bmatrix} \tag{9.4}$$

A PSS with the structure in Figure 9.3 is used to produce a supplementary signal V_s for the exciter.

Its dynamics are described in the following equation, through state variables $pss1$, $pss2$, and $pss3$:

$$\begin{bmatrix} \dfrac{dpss1}{dt} \\ \dfrac{dpss2}{dt} \\ \dfrac{dpss3}{dt} \end{bmatrix} = \begin{bmatrix} -\dfrac{1}{T_w} & & \\ \dfrac{G_{pss}(T_{n1} - T_{d1})}{T_{d1}^2} & -\dfrac{1}{T_{d1}} & \\ \dfrac{G_{pss}T_{n1}(T_{n2} - T_{d2})}{T_{d1}T_{d2}^2} & \dfrac{T_{d2} - T_{n2}}{T_{d2}^2} & -\dfrac{1}{T_{d2}} \end{bmatrix} \begin{bmatrix} pss1 \\ pss2 \\ pss3 \end{bmatrix}$$

$$+ \begin{bmatrix} \dfrac{1}{T_w} \\ \dfrac{G_{pss}(T_{d1} - T_{n1})}{T_{d1}^2} \\ \dfrac{G_{pss}T_{n1}(T_{d2} - T_{n2})}{T_{d1}T_{d2}^2} \end{bmatrix} S_{pss} \tag{9.5}$$

TCSC is modeled by the block diagram in Figure 9.4.

Its dynamics are determined by

$$\begin{bmatrix} \dfrac{dB_{t\,csc\,0}}{dt} \\ \dfrac{dB_{tcsc}}{dt} \end{bmatrix} = \begin{bmatrix} -\dfrac{1}{T_b} & \\ \dfrac{k_r(T_b - T_c)}{T_r T_b} & -\dfrac{1}{T_r} \end{bmatrix} \begin{bmatrix} B_{t\,csc\,0} \\ B_{tcsc} \end{bmatrix} + \begin{bmatrix} \dfrac{1}{T_b} \\ \dfrac{T_c k_r}{T_b T_r} \end{bmatrix} S_{tcsc} \tag{9.6}$$

More details about these models are given in [19].

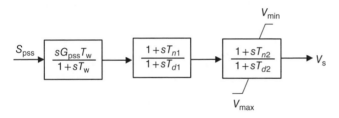

Figure 9.3 PSS block diagram.

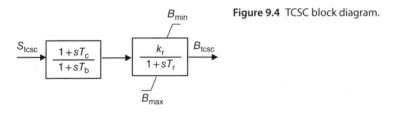

Figure 9.4 TCSC block diagram.

9.3 MPC for Damping Oscillations

9.3.1 Outline of Idea

Figure 9.5 illustrates how to apply MPC to damp oscillations of one generator. Angular speed is considered as the controlled output y. A cost function penalizes deviations of $\hat{y}[k + i|k]$ from the reference speed ω_0 over a prediction horizon H_p. At a control time k, MPC calculates a supplementary signal $\hat{u}[k|k]$ and combines it with PSS output V_{pss} as the V_s input of exciter to minimize the deviations. At subsequent control time $k+i$, the above process is repeated, and $\hat{u}[k|k]$ is continuously updated following changes in operation conditions. In this way, the angular speed $\hat{y}[k + i|k]$ is forced to return back to ω_0.

In a multi-generator system, the cost function is redefined to penalize deviations of all angular speeds. MPC signals are applied to each exciter to optimize and coordinate

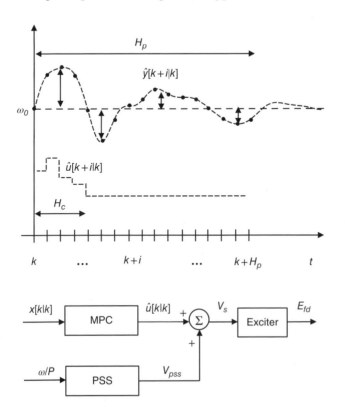

Figure 9.5 MPC for generator.

Figure 9.6 MPC for TCSC.

their damping effects. As well as exciters and PSSs, a TCSC on one inter-area tie line is also investigated in this chapter. As shown in Figure 9.6, the MPC signal $\hat{u}[k|k]$ adjusts the susceptance B_{tcsc}, and in this way changes transmission power through the tie line to reduce oscillations.

9.3.2 Mathematical Formulation

The dynamics of different devices in power systems can be represented by differential equations similar to (9.4), (9.5), and (9.6). All differential equations and network algebraic equations are combined according to network topology. So, power system dynamics can be represented by the following continuous state space model:

$$\begin{cases} \dfrac{dx(t)}{dt} = \mathbf{A}'x(t) + \mathbf{B}'u(t) \\ y(t) = \mathbf{C}'x(t) + \mathbf{D}'u(t) \end{cases} \tag{9.7}$$

Here, $\mathbf{A}'$ denotes the state matrix; $\mathbf{B}'$ is the input matrix; $\mathbf{C}'$ is the output matrix and $\mathbf{D}'$ is the feed forward matrix. $x(t)$ consists of all state variables; $u(t)$ represents a vector of control inputs; $y(t)$ is a vector of system outputs. Since there is normally no direct connection between $u(t)$ and $y(t)$, $\mathbf{D}'$ is assumed to be zero.

Based on (9.7), the transition at an initial time t_0 for a small simulation step of δ is

$$\begin{cases} x(t_0 + \delta) = (\delta\mathbf{A}' + \mathbf{I})x(t_0) + \delta\mathbf{B}'u(t_0) \\ y(t_0 + \delta) = \mathbf{C}'x(t_0 + \delta) \end{cases} \tag{9.8}$$

If t_0 is labeled as time k, (9.8) can be rewritten as

$$\begin{cases} x[k + 1|k] = \mathbf{A}x[k|k] + \mathbf{B}u[k|k] \\ y[k + 1|k] = \mathbf{C}x[k + 1|k] \end{cases} \tag{9.9}$$

where $\mathbf{A} = \delta\mathbf{A}' + \mathbf{I}, \mathbf{B} = \delta\mathbf{B}'$ and $\mathbf{C} = \mathbf{C}'$.

Based on current system state $x[k|k]$, future $x[k + i|k]$ over a time horizon H_p is calculated by iterating (9.9) N_p times ($N_p = H_p\Delta t/\delta$):

$$\begin{bmatrix} \hat{x}[k + 1|k] \\ \hat{x}[k + 2|k] \\ \vdots \\ \hat{x}[k + N_p|k] \end{bmatrix} = \mathbf{P}_x x[k|k] + \mathbf{P}_u \begin{bmatrix} \hat{u}[k|k] \\ \hat{u}[k + 1|k] \\ \vdots \\ \hat{u}[k + N_p - 1|k] \end{bmatrix} \tag{9.10}$$

where

$$\mathbf{P}_x = \begin{bmatrix} \mathbf{A} \\ \mathbf{A}^2 \\ \vdots \\ \mathbf{A}^{N_p} \end{bmatrix}, \quad \mathbf{P}_u = \begin{bmatrix} \mathbf{B} & & \cdots & \\ \mathbf{AB} & \mathbf{B} & \cdots & \\ & & \ddots & \vdots \\ \mathbf{A}^{N_p-1}\mathbf{B} & \mathbf{A}^{N_p-2}\mathbf{B} & \cdots & \mathbf{B} \end{bmatrix} \tag{9.11}$$

The corresponding $\hat{y}[k + i|k]$ is calculated by the second equation of (9.9). The MPC control objective is defined as

$$\min_{\hat{u}[\cdot]} V[k] = \min_{\hat{u}[\cdot]} \sum_{i=N_w}^{N_p} \|\hat{y}[k + i|k] - \omega_0[k + i|k]\|_{Q_i}^2 \tag{9.12}$$

In order to damp oscillations as soon as possible, (9.12) does not penalize $\hat{u}[k + i|k]$ increments. Equation 9.12 is subject to the following inequality constraints:

$$u_{\min} \leq \hat{u}[k + i|k] \leq u_{\max}$$
$$z_{\min} \leq \hat{z}[k + i|k] \leq z_{\max} \tag{9.13}$$

where z denotes a vector of constrained operation variables such as bus voltage or line current.

9.3.3 Proposed Control Schemes

9.3.3.1 Centralized MPC
A centralized MPC controller is first developed, as shown in Figure 9.7 [20]. It is assumed that a complete system model from an Energy Management System (EMS) is available, which can be refreshed from time to time following changes of load, generation, and grid topology. Every Δt seconds, the MPC controller collects current states $x[k|k]$. It then computes an open loop sequence of $\hat{u}[k + i|k]$ by solving (9.12). It applies the first control $\hat{u}[k|k]$ to exciters and TCSC, and repeats the previous process at subsequent control times.

9.3.3.2 Decentralized MPC
However, it is often not feasible to apply centralized control in large-scale systems due to control complexity and constraints on information exchange. Moreover, even if fully centralized control would be feasible, it is not necessarily desirable to do so due to considerations of reliability/vulnerability.

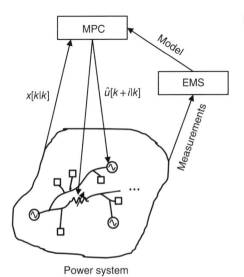

Figure 9.7 Centralized MPC.

Power system

An alternative is to break down a complex control problem into manageable subproblems that are only weakly related to each other and can be independently solved. Firstly, decentralized control uses local variables and significantly reduces communication cost. Secondly, decentralized control is more reliable than centralized control, since it can tolerate a broad range of uncertainties, both within subsystems and in their interconnections. Finally, when the information exchange among subsystems is restricted, the decentralized structure becomes an essential design constraint [21]. Therefore, the above centralized MPC is further extended into a decentralized one [22]. As with any other decentralized controls, two problems are to be solved, namely decomposition and coordination.

A large-scale control problem can be decomposed into subproblems through two kinds of approaches:

- Modeling–decomposition: construction of a global system model followed by an optimal decomposition into subsystems according to structural properties of the system and control problems under consideration [23].
- Decomposition–modeling: a decomposition of a whole system is imposed by oscillation modes and administrative areas, and hence the modeling of subsystems has to follow the decomposition already given.

Considering restrictions on information exchange in certain power grids, it is quite difficult to construct an exact system-wide model and then decompose it. So it is preferred to consider the second approach to decompose a large power system.

The decomposition adopted is demonstrated in a two-area system in Figure 9.8. This system is decomposed a priori into two areas, by replacing the tie line with an equivalent load and an equivalent generator. Then each area is modeled and an MPC controller is developed for it. Accordingly, the control objective of (9.12) is rewritten as the simultaneous and parallel resolution for the following area-wise optimization problem (subscript m refers to area m):

$$\min_{\hat{u}_m[\cdot]} V[k] = \min_{\hat{u}_m[\cdot]} \sum_{i=N_w}^{N_p} \|\hat{y}_m[k + i|k] - \omega_0[k + i|k]\|^2_{Q_{m,i}} \tag{9.14}$$

Each MPC controller solves its optimization subproblem using a detailed model of its own area and a possibly very rough model of the remaining areas (typically a black box

Figure 9.8 Decomposition of a two-area system.

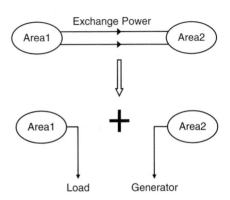

model). It then applies the first element of the optimal control sequence to its exciters or TCSC and observes the resulting effects to proceed.

Due to the interaction among subsystems, it is necessary to coordinate area-wise MPC controllers for damping system-wide oscillations. First, MPC controllers could negotiate or exchange useful information. For example, they can inform their neighbors about what they intend to do and pass along measurements that others may not be able to sense directly [24]. In [25], Lagrange penalties on common variables between subsystems are exploited to judge the coherency of MPC computations, and any controls are not applied before Lagrange penalties converge to a pre-defined small positive constant.

However, communicating and negotiating among different areas requires considerable communications infrastructure. In today's large interconnected systems, lack of communication infrastructure is a main obstacle for implementing advanced control schemes. Upgrading this infrastructure is costly and will remain an issue at least in the near future. In addition, negotiating maybe misses good control opportunities since oscillations evolve continuously and fast. Consequently, instead of explicit communication/negotiation, an implicit coordination scheme is used in this context. Specifically, each subsystem solves its own oscillation problem, and overall system stability emerges from these area-wise controls [26], with the help of a common control objective that makes all angular speeds return to the reference speed.

9.3.3.3 Hierarchical MPC

Hierarchical control is a mixture of centralized control and decentralized control. It not only reduces the computation and communication needed by a centralized controller, but also coordinates individual decentralized controllers. In [27], a wide area central controller is responsible for decoupling subsystem dynamics and calculating the interactions among them for lower-level local controllers. Reference [28] proposes a two-loop hierarchical structure: a local loop based on a machine speed signal and a global loop based on a differential frequency between two remote areas. The total PSS signal is the sum of control components generated by these two loops.

In this section, the previous area-wise MPC is further reinforced by adding a group of device-wise MPC controllers that works on a generator in order to further improve control robustness, as shown in Figure 9.9 [29]. Compared with area-wise MPC controllers in the upper level, device-wise MPC controllers in the lower level only consider the dynamic behaviors of one generator. They calculate supplementary signals for exciters in order to make generators run at the reference speed, as described in (9.15) (subscript n refers to generator n):

$$\min_{\hat{u}_n[\cdot]} V[k] = \min_{\hat{u}_n[\cdot]} \sum_{i=N_w}^{N_p} \|\hat{y}_n[k+i|k] - \omega_0[k+i|k]\|^2_{Q_{n,i}} \tag{9.15}$$

Device-wise MPC controllers need less time to measure, compute, and apply their controls, so that they can update their controls more frequently following changes of system states, and thus approach their control targets in a possibly better way. In addition, when area-wise MPC controllers can't work normally, device-wise MPC controllers are designed to work independently with the aid of internal control objectives.

Figure 9.9 Hierarchical MPC.

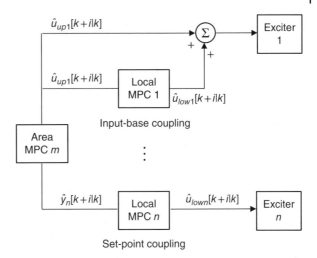

Two ways of coupling area-wise MPC controllers and device-wise MPC controllers are investigated:

- Input-base coupling (named hierarchical MPC I and indicated by the upper part of Figure 9.9): every Δt seconds, area-wise MPC controller m collects subsystem states and calculates $\hat{u}_{up}[k + i|k]$ for exciters under its authority. It sends $\hat{u}_{up}[k + i|k]$ to device-level MPC controllers as their decision bases. Every Δt_{low} seconds, the latter compute a correction $\hat{u}_{low}[k + i|k]$ depending on their local measurements, and combine $\hat{u}_{low}[k + i|k]$ with $\hat{u}_{up}[k + i|k]$ as the supplementary input for the exciter.
- Set-point coupling (named hierarchical MPC II and indicated by the lower part of Figure 9.9): area-wise MPC controller m solves an optimization problem in (9.14) and sends the predicted $\hat{y}_n[k + i|k]$ during next Δt seconds to device-wise MPC controller n as its set-points. The latter calculates supplementary input $\hat{u}_{lown}[k + i|k]$ in order to drive the speed of the controlled generator to reach the given set-points.

Control effects of a device-wise MPC controller depend not only on its controls, but also on states and controls of other MPC controllers existing in a same area. Therefore, these controllers are further coordinated with the help of area-wise MPC controllers. For a device-wise MPC controller, subsystem states and controls can be divided into two categories: its internal states and controls, and external states and controls related to other MPC controllers. Every Δt seconds, an area-wise MPC controller sends predicted states and controls to device-wise MPC controllers. When the latter calculate their controls, they take predicted external states and controls as simulation scenarios reflecting interactions among different controllers. The model of a device-wise MPC controller n is thus represented as follows:

$$x_n[k + 1|k] = \mathbf{A}_n x[k|k] + \mathbf{B}_n u[k|k]$$

$$= [\mathbf{A}_{n.n}, \mathbf{A}_{n.ext}] \begin{bmatrix} x_n[k|k] \\ x_{n.ext}[k|k] \end{bmatrix} + [\mathbf{B}_{n.n}, \mathbf{B}_{n.ext}] \begin{bmatrix} u_n[k|k] \\ u_{n.ext}[k|k] \end{bmatrix} \qquad (9.16)$$

$$= \mathbf{A}_{n.n} x_n[k|k] + \mathbf{B}_{n.n} u_n[k|k] + \text{constant}$$

where constant $= \mathbf{A}_{n.ext}x_{n.ext}[k|k] + \mathbf{B}_{n.ext}u_{n.ext}[k|k]$; x_n and u_n include the internal states and controls of MPC controllers n; A_n and B_n are the parts of A and B that are relative to x_n; $x_{n.ext}$ and $u_{n.ext}$ refer to external states and controls. The external states and controls are considered constant during one period of Δt, and hence the item "constant" is unchanged.

9.4   Test System & Simulation Setting

The developed MPC controllers are investigated on a 16-generator, 70-bus reduced-order representation of New England and New York systems in Figure 9.10 [5]. It contains five coherent areas: areas A1–A3 respectively contain an equivalent generator representing an external equivalent system, namely generators 14, 15, and 16. Area A4 consists of generators 10–13, and Area A5 includes generators 1–9. All generators are represented by sub-transient models. They are equipped with exciters as well as Turbine Governors (TGs). PSS controllers are also installed on the generators. The loads consist of 50% constant current loads and 50% constant impedance loads. Simulations are performed on a MATLAB based software platform, Power System Toolbox (PST) [19]. The work of this chapter focuses on the inter-area oscillations between A4 and A5, which are connected to each other by the tie lines 1-2, 1-27, and 8-9. The last line is equipped with a TCSC.

x includes state variables of generators, exciters, PSSs, TGs, and TCSC. y consists of angular speeds. u is a vector of supplementary controls for exciters and TCSC, which is constrained to [-0.1 0.1]. A simulation step δ=0.005 s is used to formulate system dynamics in (9.8). Every Δt=0.1 s, the centralized MPC controller and area-wise decentralized controllers refresh their controls. The prediction horizon is set to 1.5 s (namely H_p=15) and the control horizon H_c is 3. In (9.12), (9.14), and (9.15), deviations of predicted outputs from the reference are weighted uniformly and independently, that is, all Q are identity matrices. A temporary three-phase short-circuit to ground at bus 1 (cleared by opening the tie line 1-2 followed by its reconnection after a short delay) causes poorly damped oscillations, which can be described by the temporal evolution of power flow through tie line 1-2 $P_{1\text{-}2}$ and the angular speed of generator 1 Spd_1 (dashed lines in Figures 9.11 and 9.12).

9.5   Performance Analysis of MPC Schemes

9.5.1   Centralized MPC

9.5.1.1   Basic Results in Ideal Conditions
Ideal conditions mean complete state observability and controllability, without considering SE errors and control delays. Control effects are represented by the transmission power in tie line 1-2 and the angular speed of generator 1, as shown by the solid lines in Figures 9.11 and 9.12. Compared with the system response without MPC, settling time is decreased to approximately 10 s. Figures 9.13 and 9.14 show MPC signals for the exciter on generator 1 and for the TCSC.

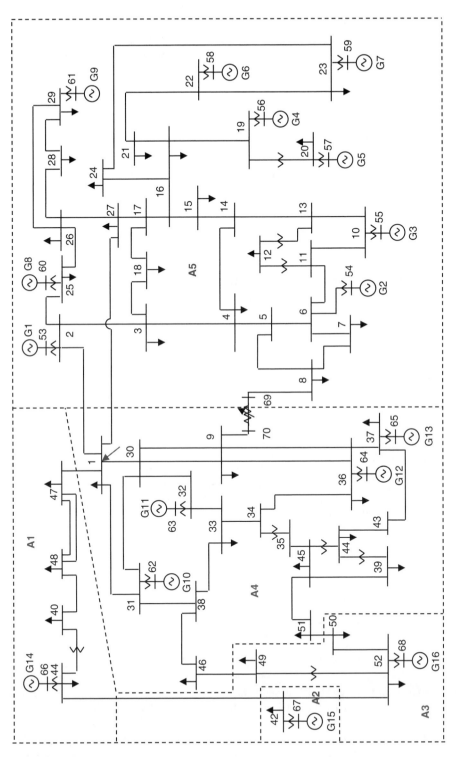

Figure 9.10 Test system.

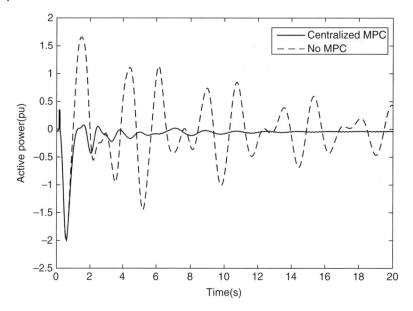

Figure 9.11 P_{1-2} in ideal conditions.

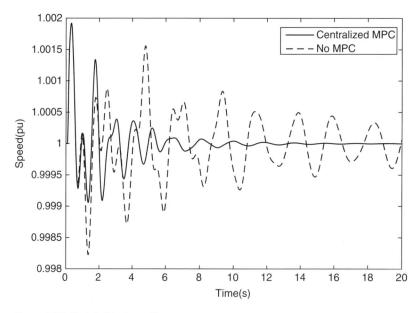

Figure 9.12 Spd_1 in ideal conditions.

9.5.1.2 Results Considering State Estimation Errors

At control time k, MPC uses $x[k|k]$ from a state estimator and WAMS measurements as an initial value for predicting future states and outputs. Consequently, $x[k|k]$ imprecision may have a detrimental effect on MPC control effects. In order to compensate for this imprecision, the difference $d[k|k]$ between $x[k|k]$ and its corresponding predicted

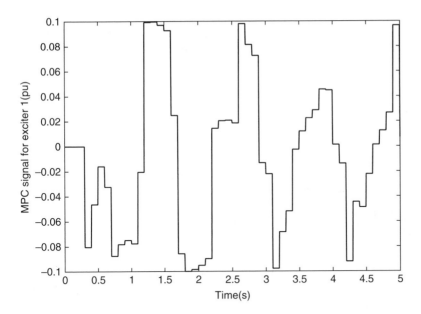

Figure 9.13 MPC signal for exciter 1.

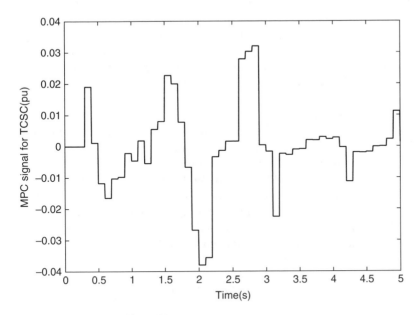

Figure 9.14 MPC signal for TCSC.

value at the previous control time k-1 is added to MPC prediction:

$$\hat{x}[k+1|k] = \mathbf{A}x[k|k] + \mathbf{B}\hat{u}[k|k] + d[k|k]$$
$$d[k|k] = x[k|k] - \hat{x}[k|k-1]$$

(9.17)

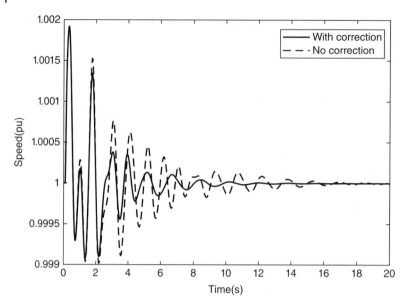

Figure 9.15 Spd_1 with SE errors.

It is assumed that $d[k|k]$ is refreshed at each control time, and then remains unchanged over the entire prediction horizon.

To simulate SE errors, $\pm10\%$ uniformly distribution pseudorandom errors are superimposed on exact states. The results are shown in Figure 9.15. As expected, SE errors affect MPC performance in terms of oscillation magnitude and settling time. On the other hand, it is observed that $d[k|k]$ correction considerably improves damping effects.

9.5.1.3 Consideration of Control Delays

The proposed MPC controller takes time to measure, compute, and apply its controls. Therefore, control delays are further added to MPC implementation. There are two possibilities to assess the impact of delays: apply controls as soon as they are available or after a common time interval. Here, assuming that all measurements are taken synchronously, a common delay of $\tau = 0.05$ s is applied to all controllers. As shown in Figure 9.16, at a control time k, the MPC controller computes $\hat{u}[k|k]$. However because of τ, $\hat{u}[k|k]$ can't be applied immediately, but from $k\Delta t + \tau$. Between $k\Delta t$ and $k\Delta t + \tau$, $\hat{u}[k|k-1]$ (which is calculated at time $k-1$) is implemented. The system response under

Figure 9.16 Input u and delay τ.

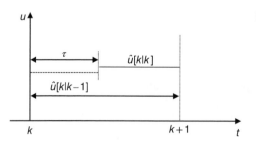

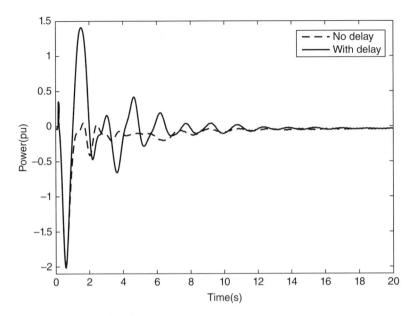

Figure 9.17 $P_{1\text{-}2}$ with delay.

this treatment is shown in Figure 9.17. It is clear that the response taking into account delays is worse than the one only considering SE corrections, but still quite superior to the response without MPC.

9.5.2 Distributed MPC

The test system is re-divided into two parts: part 1 consists of A1-A4 and part 2 is the rest. TCSC is assigned to part 1 as its control resource. One area-wise MPC controller is installed in two parts. MPC1 calculates $\hat{u}[k|k]$ for exciters 10-16 and TCSC; MPC 2 is responsible for exciters 1-9. System response is shown in Figure 9.18.

It is observed that the distributed MPC yields similar control effects to the centralized MPC. This is due to the chosen decomposition method. The system is decoupled by replacing the exchange power with one load and one generator whose values are equal to the exchange power at steady state. Hence, the subsystem models obtained in this way contain the information about this power. In addition to the explicit control objective of (9.14), the area-wise MPC controllers have another implicit objective, namely restoring the exchange power to its steady state value. This actually further strengthens the coordination between the two area-wise MPC controllers.

9.5.3 Hierarchical MPC

Device-wise MPC controllers are installed on each generator. The area-wise MPC controllers use the same simulation parameters as the distributed MPC. The device-wise MPC controllers refresh their controls every $\Delta t_{low} = 0.01$ s and use a control horizon $H_{cl} = 5$. The prediction horizon H_{pl} is set to 60 in part 1 and 40 in part 2. Figure 9.19

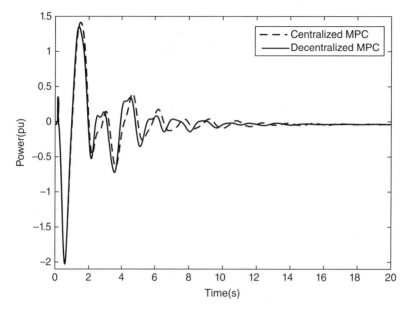

Figure 9.18 $P_{1\text{-}2}$ with decentralized MPC.

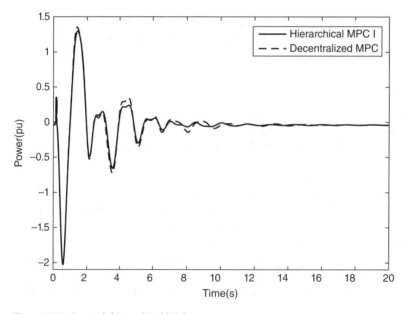

Figure 9.19 $P_{1\text{-}2}$ with hierarchical MPC1.

shows that compared with the distributed MPC, the hierarchical one further slightly improves control effects.

As explained in the Section 9.3.3.3, the advantage of hierarchical MPC is to further improve control robustness. So, next, the robustness of hierarchical MPC is verified through the following examples:

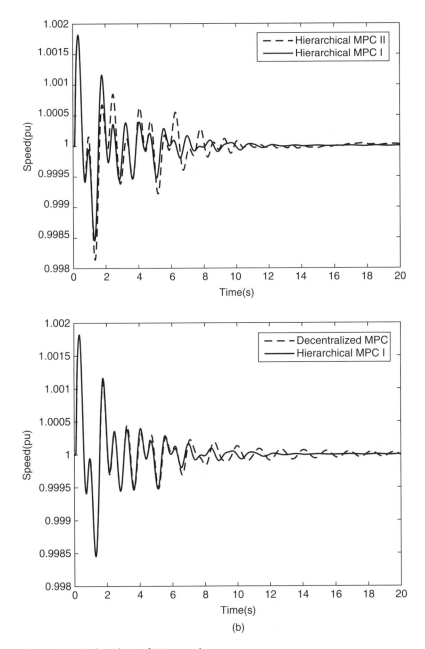

Figure 9.20 Spd_1 with Δt of 0.2 seconds.

– increasing Δt of area-wise controllers from 0.1 s to 0.2 s;
– losing area-wise MPC controller in part 1 during the first 5 s after the disturbance;
– incomplete measurements—it is assumed that only states of generators 10-12 and generators 1-4 are available for area-wise MPC controllers;

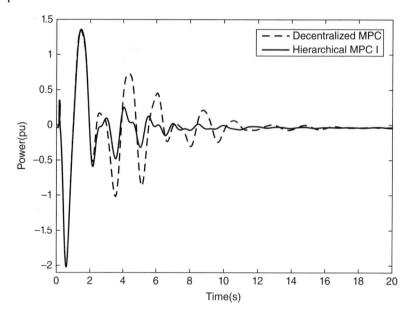

Figure 9.21 $P_{1\text{-}2}$ with a failure of area-wise MPC1.

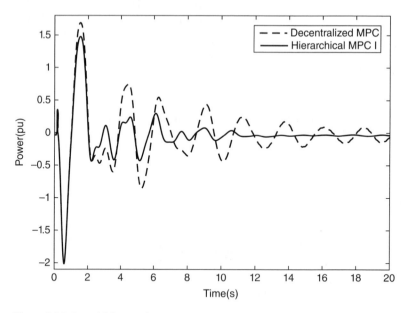

Figure 9.22 $P_{1\text{-}2}$ with incomplete measurements.

– communication failure between area-wise MPC controllers and device-wise MPC controllers—it is assumed that MPC controllers on generators 5-9 and 12-14 can't receive signals from area-wise MPC controllers.

The left part of Figure 9.20 compares two methods of coupling for hierarchical MPC. Utilizing set-point coupling, device-wise MPC controllers take worse predicted values

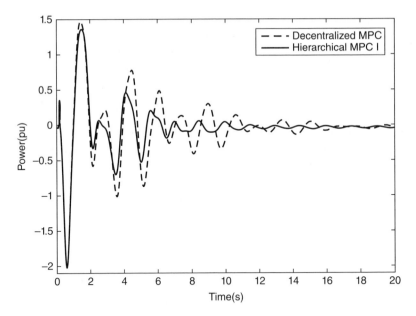

Figure 9.23 P_{1-2} with communication failure.

with larger errors as their set-points because MPC prediction precision deteriorates with increasing Δt. However, in input-base coupling, although area-wise MPC prediction becomes worse, device-wise MPC controllers can continuously correct input bases utilizing local measurements and control objectives. Therefore, the latter yields better control effects. So in other test cases, input-base coupling is used—namely hierarchical MPC I.

Figures 9.20–9.23 give the response of transmission power P_{1-2} for each test case. It is shown that the hierarchical MPC is more robust than the distributed one because of its better performances in different test cases.

9.6 Conclusions and Discussions

In contrast to traditional damping control methods that focus on oscillation modes and damping ratios, this chapter interprets damping control as a dynamic process control problem over a future time horizon. This problem penalizes future deviations of generator speeds from the reference speed, and solving it produces supplementary inputs for existing damping controllers. These inputs are refreshed at each control time, which is actually equal to adaptively adjust damping controllers' parameters following changes in operation condition, in order to improve and coordinate their damping effects.

MPC is a good solution for dynamic process control. Therefore, this chapter attempts to apply it to produce supplementary inputs for existing damping controllers. First, a centralized MPC is developed to demonstrate this idea. Next, it is extended into a decentralized MPC for viability in large power systems. Finally, a hierarchical MPC scheme is proposed with the aim of enhancing control robustness. The performances of the three MPC schemes are tested on a 70-bus system. It is observed that the distributed scheme appears to be a feasible control strategy for large systems while the hierarchical MPC further improves control effects, and at the same time offers better robustness.

There are still many improvements to be made for the proposed MPC schemes before they can be applied in real power systems. First, an exact model is a prerequisite for MPC to produce good damping effects. But even in a subsystem, it is not easy to build a complete and exact model. And such a model involves considerable communication when MPC is active. So an alternative is to build reduced-order models that can describe oscillation dynamics and need fewer measurements. The area-wise controllers in the decentralized MPC are coordinated in an implicit way, through a common control objective for subsystems. This coordination should be further verified in multi-area systems. Finally, in the hierarchical MPC, corrections from device-wise MPC controllers can be considered as disturbances for area-wise MPC controllers. In this chapter, the limits of corrections are determined through repeated simulation trials. Further work needs to be done on coordinating the two kinds of MPC controllers.

References

1 E. Grebe, J. Kabouris, S. Lopez Barba, W. Sattinger, et al, "Low Frequency Oscillations in the Interconnected System of Continental Europe", Proceedings of the IEEE Power and Energy Society General Meeting 2010, July 25–29, 2010, Minneapolis, MN, United States.

2 P. Korba and K. Uhlen, "Wide-area Monitoring of Electromechanical Oscillations in the Nordic Power System: Practical Experience", *IET Generation, Transmission, Distribution*, Vol. 4, No. 10, May 2010, pp. 1116–1126.

3 Y. Liu, J. R. Gracia, T. J. King and Y. L. Liu, "Frequency Regulation and Oscillation Damping Contributions of Variable-Speed Wind Generators in the U.S. Eastern Interconnections (EI)", *IEEE Transactions on Sustainable Energy*, Vol. 6, No. 3, July 2015, pp. 951–958.

4 D. N. Kosterev, C. W. Taylor and W. A. Mittelstadt, "Model Validation for the August10, 1996 WSCC System Outage", IEEE Transactions on Power Systems, Vol. 14, No. 3, Aug. 1999, pp. 967–979.

5 G. Rogers, *Power System Oscillations*. Norwell, Massachusets, USA: Kluwer Academic Publishers, 2000.

6 R. Preece, J. V. Milanović, A. M. Almutairi and O. Marjanovic, "Damping of Inter-area Oscillations in Mixed AC/DC Networks using WAMS based Supplementary Controller", *IEEE Transactions on Power Systems*, Vol. 28, No. 2, May 2013, pp. 1160–1169.

7 R. Majumder, B. Chaudhuri and B. C. Pal, "A Probabilistic Approach to Model-based Adaptive Control Damping of Interarea Oscillations in Power Systems", *IEEE Transactions on Power Systems*, Vol. 20, No. 1, Feb. 2005, pp. 367–374.

8 S. J. Qin and T. A. Badgwell, "A Survey of Industrial Model Predictive Control Technology", *Control Engineering Practice*, Vol. 11, No. 7, July 2003, pp. 733–764.

9 R. Scattolini, "Architectures for Distributed and Hierarchical Model Predictive Control - A Review", *Journal of Process Control*, Vol. 19, No. 5, May 2009, pp. 723-731.

10 F. Capitanescu and L. Wehenkel, "A New Iterative Approach to the Corrective Security-Constrained Optimal Power Flow Problem", *IEEE Transactions on Power Systems*, Vol. 23, No. 4, Nov. 2008, pp. 1533–1541.

11 B. Otomega, A. Marinakis, M. Glavic and T. V. Cutsem, "Model Predictive Control to Alleviate Thermal Over Loads", *IEEE Transactions on Power Systems*, Vol. 22, No. 3, Aug. 2007, pp. 1384–1385.

12 J. J. Ford, G. Ledwich and Z. Y. Dong, "Efficient and Robust Model Predictive Control for First Swing Transient Stability of Power Systems using Flexible AC Transmission Systems Devices", *IET Generation, Transmission, Distribution*, Vol. 2, No. 5, Sept. 2008, pp. 731–742.

13 T. A. Short, D. A. Pierre and J. R. Smith, "Self-tuning Generalized Predictive Control for Switched Capacitor Damping of Power System Oscillations", Proceedings of the 28th conference on decision and control, Dec. 1989, Florida, USA.

14 N. P. Johansson, H. P. Nee and L. Angquist, "An Adaptive Model Predictive Approach to Power Oscillation Damping Utilizing Variable Series Reactance FACTS Devices", Proceedings of the 41st International Universities Power Engineering Conference, Sept. 2006, Newcastle, UK.

15 L. Wang, H. Cheung, A. Hamlyn and R. Cheung, "Model Prediction Adaptive Control of Inter-area Oscillations in Multi-generators Power Systems", Proceedings of IEEE Power and Energy Society General Meeting, July 2009, Calgary, Canada.

16 R. Majumder, B. Chaudhuri and B. C. Pal, "A Probabilistic Approach to Model-based Adaptive Control for Damping of Interarea Oscillations", *IEEE Transactions on Power Systems*, Vol. 20, No. 1, Feb. 2005, pp. 367–374.

17 B. Chaudhuri, R. Majumder and B. C. Pal, "Application of Multiple-model Adaptive Control Strategy for Robust Damping of Interarea Oscillations in Power System", *IEEE Transactions on Control Systems Technology*, Vol. 12, No. 5, Sept. 2004, pp. 727–736.

18 J. Maciejowski, *Predictive Control with Constraints*, Prentice Hall, Harlow, England, 2002.

19 J. Chow, and G. Rogers, "Power System Toolbox Version 3.0", http://www.eps.ee.kth.se/personal/vanfretti/pst/Power_System_Toolbox_Webpage/PST.html, 2016.

20 D. Wang, M. Glavic and L. Wehenkel, "A New MPC Scheme for Damping Wide-Area Electromechanical Oscillations in Power Systems", Proceedings of IEEE Power and Energy Society PowerTech, June 2011, Trondheim, Norway.

21 D. D. Siljak and A. I. Zecevic, "Control of Large-Scale Systems: Beyond Decentralized Feedback", *IEEE Transactions on Power Systems*, Vol. 20, No. 1, Feb. 2005, pp. 367–374.

22 D. Wang, M. Glavic and L. Wehenkel. "Distributed MPC of Wide-Area Electromechanical Oscillations of Large-scale Power Systems", Proceedings of the 16th intelligent system applications to power systems (ISAP), Sep. 2011, Crete, Greece.

23 N. Motee and B. Sayyar-Rodsari, "Optimal Partitioning in Distributed Model Predictive Control", Proceedings of the 2003 American control conference, June 2003, Colorado. USA.

24 S. Talukdar, D. Jia, P. Hines and B. H. Krogh, "Distributed Model Predictive Control for the Mitigation of Cascading Failures", Proceedings of the 44th IEEE conference on decision and control, Dec. 2005, Seville, Spain.

25 R. R. Negenborn and B. S. Rodsai, "Multi-agent Model Predictive Control with Applications to Power Networks", PhD thesis, Delft university of Technology, 2007.

26 A. N. Venkat, I. A. Hiskens, J. B. Rawlings and S. J. Wright, "Distributed MPC Strategies with Application to Power System Automatic Generation Control", *IEEE Transactions on Control Systems Technology*, Vol. 16, No. 6, Nov. 2008, pp. 1192–1206.

27 I. Kamwa, R. Grondin and Y. Hebert, "Wide-area Measurement based Stabilizing Control of Large Power Systems - A Decentralized/Hierarchical Approach", *IEEE Transactions on Power Systems*, Vol. 16, No. 6, Feb. 2001, pp. 136–153.

28 F. Okou, L. A. Dessaint and O. Akhrif, "Power Systems Stability Enhancement Using a Wide-area Signals based Hierarchical Controller", *IEEE Transactions on Power Systems*, Vol. 20, No. 3, Aug. 2005, pp. 1465–1477.

29 D. Wang, M. Glavic and L. Wehenkel, "Considerations of Model Predictive Control for Electromechanical Oscillations Damping in Large-scale Power Systems", *International Journal of Electric Power and Energy Systems*, Vol. 58, 2014, pp. 32–41.

10

Voltage Stability Enhancement by Computational Intelligence Methods

Worawat Nakawiro

Lecturer in Electrical Power Systems, Department of Electrical Engineering, Faculty of Engineering, King Mongkut's Institute of Technology Ladkrabang, Bangkok, Thailand

10.1 Introduction

Present day power systems are highly stressed conditions due to several factors such as stronger competition in generation sectors, inadequate transmission expansion [2], intermittent renewable sources and heavy use of induction machines. With these factors, a power system becomes more vulnerable to various types of unstable behaviour among which is voltage stability. This has been reported as the cause of various incidents throughout the world over the last decades, such as those in the western region (WECC) of the United States in 1996, the Chilean power system in 1997 (accounting for a loss of 80% of its total load) and the Greek system covering the whole of Athens and the neighbouring area in 2004 [1].

Voltage stability is defined as the ability to maintain an acceptable voltage profile in many parts of the system after being subject to disturbances such as gradual load increases or outages of critical lines or generating units [6]. After these events, some loads may try to restore power consumption according to their load dynamics. This may drive the weakened power system closer to the technical limits, which can lead to voltage collapse.

Several voltage stability indices have been proposed in the literature to detect proximity of the current operating condition to the voltage collapse point. These indices can be broadly categorised into model-based methods [8] and measurement-based methods [9, 10]. The former relies on approximate data and modelling of the processes of the power system. By contrast, the latter depends solely on the actual response of the power system. Reference [5] provides an extensive review of voltage stability indices and other related operational and planning aspects. In some critical situations, voltage stability has to be enhanced in order to prevent any adverse impact that could eventually lead to blackout. Preventive and corrective actions (such as in [4, 7]) can be used according to the severity of the disturbance and present state of the power system. Voltage stability constrained optimal power flow (VSCOPF) [12, 13] is an efficient tool for determining the optimal control action for the power system while ensuring sufficient stability margin in normal operation and restoring stability in critical situations. Due to the increasing complexity of the optimisation problem such as a non-convex and discontinuous search

Dynamic Vulnerability Assessment and Intelligent Control for Sustainable Power Systems, First Edition.
Edited by José Luis Rueda-Torres and Francisco González-Longatt.

space, computational intelligence (CI) methods [14, 15] becomes an attractive tool for solving VSCOPF.

This chapter presents an application of CI methods to solve VSCOPF for preventive and corrective control actions. Ant colony optimisation (ACO) is used as the search engine in both examples. An artificial neural network (ANN) is used to provide fast estimation of the voltage stability margin (VSM) during the optimisation process in order to save computing time.

The rest of this chapter is organised as follows. Section 2 introduces voltage stability problems including key techniques for finding stability margin, namely the direct method and the continuation method [3]. In addition, the concepts of optimal power flow (OPF) and associated CI methods (namely ANN and ACO) are discussed in this section. Section 3 gives details of the IEEE-30 bus test system used in all simulations in this chapter. Section 4 presents a two-stage CI approach that guarantees minimum VSM during normal operation of the system. Section 5 explains an optimal load shedding scheme as the corrective measure to restore system stability. Corresponding simulation results are given in the respective sections. Finally the chapter is concluded and future outlook is given in Section 6.

10.2 Theoretical Background

This section reviews important theoretical concepts relevant to the method proposed in this chapter.

10.2.1 Voltage Stability Assessment

The voltage stability margin (VSM) is defined as the distance from the current operating point to the collapse point. In quasi-steady state, a power system can be modelled by a set of differential and algebraic equations

$$\dot{x} = f(\mathbf{x}, \mathbf{y}, \lambda)$$
$$0 = g(\mathbf{x}, \mathbf{y}, \lambda) \tag{10.1}$$

where $\mathbf{x}$ is the vector of state variables; $\mathbf{y}$ is the vector of algebraic variables and λ is a parameter that changes over time so that the power system moves from an equilibrium point to another. A vectorised form can be written as

$$\begin{bmatrix} \dot{x} \\ 0 \end{bmatrix} = F(\mathbf{z}, \lambda) \tag{10.2}$$

where F is the set of differential and algebraic equations and $\mathbf{z}$ is the vector of state and algebraic variables. It is known that the voltage collapse point corresponds to the singular bifurcation point with the following conditions:

$$F(\mathbf{z}, \lambda) = 0$$
$$D_z F(\mathbf{z}, \lambda)^T \mathbf{w} = 0$$
$$\|\mathbf{w}\|_\infty = 1 \tag{10.3}$$

where $\mathbf{w}$ is the left eigenvector and D_z is the differential operator with respect to z. It should be noted that a nonlinear technique such as Newton–Raphson can be applied to solve (10.3). This method is very fast and accurate but the solution depends heavily on a good initial guess. This procedure is known as the direct method. Alternatively, one could also apply optimisation to solve (10.3) as follows:

$$\text{Maximize} \quad \Delta_{mis} = \sum \left[\frac{D_z F(\mathbf{z}, \lambda)^T \mathbf{w}}{\|\mathbf{w}\|_\infty - 1} \right]$$

$$\text{subject to} \quad F(\mathbf{z}, \lambda) = 0 \tag{10.4}$$

where Δ_{mis} is the mismatch of the singularity condition and the eigenvector requirement at the collapse point.

The shortcomings of the direct method, such as a singularity near the point of collapse (POC), are overcome by the continuation method. The complete voltage profile can be generated by automatic changes of the loading parameter λ.

Figure 10.1 illustrates basic concepts of the continuation method. Assume an initial operating point $(\mathbf{z}_1, \lambda_1)$ lies on the curve. The so-called predictor step computes the direction vector $\Delta \mathbf{z}_1$ and the parametric change $\Delta \lambda_1$. This gives a new point $(\mathbf{z}_1 + \Delta \mathbf{z}_1, \lambda_1 + \Delta \lambda_1)$ which may not lie on the trajectory. The so-called corrector step is used to ensure that the new point $(\mathbf{z}_2, \lambda_2)$ will lie on the trajectory. This process continues until the complete profile is obtained and the point of collapse (POC) can be located.

10.2.2 Sensitivity Analysis

Sensitivity analysis [1] is a method used to analyse the effect of changing system parameters such as load or reactive power compensation on the stability margin of the power system. Typically change of a system parameter will cause changes to other variables. Therefore the model in (10.1) can be parameterised as

$$\dot{\mathbf{x}} = f(\mathbf{x}(\lambda(\mathbf{p}), \mathbf{p}), \mathbf{y}(\lambda(\mathbf{p}), \mathbf{p}), \lambda(\mathbf{p}), \mathbf{p})$$
$$0 = g(\mathbf{x}(\lambda(\mathbf{p}), \mathbf{p}), \mathbf{y}(\lambda(\mathbf{p}), \mathbf{p}), \lambda(\mathbf{p}), \mathbf{p}) \tag{10.5}$$

Figure 10.1 Illustration of continuation method.

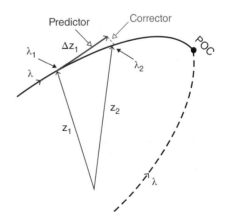

where **p** is the vector of changing parameters. The partial derivative due to the change of **p** can be written as

$$\underbrace{\begin{pmatrix} f_x & f_y \\ g_x & g_y \end{pmatrix}}_{A_{sys}} \begin{pmatrix} \dfrac{\partial x}{\partial \lambda}\dfrac{\partial \lambda}{\partial p} + \dfrac{\partial x}{\partial p} \\ \dfrac{\partial y}{\partial \lambda}\dfrac{\partial \lambda}{\partial p} + \dfrac{\partial y}{\partial p} \end{pmatrix} + \begin{pmatrix} f_\lambda \\ g_\lambda \end{pmatrix}\dfrac{\partial \lambda}{\partial p} + \begin{pmatrix} f_p \\ g_p \end{pmatrix} = 0 \tag{10.6}$$

At the collapse (bifurcation point), one can find the left eigenvector (v_f^T, v_g^T) corresponding to the zero eigenvalue of A_{sys} and multiply (10.6) by this. It is obvious that the first term will vanish, leaving only

$$(v_f^T, v_g^T)\begin{pmatrix} f_\lambda \\ g_\lambda \end{pmatrix}\dfrac{\partial \lambda}{\partial p} + (v_f^T, v_g^T)\begin{pmatrix} f_p \\ g_p \end{pmatrix} = 0 \tag{10.7}$$

Therefore, the sensitivity of λ with respect to parameter changes is

$$\dfrac{\partial \lambda}{\partial p} = \dfrac{-(v_f^T, v_g^T)\begin{pmatrix} f_p \\ g_p \end{pmatrix}}{(v_f^T, v_g^T)\begin{pmatrix} f_\lambda \\ g_\lambda \end{pmatrix}} \tag{10.8}$$

The total change of loading margin λ due to k parameter changes can be found from the sum

$$\Delta\lambda = \dfrac{\partial \lambda}{\partial p_1}\Delta p_1 + \dfrac{\partial \lambda}{\partial p_2}\Delta p_2 + \dots + \dfrac{\partial \lambda}{\partial p_k}\Delta p_k \tag{10.9}$$

This gives a linear approximation of the margin λ after the change of k system parameters.

10.2.3 Optimal Power Flow

Optimal power flow (OPF) becomes one of the important routines of modern power systems. Its objective is to achieve minimum operating cost while ensuring that constraints are within the respective limits. These limits consist of the capacity of equipment such as generators and compensating devices, security limits and stability limits. The set of optimal decision variables is determined by solving an optimisation problem with various constraints. The objective of this problem can be economic costs, system security or any other objectives. Since the early stages [20], many researchers have studied OPF with different objectives and various methods including computational intelligence [21].

Besides general objective functions of OPF such as minimisation of costs or losses, the system operator may have to assess stability of the OPF solution for all credible contingencies. If the system is unstable, the OPF solution can modified by the engineering judgement of the system operator. To guarantee optimality of the solution, the stability limit can be incorporated into the OPF software [26, 27].

10.2.4 Artificial Neural Network

An artificial neural network (ANN) is the electrical analogue of the biological neuronal system. An ANN is trained to learn the underlying relationship and perform a

Figure 10.2 An artificial neural model.

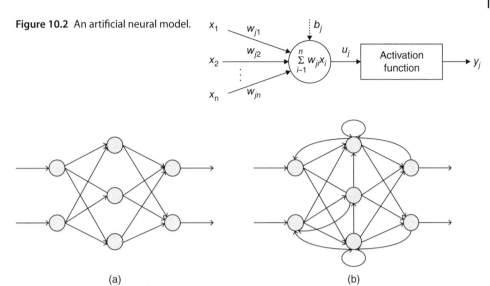

Figure 10.3 Common ANN configurations (a) feed-forward (b) recurrent networks.

mapping function between the selected inputs and the outputs [19]. A typical artificial neural model is shown in Figure 10.2. A neuron is activated by the weighted sum of n inputs x_1 to x_n and additional bias. The intermediate output u is computed before passing through the activation function which will define the range of neural output y. Multiple neurons can be connected as a network (known as an ANN) in various configurations. Figure 10.3 shows an example of two common ANN configurations, namely feed-forward and recurrent networks. The former has no feedback from one neuron to another. This network is generally sufficient for most engineering problems. The latter is capable of handling temporal information and very suitable for predicting time series such as long-term wind speed and power [22] and solar radiation [23].

10.2.5 Ant Colony Optimisation

Ant colony optimisation (ACO) is an algorithm inspired by the foraging behaviour of ants. ACO was initially developed for combinatorial optimisation problems [16]. Later, several variants of ACO emerged adapted for real-space problems, among which ACO_R shows promising performance in solving test problems for unconstrained optimisation [17]. In this chapter, we have modified the ACO_R algorithm to handle VSCOPF.

ACO_R defines the entire search space by the weighted sum of multiple normal distributions as follows:

$$G^i(x) = \sum_{l=1}^{k} \omega_l g_l^i(x) = \sum_{l=1}^{k} \omega_l \frac{1}{\sigma_l^i \sqrt{2\pi}} e^{-(x-\mu_i^l)^2/2(\sigma_i^l)^2} \tag{10.10}$$

where k is the number of single Gaussian PDFs at the construction step i; ω, μ and σ are vectors of size k defining the weights, means and standard deviations associated with every individual Gaussian PDF at the construction step i.

The location of ants at each construction step represents the value of variables in real space. It is updated using a random sampling technique based on the parameters of

	x_1^1	x_1^2	$\cdots$	x_1^i	$\cdots$	x_1^D	$f(\mathbf{x}_1)$	p_1	1
$\mathbf{x}_1$	x_1^1	x_1^2	$\cdots$	x_1^i	$\cdots$	x_1^D	$f(\mathbf{x}_1)$	p_1	1
$\mathbf{x}_2$	x_2^1	x_2^2	$\cdots$	x_2^i	$\cdots$	x_2^D	$f(\mathbf{x}_2)$	p_2	1
$\vdots$	$\vdots$	$\vdots$	$\vdots$	$\vdots$	$\vdots$	$\vdots$	$\vdots$	$\vdots$	$\vdots$
$\mathbf{x}_j$	x_j^1	x_j^2	$\cdots$	x_j^i	$\cdots$	x_j^D	$f(\mathbf{x}_j)$	p_j	0
$\vdots$	$\vdots$	$\vdots$	$\vdots$	$\vdots$	$\vdots$	$\vdots$	$\vdots$	$\vdots$	$\vdots$
$\mathbf{x}_n$	x_n^1	x_n^2	$\cdots$	x_n^i	$\cdots$	x_n^D	$f(\mathbf{x}_n)$	p_n	0
	Candidate solution						Fitness	Probability	Feasibility

Figure 10.4 Data structure of the solution archive.

the PDF. Unlike the pheromone trail of the classical ACO, ACO_R stores the knowledge gained from previous searches in a tabular format, namely the archive $\mathbf{X}_T$.

Figure 10.4 shows the solution archive $\mathbf{X}_T$ used to store n good candidate solutions that the algorithm has found so far. Since it is meant to handle the constrained problem, the quality of each solution is measured by the fitness value. Feasibility status of each solution also indicates whether the solution is desired or not. Each solution is assigned a distinct probability that the solution is chosen for the solution updating procedure of the next construction step.

For each solution j, the weight ω_j is computed from the Gaussian PDF value with mean of 1 and standard deviation of qn as

$$\omega_j = \frac{1}{qn\sqrt{2\pi}} e^{-(j-1)^2/2q^2n^2} \tag{10.11}$$

where q is a parameter of the algorithm and n is the size of solution archive. If q is small, the solutions with lower ranks in the archive have very strong influences in guiding new search directions whereby a larger q allows the wider search diversification over the entire space. For each archive solution of rank j in $\mathbf{X}_T$, the corresponding probability is calculated by

$$p_j = \frac{\omega_j}{\sum_{r=1}^{n} \omega_r} \; ; \; \forall j = 1, 2, \ldots, n \tag{10.12}$$

The new position of ant k is randomly chosen from a normal distribution again. Once the distinct selection probability is assigned to each solution, roulette wheel selection is used to select the solution which will be the mean of the distribution. The standard deviation is also computed using the average distance from the chosen solution $\mathbf{s}_j$ to the other solutions in $\mathbf{X}_T$ according to

$$\sigma_k^i = \xi \sum_{e=1}^{n} \frac{\left| x_e^i - x_j^i \right|}{n-1}; \; \forall i = 1, 2, \ldots, D \tag{10.13}$$

where ξ is the pheromone evaporation coefficient and D is the number of variables. The position of x_a becomes

$$x_{a,k}^i = x_j^i + \sigma_k^i N(0, 1) \tag{10.14}$$

where $N(0,1)$ is the normal distribution with zero mean and unity standard deviation.

10.3 Test Power System

The IEEE 30 bus is used for all simulations in this chapter. The single line diagram is depicted in Figure 10.5.

All generators are assumed to be able to operate in under- and over-excited modes where the reactive power limits are listed in Table 10.1. Shunt reactive power sources are installed throughout the system. The limits of these sources are given in Table 10.2

Figure 10.5 IEEE 30 bus test system.

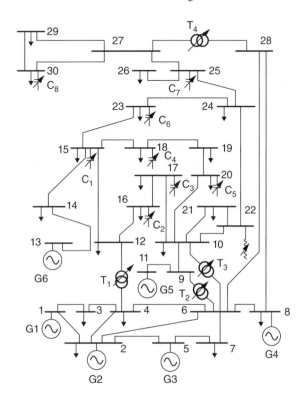

Table 10.1 Generator reactive power limit (pu).

Bus no.	2	5	8	11	13
q_{Gi}^{min}	−0.4	−0.4	−0.1	−0.06	−0.06
q_{Gi}^{max}	0.5	0.4	0.4	0.24	0.24

Table 10.2 Reactive power source limits (pu).

Bus no.	22	15	16	17	18	20	23	25	30
q_i^{min}	−0.1	0	0	0	0	0	0	0	0
q_i^{max}	0	0.2	0.15	0.1	0.2	0.1	0.3	0.25	0.2

Power system simulations are carried out using the PSAT software package [24]. The MATLAB ANN toolbox [25] is used to classify stability of operating conditions and to estimate the voltage stability margin. ACO was developed as a MATLAB function.

10.4 Example 1: Preventive Measure

10.4.1 Problem Statement

In this example, a two-stage approach is developed as a preventive measure to enhance the stability margin. The first stage is to assess the stability margin of the system. If the margin is insufficient, the second stage will propose a suitable preventive control measure. Figure 10.6 shows the conceptual diagram of the proposed design. An intelligent classifier (i.e. an ANN) is trained offline so that it can inform whether or not the system has sufficient margin. For the operating conditions with insufficient margin, the optimal control parameters will be determined by solving a VSCOPF problem. The result is sent to the system operator to make the final decision.

The objective of VSCOPF is to minimise transmission power loss and ensure that all security and stability parameters are within their allowable limits. Here we deal with reactive power-related variables. Therefore, the problem in this example is called voltage stability constrained optimal reactive power (VSCORPD) with the following mathematical formulation:

$$\text{Minimize } f(\mathbf{x}) = P_{loss}(\mathbf{x}, \mathbf{d}) \tag{10.15}$$

The constraints consist of

a) Generator bus voltage limits

$$u_{Gi}^{\min} \leq u_{Gi} \leq u_{Gi}^{\max} \quad \forall i \in \mathbf{s}_{PV} \tag{10.16}$$

b) Shunt compensator limits

$$q_{Ci}^{\min} \leq q_{Ci} \leq q_{Ci}^{\max} \quad \forall i \in \mathbf{s}_{QC} \tag{10.17}$$

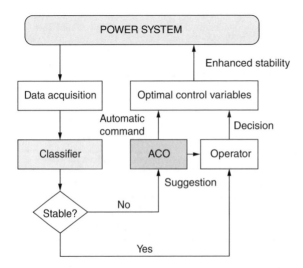

Figure 10.6 Conceptual diagram of the proposed two-stage design.

c) Transformer tap setting limits

$$a_i^{\min} \leq a_i \leq a_i^{\max} \qquad \forall i \in \mathbf{s}_T \tag{10.18}$$

d) Load bus voltage limits

$$u_{Li}^{\min} \leq u_{Li} \leq u_{Li}^{\max} \qquad \forall i \in \mathbf{s}_{PV} \tag{10.19}$$

e) Line flow limits

$$s_{Fi} \leq s_{Fi}^{\max} \qquad \forall i \in \mathbf{s}_L \tag{10.20}$$

f) Voltage stability margin limit

$$\lambda \geq \lambda^{\min} \tag{10.21}$$

where P_{loss} is the total active power losses in the transmission system; $\mathbf{s}_{PV}$ is the set of generator (PV) buses; $\mathbf{s}_{QC}$ is the set of shunt compensators; $\mathbf{s}_T$ is the set of transformers; $\mathbf{s}_{PQ}$ is the set of load (PQ) buses; $\mathbf{s}_L$ is the set of transmission lines. The vector $\mathbf{x}$ contains control variables listed in (10.16)–(10.18) where (10.16) is treated as a continuous variable and the rest are treated as discrete variables. The vector $\mathbf{d}$ describes the operating condition of the power system represented by active and reactive power demand and active power generation. Note that after changing the control variables, the VSM is approximated by an offline trained ANN in order to bypass the continuation power flow procedure.

10.4.2 Simulation Results

A classifier, namely learning vector quantisation (LVQ) [18], a special type of ANN, was used to identify if the system has sufficient VSM. A dataset of 5000 samples as shown in Table 10.3 was used for training and testing. The partitioning of this dataset is given in Table 10.3 where class A contains operating points with sufficient VSM while operating points in class B have insufficient VSM and need improvement. The developed classifier shows acceptable performance on the testing dataset as shown in Table 10.4.

Ten operating conditions which were correctly classified as in class B are considered for voltage stability enhancement. Figure 10.7 shows that as a result of optimisation, transmission power loss is reduced significantly. Importantly, the VSM of these 10 operating points below the threshold of 25% are enhanced after VSCORPD as shown in Figure 10.8. Note that the VSM of 25% means that the system is stable with extra loadability of 25% before reaching collapse.

Table 10.3 Partitioning of dataset.

Database	Training set	Testing set
5000	4000	1000
	Two-class partition	
Class A	1623	390
Class B	2377	610

Table 10.4 Performance of the classifier on testing dataset.

	Class A	Class B
Samples of class A (390)	349	41
Samples of class B (610)	156	454
Classification performance evaluation		
Success rate	80.3%	(803/1000)

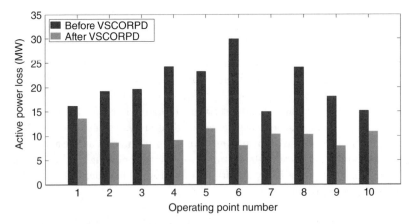

Figure 10.7 Transmission power losses.

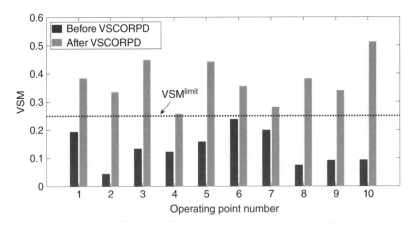

Figure 10.8 Voltage stability margin.

10.5 Example 2: Corrective Measure

10.5.1 Problem Statement

In some serious cases, adjusting the reactive power sources is not adequate to restore system stability. Load shedding becomes an unavoidable but yet powerful option. The solution of optimal load shedding involves the determination of the effective locations and optimal amount of load to be curtailed. The former is achieved by analysing the sensitivity of load on all buses to the stability margin. However this may result in violation of some security constraints. Therefore this chapter proposes an OPF for load shedding with the objective as

$$f(\Delta p_d) = \sum_{i \in n_s} C_i \cdot \left(\frac{\Delta p_{di}}{\partial \lambda / \partial p_{di}} \right) \tag{10.22}$$

where C_i is the power interruption cost at bus i ($/kW); Δp_d represents the active power load curtailment at effective buses; $\partial \lambda / \partial p_{di}$ is the change of voltage stability margin due to the change of power demand at bus i. The constraints consist of

a) Load bus voltage limits

$$u_{Li,b}^{min} \le u_{Li,b} \le u_{Li,b}^{max} \text{ and } u_{Li,m}^{min} \le u_{Li,m} \le u_{Li,m}^{max} \quad \forall i \in \mathbf{s}_{PQ} \tag{10.23}$$

b) Line power flow limits

$$s_{Li,b}^{min} \le s_{Li,b} \le s_{Li,b}^{max} \text{ and } s_{Li,m}^{min} \le s_{Li,m} \le s_{Li,m}^{max} \quad \forall i \in \mathbf{s}_{PQ} \tag{10.24}$$

where the subscripts b and m denotes the base- and maximum-loading conditions and $\mathbf{s}_{PQ}$ is the set of load (PQ) buses.

c) Fixed power factor

$$\frac{\Delta p_{di}}{p_{di}^0} = \frac{\Delta q_{di}}{q_{di}^0} \quad \forall i \in \mathbf{s}_S \tag{10.25}$$

where p_{di}^0 and q_{di}^0 are initial active and reactive power demand of the bus i, respectively, and $\mathbf{s}_S$ is the set of effective load buses selected for load shedding. This constraint assumed that the power factor at the selected buses remains unchanged after load shedding.

d) Allowable load curtailment

$$\Delta p_{di}^{min} \le \Delta p_{di} \le \Delta p_{di}^{max} \quad \forall i \in \mathbf{s}_S \tag{10.26}$$

e) Voltage stability margin

$$1 \le \lambda_0 + \sum_{i=1}^{N} \frac{\partial \lambda}{\partial p_{di}} \Delta p_{di} + \sum_{i=1}^{N} \frac{\partial \lambda}{\partial q_{di}} \Delta q_{di} \le 1.06 \tag{10.27}$$

This is to ensure that the stability of the power system ($\lambda \ge 1$) can be restored after the load shedding. It should be noted that a high stability margin may involve excessive load shedding. Therefore the maximum stability margin is chosen at 6%.

Table 10.5 gives the costs due to power interruption incurred by power consumers in different sectors according to [20].

10.5.2 Simulation Results

In this example, the power system is initially voltage unstable, represented by a voltage stability margin (VSM) less than 1. Load shedding is carried out at the effective locations identified by the sensitivity analysis discussed earlier. The sensitivity of VSM due to the change of active and reactive power at load buses is shown in Figure 10.9.

Table 10.5 Power interruption cost in different sectors.

	Interruption cost ($/kW)		
Transportation (t)	Industrial (i)	Commercial (c)	Residential (r)
16.42	13.93	12.87	0.15

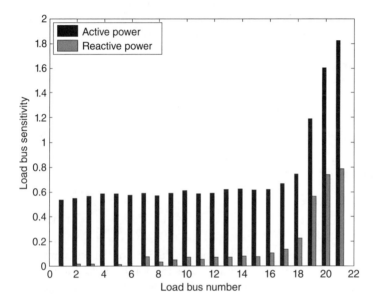

Figure 10.9 Sensitivity due to load change at different buses.

Table 10.6 Information for load shedding at selected locations.

Bus	$\Delta p_{di,min}$ (pu)	$\Delta p_{di,max}$ (pu)	Configuration	Cost ($/100 kW)
23	0	0.032	0.5c+0.5r	651
24	0	0.087	0.3t+0.7i	1467.7
26	0	0.035	1r	15
29	0	0.024	0.4i+0.2c+0.4r	820.6
30	0	0.070	0.6t+0.2i+0.2r	1266.8

Five locations are selected and their allowable ranges of load shedding are given in Table 10.6. Note that load bus numbers 17–21 correspond to buses 23, 24, 26, 29 and 30, respectively. These buses are geographically located far from the generating units and lack reactive power supports. Loads at these buses are assumed to have different configurations according to the sectors in Table 10.5. For example, the load bus 30 is of the 0.6t+0.2i+0.2r composition. This means 60% of total demand of this bus comes from the transportation (t) sector, 20% from the industrial (i) sector and 20% from the

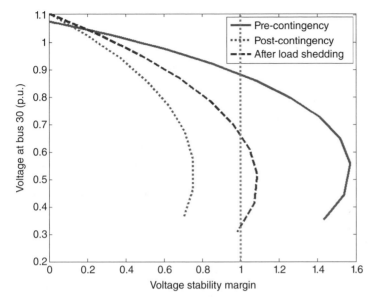

Figure 10.10 PV profile at bus 30.

residential (r) sector. The cost of shedding 100 kW at bus 30 is \$1266.8, for example. It is noted that the interruption cost at these locations is different.

Initial analysis reveals that bus 30 is the most critical one. Outage of the line between buses 28 and 27 causes a severe contingency. Figure 10.10 shows the PV profile of bus 30 after load shedding, against pre- and post-contigency (with no control actions) conditions. It is demonstrated that the proposed method is able to restore the voltage stability of the system while maintaining a number of constraints within their limits.

10.6 Conclusions

This chapter presents two examples that apply CI techniques to determine optimal preventive and corrective control measures for enhancing voltage stability of the IEEE 30-bus test system. In the first example, a combined LVQ and ACO method is proposed. The operating conditions identified to have insufficient VSM by LVQ are further analysed. Then ACO is used to determine the optimal control action that will guarantee minimum VSM. In the second example, an optimal load shedding scheme for restoring unstable operating conditions is proposed. The change in VSM due to load curtailment at effective locations is approximated by the sensitivity method in order to save computing time. Simulation results show the capability of the proposed method to improve VSM in normal operation and to restore stable operation in critical situations. Given the limited learning capability of classical ANNs when dealing with changes such as topology in power systems, further investigation on deep learning ANN is needed. Moreover, the test system used in this chapter is still not of a practical size. Selection of suitable information from a large-scale system as input features for learning purposes is of great importance.

References

1 V. Ajjarapu, *Computational Techniques for Voltage Stability Assessment and Control.* Springer Verlag, 2006.

2 G. Reed, J. Paserba, and P. Salavantis, "The key to resolving transmission gridlock: the case for implementing power electronics control technologies" In: *Conference on electricity transmission in deregulated markets: challenges,* opportunities, and necessary R&D Agenda, Pittsburgh PA, USA, pp 1–5, 2004.

3 C. Canizares (ed.), "Voltage stability assessment: Concepts, practices and tools" Technical Report, Special Publication of IEEE Power System Stability Subcommittee, 2002.

4 T. Van Cutsem and C. Vournas, *Voltage Stability of Electric Power Systems.* Kluwer Academic, 1998.

5 W. Nakawiro, "Voltage Stability Assessment and Control of Power Systems using Computational Intelligence," Ph.D. Thesis, Faculty of Engineering, University of Duisburg-Essen, Duisburg, Germany, 2011

6 P. Kundur, J. Paserba, V. Ajjarapu, G. Andersson, A. Bose, C. A. Canizares, N. Hatziargyriou, D. J. Hill, A. Stankoviv, C. W. Taylor, T. Van Cutsem, and V. Vittal, "Definition and Classification of Power System Stability," *IEEE Trans. Power Syst.,* vol. 19, no. 2, pp. 1387–1401, May 2004.

7 P. Kundur, *Power System Stability and Control.* McGraw-Hill, 1994.

8 P. Kessel and H. Glavitsch, "Estimating the Voltage Stability of a Power System," *IEEE Trans. Power Delivery,* vol. PWRD-1, no. 3, pp. 346–352, July 1986.

9 K. Vu, M. Begovic, D. Novosel, and M. M. Saha, "Use of Local Measurements to Estimate Voltage Stability Margin," *IEEE Trans. Power Syst.,* vol. 14, no. 3, pp. 1029–1035, Aug. 1999.

10 B. Milosevic and M. Begovic, "Voltage Stability Protection and Control using a Wide-Area Network of Phasor Measurements," *IEEE Trans. Power Syst.,* vol. 18, no. 1, pp. 121–127, Feb. 2003.

11 I. Simon, G. Verbic, and F. Gubina, "Local Voltage-Stability Index Using Tellegen's Theorem," *IEEE Trans. Power Syst.,* vol. 21, no. 3, pp. 1267–1275, Aug. 2006.

12 S. Kim et al., "*Development of voltage stability constrained optimal power flow (VSCOPF),*" Power Engineering Society Summer Meeting, 2001, Vancouver, BC, Canada, 2001, pp. 1664–1669 vol. 3.

13 W. Zhang, F. Li, and L. M. Tolbert, "*Voltage stability constrained optimal power flow (VSCOPF) with two sets of variables (TSV) for reactive power planning,*" Transmission and Distribution Conference and Exposition, 2008. T&D. IEEE/PES, Chicago, IL, 2008, pp. 1–6.

14 R. Eberhart and Y. Shi, *Computational Intelligence.* MA, USA: Morgan Kaufmann, 2007.

15 A. Konar, *Computational Intelligence: Principles, Techniques and Applications.* Heidelberg: Springer Verlag, 2005.

16 M. Dorigo, V. Maniezzo, and A. Colorni, "The Ant System: Optimization by a Colony of Cooperating Agents," *IEEE Trans. Syst. Man and Cybern. Part B-Cybernatics,* vol. 26, no. 1, pp. 1–13, 1996.

17 K. Socha and M. Dorigo, "Ant Colony Optimization for Continuous Domains," *Euro. J. of Oper. Res.,* vol. 185, pp. 1155–1173, 2008.

18 T. Kohonen, *Self-Ogranization and Associative Memory*, 2nd edn. Berlin: Springer, 1988.

19 C. M. Bishop, *Neural Networks for Pattern Recognition*. Oxford, UK: Oxford University Press, 2007.

20 H. W. Dommel and W. F. Tinney, "Optimal Power Flow Solutions," *IEEE Trans. Power App. Syst.*, vol. PAS-87, no. 10, pp. 1866–1876, Oct. 1968.

21 M. R. AlRashidi and M. E. El-Hawary, "Applications of Computational Intelligence Techniques for Solving the Revived Optimal Power Flow Problem," *Elect. Power Syst. Res.*, vol. 79, pp. 694–702, 2009.

22 T. G. Barbounis, J. B. Theocharis, M. C. Alexiadis and P. S. Dokopoulos, "Long-Term Wind Speed and Power Forecasting Using Local Recurrent Neural Network Models," in *IEEE Transactions on Energy Conversion*, vol. 21, no. 1, pp. 273–284, March 2006.

23 G. Capizzi, C. Napoli, and F. Bonanno, "Innovative Second-Generation Wavelets Construction With Recurrent Neural Networks for Solar Radiation Forecasting," in *IEEE Transactions on Neural Networks and Learning Systems*, vol. 23, no. 11, pp. 1805–1815, Nov. 2012.

24 F. Milano, L. Vanfrett, and J. C. Morataya, "An Open Source Power System Virtual Laboratory: The PSAT Case and Experience," *IEEE Trans on Education*, vol. 51, no. 1, pp. 17–23, Feb. 2008.

25 Mathwork Inc., "Neural Network Toolbox", R2015a

26 H. R. Cai, C. Y. Chung, and K. P. Wong, "Application of Differential Evolution Algorithm for Transient Stability Constrained Optimal Power Flow," *IEEE Trans. Power Syst.*, vol. 23, no. 2, pp. 719–728, May 2008.

27 D. Gan, R. J. Thomas, and R. D. Zimmermann, "Stability-Constrained Optimal Power Flow," *IEEE Trans. Power Syst.*, vol. 15, no. 2, pp. 535–540, May 2000.

28 P. J. Balducci, J. M. Roop, L. A. Schienbein, J. G. Desteese, and M. R. Weimar, "Electrical power interruption cost estimates for individual industries, sectors and U.S. economy," Pacific Northwest National Lab Feb. 2002.

11

Knowledge-Based Primary and Optimization-Based Secondary Control of Multi-terminal HVDC Grids

Adedotun J. Agbemuko[1], Mario Ndreko[2], Marjan Popov[3], José Luis Rueda-Torres[3], and Mart A.M.M van der Meijden[3,4]

[1] *Institut de Recerca en Energia de Catalunya (IREC), Barcelona, Spain*
[2] *TenneT TSO GmbH, Bayreuth, Germany*
[3] *Delft University of Technology - Department of Electrical Sustainable Energy, Delft, The Netherlands*
[4] *TenneT TSO BV, Arnhem, The Netherlands*

Nomenclature

P_e^i	Active power error at node i
V_e^i	DC voltage error at node i
E_{grid}^i	AC grid voltage at node i
dP_*^i	Rate of change of active power reference at node i
$\mu(x)$	Membership degree of a variable x in fuzzy domain
β_i	Degree of fulfillment of rule i
$\min\limits_{x} f$	Minimization of a function f
$\tilde{\mathbf{X}}$	Vector of individual chromosomes to be optimized
$\overline{P}_{loss}$	Total DC power losses of the system
P_i	Active power at node i
U_{dci}	DC voltage at node i
$\underline{\mathbf{L}}$	Lower bound on all solutions of optimization
$\overline{\mathbf{U}}$	Upper bound on all solutions of optimization
$\hat{P}$	Pre-defined power at a node

This chapter presents a novel control scheme for voltage and power control in high voltage multi-terminal DC (HV-MTDC) grids used for the grid connection of large offshore wind power plants. The proposed control scheme employs a computational intelligence technique in the form of a fuzzy controller for primary voltage control and genetic algorithms at the secondary control level. At the primary level, the fuzzy approach does not necessarily change the conventional structures proposed in the literature, but instead combines the best of several conventional strategies in a single "box" whilst at the same time eliminating their drawbacks. At the secondary level, the solution of the optimal HVDC grid load flow via genetic algorithms provides

Dynamic Vulnerability Assessment and Intelligent Control for Sustainable Power Systems, First Edition.
Edited by José Luis Rueda-Torres and Francisco González-Longatt.
© 2018 John Wiley & Sons Ltd. Published 2018 by John Wiley & Sons Ltd.

optimal set-points which are passed on to the primary level control. Time domain dynamic simulations on a three terminal HVDC grid benchmark are presented. It is demonstrated that the proposed method ensures optimal power dispatch in the HVDC grid ensuring robustness and flexibility in operation.

11.1 Introduction

An ever increasing demand for reduction of the CO_2 emissions partly responsible for climate change has led to the proliferation of renewable energy sources (RES) into conventional power systems. Challenges attributed to RES, particularly their intermittent and distributed nature, has called for radical intelligent control techniques. Intelligent control techniques provide seemingly improved flexibility, extended capabilities, self-monitoring ability, self-diagnostics, and self-recovery, to mention a few, which are near impossible tasks with conventional control methods.

This chapter discusses the application of intelligent control techniques to offshore wind plants with high voltage DC (HVDC) transmission. More specifically, the focus is placed on multi-terminal high voltage direct current (HV-MTDC) grids based on voltage source converter (VSC) technology. Typically, three control levels can be identified—primary, secondary, and tertiary control. The subject of this chapter is mainly the primary and secondary levels. The tertiary level, while it exists in the HV-MTDC literature, goes beyond the scope of this chapter. Prior to discussing the intelligent control methods, a brief overview of conventional control strategies applied in an HV-MTDC grid is provided. The latter introduces the reader to the problem and justifies the need for applied intelligent control.

11.2 Conventional Control Schemes in HV-MTDC Grids

In HV-MTDC literature, two broad categories of control strategies have been proposed, *viz.* PI (proportional integral) based strategies—constant power, constant voltage—and droop (proportional control) strategies. All other strategies/modification schemes are related to one of these either singly or in combination. Such schemes include voltage margin, piecewise droop, dead/un-dead band droop, and numerous other schemes [1–13].

Constant voltage and power control, whilst very simple and straightforward in use, have a drawback of inherently assuming continuing normal operation of the grid [7]. Hence, they are incapable of online power compensation in the event of contingencies without applying communication resources. Power compensation during emergencies requires timeliness, for which reliability issues and the possible unavailability of communication channels or communication resources may prove a hindrance to the security of supply objective, not to mention possible damage to sensitive devices. However, they do have the advantage of reaching the desired set points without any deviation. That is, whenever a new power flow pattern is established and set points are sent to the respective converters, with PI strategies, the pre-determined power flow pattern will be reached (assuming the choice of other parameters is well conditioned).

In contrast, the droop control strategy similar to that employed in HVAC (High Voltage Alternating Current) grids allows for several capable terminals to contribute to power balancing during contingency, in order to keep the grid voltage within acceptable limits without the need for communication resources. This offers advantages in terms of security, robustness, stability, and safe operation. Nevertheless, droop strategies have a drawback of not reaching desired set points when a new power flow pattern is required. This is as a result of the inherent architectural structure of the droop controller.

As a result of the drawbacks attributed to these two broad categories, modifications have been proposed. Voltage margin is a modified form of constant voltage that whilst still a PI strategy, allows several converters to control the voltage in a coordinated manner in the event of an outage of the voltage priority control terminal [5, 14]. However, as the MTDC grid grows in size, there may be a need for a larger margin (which may not be feasible or realistic); this puts a limitation on the size of the grid, besides being not suitable in the event of cascaded outages where communication may have to be relied upon. Therefore, the droop strategy is still a better alternative notwithstanding its drawbacks.

Dead and un-dead band strategies have been proposed as acknowledged in [13, 15, 16]. The dead band approach combines the capabilities of voltage margin and droop voltage schemes, while the un-dead band scheme can be viewed as a combination of several droop characteristics (different gains in different regions). The principles of operation remain the same as that of the basic strategies. A huge drawback of both strategies are the sharp edges between transition points as acknowledged in [16]. Thus, smoothed nonlinear edges at transition points have been proposed. However, the drawbacks of conventional droop still presents.

Voltage and power deviations of the droop voltage strategy can be attributed to a number of factors. The voltage and power control blocks basically fight each other for control, leading to steady state deviations. Also, the fact that a voltage difference at terminals is required to establish a power flow contributes to the deviations. Converter losses, uneven DC line voltage drops, and topology all contribute to deviations. Deviations result to inaccurate power flows, poor loss profile, and inaccurate instantaneous power balancing [7, 9].

Some research effort has been dedicated to solving the problem of voltage and power deviations. Proposed methods include the influence of fixed droop constants analytically in power flow equations, exploiting the role of the Jacobian matrix of the traditional Newton–Raphson (N–R) method as reported in [7, 9]. According to [9], some percentage of deviation is still incurred. Besides, these methods require an elaborate mathematical framework that is tedious and not entirely clear.

This chapter proposes a new methodology with zero voltage and power deviation that does not include the influence of the droop constant in power flow equations, nor does it employ the traditional N–R method. Instead, an optimization problem is solved using genetic algorithms at the secondary level to calculate the optimal voltage references that minimize the losses and establish a pre-defined power flow pattern in the same way as N–R. These optimal voltage and power references are then passed on to a knowledge-based fuzzy-droop controller at the primary level that combines the capabilities of both constant power and droop voltage strategies.

Upon receiving the new set points, the fuzzy controller transitions the system to a constant power strategy for which set points will be reached and remains in this strategy until errors start to reduce. As errors starts to reduce below a given boundary, the strategy senses this and very smoothly transitions from constant power back to droop voltage control, and when steady state conditions are fulfilled, the system remains fully in droop mode. The MTDC stays in droop voltage control until new set points are received, irrespective of any contingency. The methodology is in retrospect a combination of the aforementioned broad strategies. This combination is made possible by fuzzy logic, which introduces inherent knowledge into the system control.

11.3 Principles of Fuzzy-Based Control

The objective of this section is to highlight the implementation of fuzzy logic and control as an intelligent technique and not to discuss the basics of fuzzy logic control. Readers are referred to [17] for a detailed treatment of the concept of fuzzy logic and control. Notwithstanding, a short overview is given.

Fuzzy logic and control is classified as one of the knowledge-based computational intelligence methods that mimic human behavior and actions. Fuzzy logic was formulated in 1965 by L. Zadeh who put forward the mathematical formulation as we know it today [18]. Fuzzy control can also be viewed as a natural language processing tool that allows for the use of human knowledge and experience about a system behavior, coded as rules, in order to take control actions without an explicit mathematical formulation of the system. Thus, fuzzy control is suitable for problems in which mathematical formulation is too expensive or for which a single control law is not applicable. A particular class of fuzzy systems is applicable to this work—*rule-based fuzzy systems*. Rule-based fuzzy systems are one of the most popular classes of fuzzy systems and most relevant to electrical power systems application. Rule based fuzzy system take the form:

IF *antecedent* proposition **THEN** *consequent* proposition

Both antecedents and consequents can combine variables with connectives such as "AND," "OR," and "NOT." The form of the *consequent* determines the sub-class of rule-based systems. For this chapter, the linguistic model with special case of singleton consequents is utilized.

11.4 Implementation of the Knowledge-Based Power-Voltage Droop Control Strategy

As previously described, the broad categories of conventional strategies are in a way complementary to each other. When each broad category is used exclusively with and without modifications, several drawbacks exist, thus restricting practical use of these strategies. Therefore, the proposed strategy is capable of combining several complementary conventional strategies in such a way that it is easy to transition from one to the other very smoothly without causing instabilities. This combination eliminates the most important drawbacks of each while combining their advantages. In this chapter, the proposed strategy combines constant active power and droop voltage control strategies. The fuzzy controller acts as an autonomous supervisor in transitioning from one strategy to the other when the pre-defined conditions are fulfilled. A block diagram of the proposed control strategy is depicted in Figure 11.1. As can be seen, the droop gain is scheduled

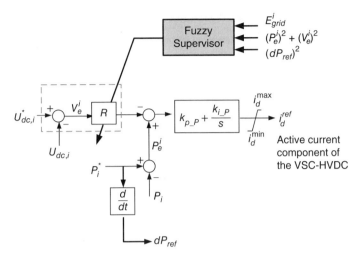

Figure 11.1 Generic fuzzy-based droop controller.

by the output of the fuzzy controller. The scheduling is performed online, based on local measurements as will be given in following sections. The main assumption for practical implementation is that the inner control loop of the converter where the fuzzy logic is applied is well tuned such that it is stable for the whole range of droop values. Also, we assume that the converters are connected to stiff grids with high short circuit ratios.

11.4.1 Control Scheme for Primary and Secondary Power-Voltage Control

A schematic diagram of the proposed primary and secondary control schemes is given in Figure 11.2. A three-terminal HVDC grid test system is utilized, connecting a far-from-shore wind power farm. An optimal power flow using genetic algorithms is solved at the secondary level. This defines the power and voltage set points for the given wind power generation and power dispatch at the onshore converters. Since the fuzzy controller schedules the gains such that the onshore converter is in constant power control when new power set points are ordered and achieved, droop gains do not affect the transition. As a result, the effect of the droop gain is not included in the optimal HVDC grid load flow. This simplifies the optimal load flow calculations.

The calculated optimal power and voltage set points, in combination with information obtained locally by the fuzzy controller, are used to determine if conditions are fulfilled for the transition from one strategy (constant power) to another (droop) and vice versa. By default, the system is designed to remain in droop strategy during normal operation. Hence, when set points are received, the transition occurs from the droop strategy to the constant active power strategy.

Hence, as soon as the the power set points are reached and the error gradually reduces to zero, the fuzzy controller smoothly transitions the system from constant active power back to droop control by increasing the droop gain. It completes the transition to droop and remains in droop until new set points are ordered. As long as new set points are not received (even if communication fails), the converter remains in droop strategy, ensuring that terminal voltages will be kept within bounds.

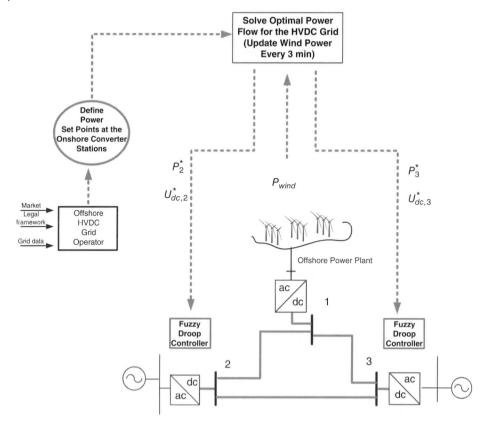

Figure 11.2 Pictorial overview of the complete system and control modules.

11.4.2 Input/Output Variables

Knowledge gained from system operation and desired control objectives determines the fuzzy supervisor inputs and outputs that would give sufficient information to take control actions on the main system. The distinct input variables for this strategy are: P_e^i, V_e^i, E_{grid}^i, dP_{ref} with $1 \leq i \leq (N_{dc} - k)$, N_{dc} the number of DC nodes in the DC grid, and k the number of the offshore nodes. Furthermore, we define the following variables: $P_e^i = P_*^i - P^i$ and $V_e^i = U_{dci}^* - U_{dci}$. Finally, E_{grid}^i is the AC grid voltage amplitude at the point of common coupling (PCC) of the onshore station i. The error variable that is used in the fuzzy rules is defined as $(P_e^i)^2 + (V_e^i)^2$. Finally, dP_{ref} is the rate of change of the power reference set point.

A combination of P_e^i, V_e^i, E_{grid}^i, and dP_{ref} determines the current and future state of the fuzzy controller. The signal dP_{ref} in particular signals to the controller whether a reference change has occurred or not. Without this signal, the controller will always initiate a transition to constant power for any non-zero P_e^* and V_e^*. That is, even for a contingency scenario where both P_e^i and V_e^i signals are non-zero, the controller will try to switch over to constant power, which is quite the opposite of the control objectives. Thus, dP_{ref} acts as a discriminatory signal to prevent controller from changing unless there is a reference change for which dP_{ref} will be non-zero. E_{grid}^i acts to give information about fault scenarios during which we want priority to switch to droop mode to ensure voltage is kept within a tolerable range.

The output of the fuzzy controller is a singleton with gain bounded between the values 0 and 20 (VLOW or HIGH respectively) at the extremes. During the transitions, the gain (output of the fuzzy algorithm) varies smoothly between 0 and 20 as dictated by the sigmoidal MF of the error signal as shown in Figure 11.3.

11.4.2.1 Membership Functions and Linguistic Terms

There is no standard method for the choice of membership function (MF) type. The choice relies upon the developers intuition and expected performance, and is application-specific. There are several typical MFs to choose from, and more complicated MFs can be obtained from basic MFs. In very special circumstances depending on the application, there might be a need to develop customized MFs. Alternatively, different MFs can also be employed for different variables, again application specific. In this chapter, two major MFs, trapezoidal and sigmoidal, were employed to map variables from their natural domain onto the fuzzy domain [0, 1]. The simplest MFs, trapezoidal and triangular, can be employed at the development phase, and as experience is gained, more complicated MFs can be employed. Parameters of the MFs can be chosen based on intuition of the underlying dynamics, optimization based on expected performance, or other computational methods available.

In this chapter, the choice of parameters for the trapezoidal MF was straightforward. However, the sigmoidal MFs requires the solution of a system of nonlinear sigmoidal MF equations equal to the number of linguistic terms of the sigmoidal MF. Parameters of the sigmoidal MF are the unknowns of the equation. The sigmoidal MF is dictated by the equation:

$$\mu(x) = \frac{1}{1 + e^{-a(x-c)}} \tag{11.1}$$

where a and c are parameters of the sigmoidal curve.

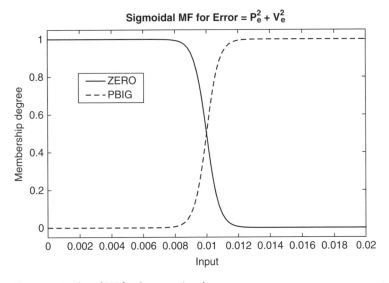

Figure 11.3 Plot of MF for the error signal.

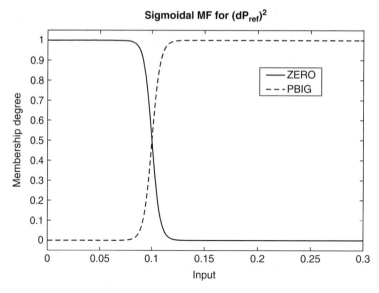

Figure 11.4 Plot of MF for the square of rate of change of active power signal.

The sigmoidal MF with two linguistic terms, "*ZERO*" and "*PBIG*," were chosen for each of $(P_e^i)^2 + (V_e^i)^2$ and $(dP_{ref})^2$, as shown in Figs. 11.3 and 11.4. PBIG implies "*positive big*," and is just a term that is highly subjective. The sigmoidal MF was chosen for its smoothed curved surface. The trapezoidal MF was sufficient for AC grid voltage with two linguistic terms, "*Faulty*" and "*OK*," following discretization of fault values, as shown in Figure 11.5.

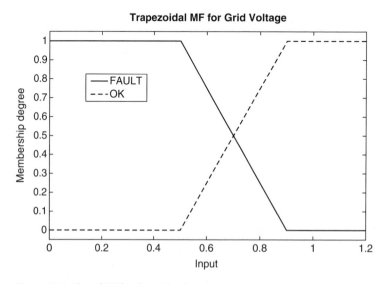

Figure 11.5 Plot of MF for the grid voltage.

11.4.3 Knowledge Base and Inference Engine

The knowledge base contains the rules that encode the desired control objectives in a manner understandable by a human operator. The rules usually take the form of *IF...THEN*. For this chapter, the rules have been reduced from 27 possible rules that cover the entire universe of discourse to 4 rules, through mathematical manipulation and elimination of redundant rules.

The inference engine is the brain behind the implementation that provides the computational framework for transforming the rules from natural language into a meaningful mathematical framework for decision making.

The rules are highlighted below:

- IF **Error** is **ZERO** AND **AC Grid Voltage** is **OK** THEN **DroopGain** is **HIGH**
- IF **Error** is **PBIG** AND **AC Grid Voltage** is **OK** AND **dPref** is **ZERO** THEN **DroopGain** is **HIGH**
- IF **Error** is **PBIG** AND **AC Grid Voltage** is **OK** AND **dPref** is **PBIG** THEN **DroopGain** is **VLOW**
- IF **AC Grid Voltage** is **Faulty** THEN **DroopGain** is **HIGH**

11.4.4 Defuzzification and Output

Defuzzification takes the output of the inference engine from the fuzzy domain into the physical domain the system understands. The two most common defuzzification methods are the center of gravity (COG) method and the mean of max (MOM) method. In this chapter, COG with Mamdani inference was used for defuzzification for its ability to interpolate. In our case, only the extremes of 0 and 20 for droop gains were specified in the fuzzy design process. However, looking at the results in Section 11.6, it is clearly seen how interpolation occurred despite the use of discretized values. This was the intended purpose and is the rationale behind its choice. Equation (11.2) gives the general expression for the COG defuzzification method:

$$y = \frac{\sum_{i=1}^{j} \mu(y_i) y_i}{\sum_{i=1}^{j} \mu(y_i)} \tag{11.2}$$

where j is the number of elements, and $\mu(y_i)$ is the membership grade of each element in the output fuzzy set. Since the singleton model was employed for this work, Equation (11.2) is adapted to:

$$y = \frac{\sum_{i=1}^{M} \beta_i b_i}{\sum_{i=1}^{j} \beta_i} \tag{11.3}$$

Where M is the number of rules, β_i is the degree of fulfilment of each rule, and b_i is the output singletons (0 and 20). Figure 11.6 shows a 3-D nonlinear surface plot that shows clearly the transition surface. Finally, Figure 11.7 presents a state machine that describes the operating steps of the proposed fuzzy logic based adaptive droop controller.

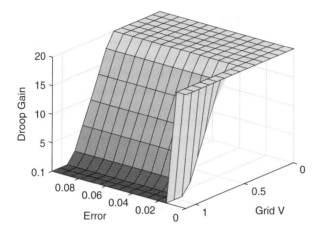

Figure 11.6 3-D surface plot.

11.5 Optimization-Based Secondary Control Strategy

The Newton–Raphson (N–R) method is the traditional iterative procedure for solving power flow equations in power systems. Though very effective in terms of convergence, its flexibility to rapidly changing conditions involving a relatively large DC grid remains to be seen. The genetic algorithm (GA), though an optimization routine, offers flexibility irrespective of topology and size of grid and can be easily integrated with communication and measurement infrastructure, allowing it to adapt to changing situations—self-healing without any need to explicitly specify the power flow equations. Comparison of both GA and N–R show strikingly similar results, with negligible differences if any. The GA uses its operators—*selection, crossover, mutation, and inheritance*—to direct solutions toward convergence or a global optimum [19].

In this proposal, the GA is employed at the secondary control level to provide the unknown optimal voltage that minimizes losses and establishes a pre-defined power flow pattern in the DC grid from which the unknown power can be calculated—in essence, an optimal power flow (OPF). Optimal references are then passed on to the fuzzy-based strategy at the primary control level, responsible for taking the final action. With communication infrastructure (as typical with secondary control) the GA can be updated periodically to reflect current conditions in a dynamic way. The GA thus effectively replaces the N–R method despite being an optimization routine.

11.5.1 Fitness Function

As with most optimization algorithms, there is an objective function—or *"fitness function"*—that determines how good a solution is relative to the set tolerance. The objective of the OPF is to find a set of nodal voltages that minimizes the losses and establishes a power flow pattern in the network. The individual chromosomes are given in Equation (11.4):

$$\tilde{X} = \{U_{dci}^*, U_{dc(i+1)}^*, ..., U_{dcN}^*\} \tag{11.4}$$

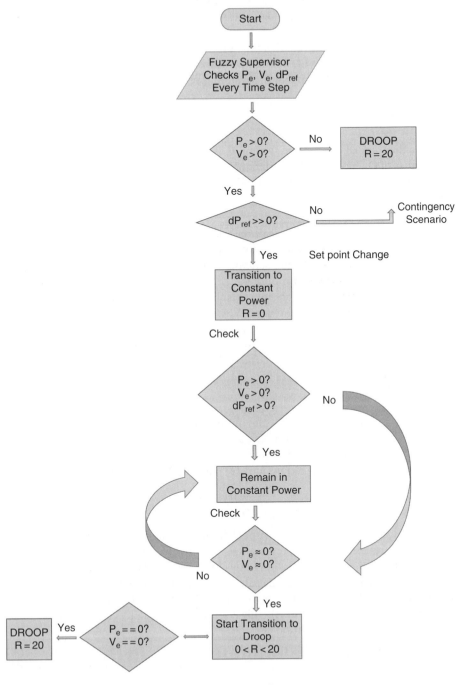

Figure 11.7 Equivalent state machine of the fuzzy controller.

where U^*_{dci} is the DC voltage reference of terminal i and N is the number of nodes in the DC grid. The nodal power at each terminal can be calculated from the voltage references by considering the topology and structure of the grid. All the information required to calculate the unknown power is in the *"incidence matrix."*

The fitness function defines mathematically what is to be minimized and takes the form described by Equation (11.5). The fitness function (objective function) employed in this work is the total DC losses of the system given by the sum of DC nodal power.

$$min\ \overline{P} = \sum_{i=1}^{N_{dc}} P_i \tag{11.5}$$

where, $\overline{P}$ is the total DC loss in the system, P_i is the DC nodal power at node i, and N_{dc} is the number of DC nodes in the grid. Equation (11.5) is thus expressed in terms of the chromosomes contained in Equation (11.4), that is, the nodal voltages.

11.5.2 Constraints

There are certain technical and physical/material limits that cannot be violated irrespective of any other conditions. In the present work, the consideration of the constraints is performed based on a static penalty scheme. Constraints can take the form of bound, linear, or nonlinear constraints, for both equality and inequality constraints.

Bounds constraints ensure that acceptable solutions are within a technically allowable range. With the current state-of-the-art of VSCs, the IGBT (insulated gate bipolar transistor) switches that make up the VSCs are still very sensitive devices. Over-voltages cause irreparable damage to switches; under-voltage in general lead to loss of controllability. Hence there must be a bound on the solution of GA. Bounds constraints take the form of Equation (11.6):

$$\underline{\mathbf{L}} \leq \mathbf{x} \leq \overline{\mathbf{U}} \tag{11.6}$$

where $\underline{\mathbf{L}}$ is the lower bound and $\overline{\mathbf{U}}$ is the upper bound on all solutions. For this work, the bounds constraints on the references obtained are as described in Equation (11.7). A slight modification of Equation (11.7) was employed, as shown in Equation (11.8), where $\underline{c}_1$, and $\overline{c}_2$ define allowable relaxations and are equal to 0.05 in this work:

$$0.9pu \leq \mathbf{U}_{dci} \leq 1.1pu \qquad i = 1, 2, ..., N_{dc} \tag{11.7}$$

$$0.9pu + \underline{c}_1 \leq \mathbf{U}_{dci} \leq 1.1pu + \overline{c}_2 \qquad i = 1, 2, ..., N_{dc} \tag{11.8}$$

Nonlinear equality constraints are especially important from an operational point of view. There are certain common instances that involve market, legal, or associated conditions, which stipulate that certain nodes in the DC grid get a fixed or pre-defined amount of power irrespective of changes in operating conditions, especially the amount of wind power (excluding any disruptive changes that may give priority to security instead of market or legal conditions). This is a nonlinear equality constraint, and these have the form described in Equation (11.9):

$$-P_{max} \leq \mathbf{P}_i \leq P_{max} \qquad i = 1, 2, ..., N_{dc} \tag{11.9}$$

where P_i is the nodal power of node i.

$$\mathbf{P}_i = \hat{P} \qquad i = 1, 2, ..., N_{dc} \tag{11.10}$$

where $\hat{P}$ is the pre-defined power at a node.

11.6   Simulation Results

This section presents the time domain simulation results carried out with the proposed intelligent control strategy both at the primary and secondary levels. The results are meant to demonstrate the capabilities, robustness, and efficacy of the strategies. The strategy was implemented on the three terminal grid described in Figure 11.2. Several realistic scenarios were simulated. The proposed fuzzy logic based adaptive droop is installed at VSC 2 while VSC 3 applies constant voltage control operating as a slack DC bus.

11.6.1   Set Point Change

Prior to this simulation, the network is first initialized at equilibrium operating point. Wind power at VSC 1 is 1000 MW, VSC 2 is set at 300 MW (inversion). VSC 3 takes the difference after losses are accounted for.

At t = 1 s, the power set point of VSC2 changes from 300 MW to 500 MW (inversion). All the conditions were thus fulfilled for a transition from droop strategy to constant power mode dictated by a droop gain of *zero*. Figure 11.8 presents the transition from droop control (R = 20) to constant power (R = 0). At about t = 4 s when the steady state power and voltage error reduces to zero, the fuzzy supervisor gradually transitions back to droop mode (dictated by the droop gain) in a very smooth manner without sharp edges. A time delay block is added to the reference change in order to prevent oscillations as both power and voltage are dependent on each other.

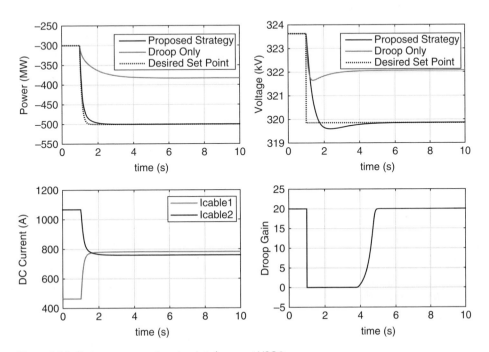

Figure 11.8 System response to set point change at VSC 2.

11.6.2 Constantly Changing Reference Set Points

With the goal to test the efficacy and robustness of the proposed strategy, consequent reference changes were sent to the converter VSC2 every t = 10 s over a 50 s range just to ensure that the expected objective will be met regardless of the the way set points are provided and for different initial conditions. For this scenario, set points were changed from inversion to rectification to prove the efficacy of the strategy. As expected, the fuzzy supervisor met the objectives as designed regardless of the initial operating point and the change applied, ensuring a robust operation over the entire range. Figure 11.9 shows the superiority of the proposed strategy over the conventional droop only strategy as the fuzzy supervisor responded to every set point change as required. Finally, is worth observing the output of the fuzzy droop gain changing from 0 to 20.

11.6.3 Sudden Disconnection of Wind Farm for Undefined Period

During contingencies, it is important that the system remains in droop mode to keep voltages at all nodes controllable within acceptable bounds. In a DC grid, any change at all will cause a non-zero error to be measured at all nodes. Thus, the proposed strategy must be able to distinguish between a measured error due to a change in reference (or other conditions that facilitate a change to constant power mode) and an error due to contingencies, to make sure it does not act or transition to another strategy. Figure 11.10 shows the system response for sudden disconnection of a complete wind power plant.

Such sudden disconnection results in a deficiency of power in the DC grid, and an obvious response is a sag in voltage at all terminals. Since VSC 2 is in droop mode, it reacts to keep the grid voltage within acceptable bounds by injecting power from the

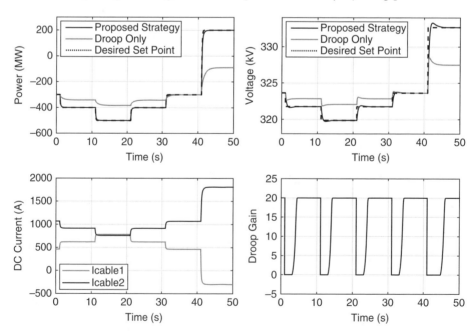

Figure 11.9 System Response to constantly changing reference set points at VSC 2.

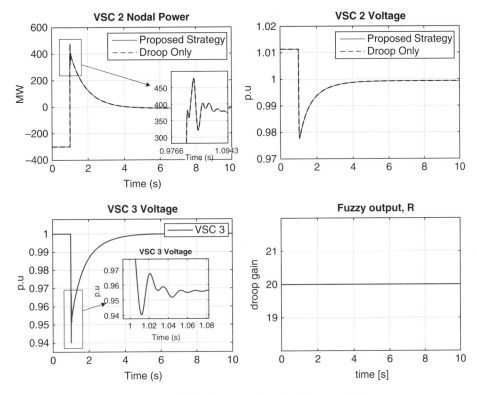

Figure 11.10 System response to sudden disconnection of wind power plant.

adjacent AC grid. It is obvious from the subplot of droop gain that the system remained in droop, and terminal voltages barely rose from their set reference. With PI based strategies, the system voltage may well dip below the minimum threshold of 0.9 pu for stable operation.

11.6.4 Permanent Outage of VSC 3

This is a rather important scenario that must be considered in any realistic operation of the grid, that is, loss of any terminal. The response of a DC grid to loss of a terminal consuming power is voltage rise, as this surplus power charges the capacitance of the network. Notwithstanding, the rise must be kept in check to prevent instability. Any rise must not exceed the maximum voltage that a grid is designed to withstand (typically 1.2 pu). Figure 11.11 depicts this rise. As can be seen from the plot, despite the rise, the nodal voltages at all terminals are much below the threshold because the system remained in droop mode. For conventional PI strategies, a chopper may need to operate when voltage exceeds the threshold. This significantly adds to the total cost of a design.

11.7 Conclusion

This chapter presents a knowledge-based primary and an optimization-based secondary control scheme for offshore multi-terminal HVDC grids. The presented results reveal

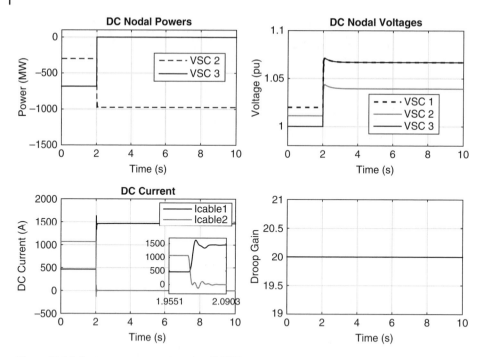

Figure 11.11 System response to outage of VSC 3.

the superiority of adaptive fuzzy droop control to solve major drawbacks presented by conventional direct voltage and power control strategies. Based on the results, there were no observed deviations in the power or voltage with fuzzy control, which is a major drawback of power based droop. As a matter of fact, there is no need for secondary corrective actions.

In addition, fuzzy control is flexible, and can be applied as a *"plug 'n' play"* device to any VSC-HVDC converter without knowledge of inner controller proprietary information (interoperability). Moreover, its capabilities can be expanded off-line and on-line as more experience is gained.

The proposed strategy reveals the capabilities of a natural language processing tool to combine strategies that would not be possible with a single control law. Besides, the implementation of the proposed strategy requires very little knowledge or expertise of power systems, let alone HVDC grid operation. Topology, market conditions, configuration, size, and so on do not influence the proposed strategy in any way detrimental to the operation of the grid and can be extended to as many terminals as necessary.

References

1 Lu, W. and Ooi, B.T. (2005) Premium quality power park based on multi-terminal HVDC. *IEEE Transactions on Power Delivery*, **20** (2), 978–983, doi: 10.1109/TPWRD.2004.838633.

2 Haileselassie, T. and Uhlen, K. (2012) Precise control of power flow in multiterminal VSC-HVDCs using DC voltage droop control, in *Power and Energy Society General Meeting, 2012 IEEE*, pp. 1–9, doi: 10.1109/PESGM.2012.6343950.

3 Lu, W. and Ooi, B. (2002) Multi-terminal HVDC as enabling technology of premium quality power park, in *IEEE Power Engineering Society Winter Meeting, 2002.*, vol. 2, pp. 719–724, doi: 10.1109/PESW.2002.985098.

4 Haileselassie, T., Uhlen, K., and Undeland, T. (2009) Control of multiterminal HVDC transmission for offshore wind energy, in *Nordic wind power conference*, pp. 10–11.

5 Ismunandar, C., Meer, v.d.A., Gibescu, M., Hendriks, R., and Kling, W. (2010) Control of multi-terminal VSC-HVDC for wind power integration using the voltage-margin method, in *in Proc. 9th International Workshop on Large Scale Integration of Wind Power into Power Systems as well as on Transmission networks for offshore Wind Power Plants, Québec, Canada*, pp. 427–434.

6 Vrana, T.K., Beerten, J., Belmans, R., and Fosso, O.B. (2013) A classification of DC node voltage control methods for HVDC grids. *Electric Power Systems Research*, **103**, 137–144, doi: http://dx.doi.org/10.1016/j.epsr.2013.05.001. URL http://www.sciencedirect.com/science/article/pii/S0378779613001193.

7 Haileselassie, T. and Uhlen, K. (2012) Impact of dc line voltage drops on power flow of mtdc using droop control. *IEEE Transactions on Power Systems*, **27** (3), 1441–1449, doi: 10.1109/TPWRS.2012.2186988.

8 Beerten, J. and Belmans, R. (2013) Analysis of power sharing and voltage deviations in droop-controlled DC grids. *IEEE Transactions on Power Systems*, **28** (4), 4588–4597, doi: 10.1109/TPWRS.2013.2272494.

9 Zhao, X. and Li, K. (2015) Droop setting design for multi-terminal HVDC grids considering voltage deviation impacts. *Electric Power Systems Research*, **123**, 67–75, doi: http://dx.doi.org/10.1016/j.epsr.2015.01.022. URL http://www.sciencedirect.com/science/article/pii/S0378779615000322.

10 Chaudhuri, N. and Chaudhuri, B. (2013) Adaptive droop control for effective power sharing in multi-terminal DC (MTDC) grids, in *Power and Energy Society General Meeting (PES), 2013 IEEE*, pp. 1–1, doi: 10.1109/PESMG.2013.6672382.

11 Egea-Alvarez, A., Beerten, J., Hertem, D.V., and Gomis-Bellmunt, O. (2015) Hierarchical power control of multiterminal HVDC grids. *Electric Power Systems Research*, **121**, 207–215, doi: http://dx.doi.org/10.1016/j.epsr.2014.12.014. URL http://www.sciencedirect.com/science/article/pii/S037877961400460X.

12 Liang, J., Jing, T., Gomis-Bellmunt, O., Ekanayake, J., and Jenkins, N. (2011) Operation and control of multiterminal HVDC transmission for offshore wind farms. *IEEE Transactions on Power Delivery*, **26** (4), 2596–2604, doi: 10.1109/TPWRD.2011.2152864.

13 Wang, W., Barnes, M., and Marjanovic, O. (2012) Droop control modelling and analysis of multi-terminal VSC-HVDC for offshore wind farms, in *International Conference on AC and DC Power Transmission (ACDC 2012), 10th IET*, pp. 1–6, doi: 10.1049/cp.2012.1963.

14 Pinto, R., Bauer, P., Rodrigues, S., Wiggelinkhuizen, E., Pierik, J., and Ferreira, B. (2013) A novel distributed direct-voltage control strategy for grid integration of offshore wind energy systems through MTDC network. *IEEE Transactions on Industrial Electronics*, **60** (6), 2429–2441, doi: 10.1109/TIE.2012.2216239.

15 Wang, W. and Barnes, M. (2014) Power flow algorithms for multi-terminal VSC-HVDC with droop control. *IEEE Transactions on Power Systems*, **29** (4), 1721–1730, doi: 10.1109/TPWRS.2013.2294198.

16 Vrana, T., Zeni, L., and Fosso, O. (2012) Dynamic active power control with improved undead-band droop for HVDC grids, in *10th IET International Conference on AC and DC Power Transmission (ACDC 2012)*, pp. 1–6, doi: 10.1049/cp.2012.1975.

17 Babuška, R. (2012) *Fuzzy modeling for control*, vol. 12, Springer Science & Business Media.

18 Zadeh, L. (1965) Fuzzy sets. *Information and Control*, **8** (3), 338–353, doi: http://dx.doi.org/10.1016/S0019-9958(65)90241-X. URL http://www.sciencedirect.com/science/article/pii/S001999586590241X.

19 Coley, D.A. (1999) *An introduction to genetic algorithms for scientists and engineers*, World scientific.

12

Model Based Voltage/Reactive Control in Sustainable Distribution Systems

Hoan Van Pham [1,2] and Sultan Nasiruddin Ahmed [3]

[1] *Power Generation Corporation 2, Vietnam Electricity*
[2] *School of Engineering and Technology, Tra Vinh University, Vietnam*
[3] *FGH GmbH, Aachen, Germany*

12.1 Introduction

Electricity demand increases continuously, hence greater amounts of power need to be transferred. System reinforcement has, however, been kept to a minimum for economic reasons. Moreover, many countries have experienced liberalization trends in electricity markets, which call for intensive variations in generation and power flows. Furthermore, increasing incentives in several countries to integrate intermittent renewable energy sources (RES) into the systems entails several challenges to system reliability and security, especially in short term operational timeframes, where a high degree of variability of power supply from RES may occur [1]. As a result, temporary over/under voltages and congested lines happen more often. This creates a demand for developing new control schemes to achieve optimal, flexible, and efficient operation for voltage regulation.

Numerous recent publications on voltage control have demonstrated its significance in system operations given several power system blackouts around the world [2, 3] and voltage violation problems [4–11]. Currently, corrective voltage control, also known as online voltage control, attracts much attention from researchers due to that fact that power systems are undergoing more uncertainties caused by increased integration of renewable energy. In [7], an online centralized controller based on a heuristic optimization algorithm is introduced for voltage control applications in wind farms. In [12], the authors propose an online voltage scheme based on a steady-state model, whose objective function is handled by using chance constrained optimization. Reference [13] describes a voltage control model in which both mitigation of voltage violations and minimization of transmission losses are taken into account. A coordinated secondary voltage control presented in [14] and based on a sensitivity model is in operation in two French control centres. In [15], an online voltage control strategy is presented to minimize the operational conflicts by prioritizing the operations of different regulating devices while maximizing the voltage regulation support by distributed generators (DGs). A novel implementation of OPF for corrective power flow management in an online operational mode, for MV distribution networks, is presented in [16].

Dynamic Vulnerability Assessment and Intelligent Control for Sustainable Power Systems, First Edition.
Edited by José Luis Rueda-Torres and Francisco González-Longatt.
© 2018 John Wiley & Sons Ltd. Published 2018 by John Wiley & Sons Ltd.

Model Predictive Control (MPC) is categorized into a class of online algorithms based on collected measurements and a system model to calculate a sequence of future control actions (control variables); however, only the first set of control actions are applied, and the process is repeated with newly received measurements in the next step. MPC is extensively used in industry because it has a simple concept and easy integration of constraints and limitations. MPC is mature technique and has been used in several applications in power systems such as frequency control, network congestion management, and integration of energy markets and weather forecasts. MPC has also been successfully applied to solving voltage-related problems [17–23]. Reference [17] describes a sensitivity analysis based controller for emergency voltage control. In [18], generator voltages, tap changers, and load shedding are considered as control variables, and they are coordinated via a tree search optimization approach. Reference [19] studies the use of MPC with a Lagrangian decomposition based solver to optimally coordinate generator voltage references and load shedding. In [20], a control switching strategy of shunt capacitors is proposed to prevent voltage collapse and maintain a desired stability margin after a contingency. Reference [21] presents a receding horizon multi-step optimization inspired by MPC, based on steady-state power flow equations, to control transmission voltages. In [22], a corrective control based on MPC is introduced to correct voltages out of limits by applying optimal changes of the control variables (mainly active and reactive power of distributed generation and LTC voltage set-point). An MPC based controller at the intermediate level in a hierarchical (cascade) three-layer structure is presented in [23] that is capable of explicitly handling constraints on the main process variables, such as the voltages along the grid or the adopted power factors.

This chapter aims to present a general philosophy of MPC-inspired voltage control schemes to correct power system voltages during normal (i.e. quasi-steady-state) conditions. These kinds of control schemes have a slow response, from – say – 10 to 60 seconds, to small operational changes and do not provide fast reactions during large disturbances in order to prevent undesirable adverse implications. Within this timeframe, long-term dynamics are of concern as well [5]. In the first part of the chapter, background theory relating to relevant knowledge to design such control schemes is presented. In the second part, a voltage control scheme in distribution networks is outlined to demonstrate the performance of MPC-inspired control schemes. The central idea of this scheme is a static optimization performed in the closed-loop mode of MPC to achieve certain control objectives.

The remainder of this chapter is structured as follows: General background behind voltage control, MPC, model analysis, and controller implementation are provided in Section 12.2. In Section 12.3, an example of designing a MPC based voltage controller is presented. Simulation results obtained using a U.K. generic distribution network are given in Section 12.4. Finally, Section 12.5 concludes the chapter.

12.2 Background Theory

12.2.1 Voltage Control

The main objective of voltage control is to maintain system voltages as close as possible to reference values. Initially, system operators will determine a target voltage for each bus in the system. This target voltage may track certain economic or security objectives, for example active power losses minimization. In distribution networks, where economic issues are often given more priority, voltage is allowed to vary more frequently within a wider band of ± 5 or $\pm 10\%$ of nominal voltage according to the grid code of each country. Therefore, operators usually take advantage of this to achieve other control targets instead of keeping voltages stable at their reference values, and voltage limitation is often treated as a constraint in an optimization problem.

There are a variety of voltage control means in power systems. In terms of system specifics, the following control means can be used: voltage set-points of Load Tap Changing (LTC) transformers, voltage set-points of advanced FACTS devices, and active and reactive power of distributed generators. It should be mentioned that generator bus voltages in transmission systems should be kept as constant as possible because the design of a power station assumes the terminal voltage will remain in a tight band around its nominal value. This implies the generator terminal voltage cannot be varied due to physical constraints.

In normal conditions, the voltage controller will use minimum control actions of the cheapest control variables, such as set-points of reactive devices – SVC, STATCOM, capacitor banks – in order to achieve the optimization targets. This is because it is not economically justifiable to use expensive control variables (e.g. load or generation curtailment). However, when voltages become unstable, the controller will use all of the available control variables (both cheap and expensive) to bring these voltages within acceptable limits.

12.2.2 Model Predictive Control

The use of MPC is motivated by its capability to compensate for system model inaccuracies by implementing controls in a closed-loop manner. Moreover, constraints on either equality (system model) or inequality (rate of change, limits on available controls, etc.) are easily integrated into a multiple time step optimization problem with a finite horizon. Another motivation for using MPC in power systems is to coordinate the various control means, which relieves system operators from the stress of properly adjusting a variety of controls.

The principle of MPC is shown in Figure 12.1. The main feature of MPC is to solve the optimal control problem in finite time horizon based on online analysis:

- At each time k, MPC computes the open-loop sequence of control actions that will make the future predicted controlled variables best follow the reference over a finite control horizon T_c, while also accounting for the control effort. Meanwhile, MPC explicitly uses the system model to predict future system behaviour over the prediction horizon, T_p.
- Only the first set of control actions are implemented and the rest disregarded.
- The process is repeated at the next sampling time $k + 1$ when the next measurements from the system are available.

It is convenient to formulate the MPC problem in the context of a discrete-time, non-linear system:

$$x_{k+1} = f(x_k, u_k)$$
$$y_k = g(x_k) \tag{12.1}$$

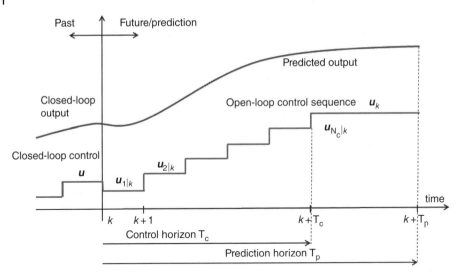

Figure 12.1 Principle of MPC.

where x_k is a vector of state variables, u_k a vector of controls, and y_k a vector of outputs or controlled variables.

MPC is formulated as a repeated online solution of a finite horizon open-loop optimal control problem. When measuring the initial system states $x_{1|k}$, a control input $u_k = u_{i|k}$ ($i = 1, \dots, N_c$) over the prediction horizon T_p can be determined by solving the following optimal control task:

$$\min_{u_k} \quad J_k = \sum_{i=1}^{N_c} l_i \left(x_{i|k}, u_{i|k} \right) \tag{12.2}$$

subject to the constraints

$$x_{i+1|k} = f(x_{i|k}, u_{i|k}), \ i = 1, \dots, N_p$$

$$y_{i+1|k} = g(x_{i+1|k}), \ i = 1, \dots, N_p$$

$$y_{min} \leq y_{i+1|k} \leq y_{max}, \ i = 1, \dots, N_p$$

$$u_{min} \leq u_{i|k} \leq u_{max}, \ i = 1, \dots, N_c$$

$$\delta_{min} \leq u_{i|k} - u_{i-1|k} \leq \delta_{max}, \ i = 1, \dots, N_c$$

where $i|k$ is the notation for prediction i steps ahead at evaluation time k, and $l_i(x, u)$ is stage cost. The last three constraints aim at making sure that controlled variables, control variables, and their changes remain in limits according to their physical limitation.

MPC is an approach based on a system model, hence the accuracy level of the model significantly influences the efficiency of MPC implementation. Also, the efficiency decreases with increasing complexity of the model, so trying to keep the model computationally tractable by using several successive system linearizations is usually considered.

The prediction horizon T_p must be chosen in order to account for expected effects of the computed control actions. In this sense, the control horizon T_c should be at most equal or less than the prediction horizon T_p. It is also worth mentioning that short horizons would require less computational expense, but may negatively affect closed-loop stability.

12.2.3 Model Analysis

As discussed, MPC explicitly uses a system model to predict future system behaviour caused by the computed control actions over the prediction horizon T_p. To be incorporated in the overall MPC optimization problem, the system model must use linear approximations. This may sometimes not be acceptable; however, it is not a problem in the sense that MPC is able to compensate for those model inaccuracies.

Sensitivities are used to synthesize the approximate behaviour that would occur for altered parameter values. Alternatively, the sensitivities can be adopted to formulate (12.1) and (12.2). Regarding sensitivities of bus voltages with respect to generation power, these can be obtained from the inverse of the Jacobian matrix extracted from an off-line power flow calculation. They can also be extracted from the solutions of two power flow calculations with different generation power, by computing the ratio of variation of the monitored bus voltage to the variation of generation power. In this section, the term sensitivity is first defined, and then we demonstrate how sensitivity analysis is carried out numerically and analytically.

12.2.3.1 Definition of Sensitivity

The sensitivity of a variable y with respect to a control variable u is a measure of the influence of changes in u on the variable y. The higher the sensitivity value, the more influence the control variable has on the controlled variable. Therefore, the sensitivity K_{y-u} at the operating point $x = x_0$, $u = u_0$ corresponds to the derivative of y with respect to u:

$$K_{y-u} = \left. \frac{dy(x, u)}{du} \right|_{x=x_0, u=u_0} \tag{12.3}$$

where y is a function of the control variables u and the state variables x. The relation between the state variables and the control variables is given by the system constraints $g(x, u) = 0$.

12.2.3.2 Computation of Sensitivity

There are two possibilities to obtain sensitivity values [25]. The straightforward option is to numerically estimate the sensitivities due to changes in the system variables between different operating points of the system. An alternative way is to calculate the derivative (12.3) analytically.

i) Numerical

The sensitivity values are numerically determined by approximating the derivative (12.3) as follows:

$$K_{y-u} = \frac{dy}{du} \approx \frac{\Delta y}{\Delta u} = \frac{y_1 - y_0}{u_1 - u_0} \tag{12.4}$$

where y_0, u_0, and y_1, u_1 are system and control variables at an initial operation point and at another operating point with a changed control setting, respectively.

In a linear system as shown in in Figure 12.2a where the system variables are linear with the control variables, the sensitivities are the same over the whole range of set

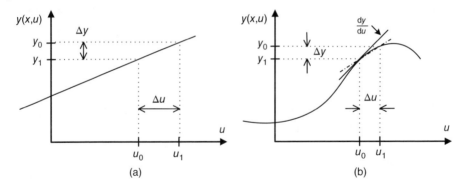

Figure 12.2 Energy Illustration of sensitivity for (a) linear and (b) nonlinear dependency.

values and are equal to (12.4). But the sensitivity becomes dependent on the setting of the control variable in a nonlinear system as presented in Figure 12.2b. In this case, the sensitivity values have to be determined based on initial operating points.

ii) Analytical

The derivative (12.3) is determined analytically using partial derivatives, leading to

$$K_{y-u} = \frac{dy}{du} = \frac{\partial y}{\partial u} + \frac{\partial y}{\partial x} \cdot \frac{\partial x}{\partial u} \tag{12.5}$$

The expression for the last term in (12.4) can be derived by utilizing the Taylor approximation of the system constraints $g(x,u) = 0$:

$$0 = g(x_0 + \Delta x, u_0 + \Delta u) \approx g(x_0, u_0) + \frac{\partial g}{\partial u} \Delta u + \frac{\partial g}{\partial x} \Delta x \tag{12.6}$$

For non-singular $\frac{\partial g}{\partial x}$, this yields

$$\frac{\partial x}{\partial u} \approx \frac{\Delta x}{\Delta u} = -\left(\frac{\partial g}{\partial x}\right)^{-1} \cdot \frac{\partial g}{\partial u} \tag{12.7}$$

The sensitivity value of the system variable y with respect to changes in the control variable u is therefore determined by

$$K_{y-u} = \frac{dy}{du} = \frac{\partial y}{\partial u} - \frac{\partial y}{\partial x} \cdot \left(\frac{\partial g}{\partial x}\right)^{-1} \cdot \frac{\partial g}{\partial u} \tag{12.8}$$

Note that in many cases in power systems, the system variable y does not depend explicitly on the control variable u, hence $\partial y / \partial u = 0$. And (12.8) can be rewritten as

$$K_{y-u} = \frac{dy}{du} = -\frac{\partial y}{\partial x} \cdot \left(\frac{\partial g}{\partial x}\right)^{-1} \cdot \frac{\partial g}{\partial u}$$

The analytical approach has some advantages, such as the fact that the calculation is carried out only once and the same formula used for every operating point. Moreover, due to no approximation being used, the estimated sensitivities of the analytical method are more accurate than the numerical. However, it requires more computational expenses, so the numerical approach is often preferred for a large system. Recall that MPC is able to compensate for model inaccuracies, so the numerical based sensitivity analysis is completely sufficient.

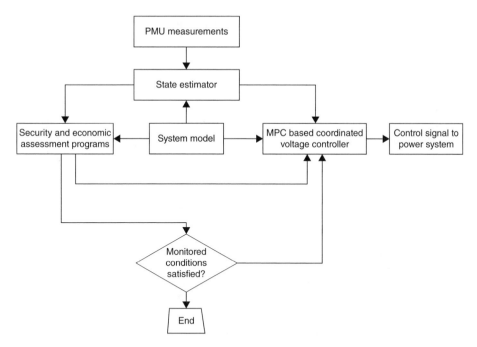

Figure 12.3 Implementation framework of MPC based voltage control.

12.2.4   Implementation

The implementation framework of MPC based voltage control is shown in Figure 12.3. Online information flows, bus voltages, and switch statuses measured by phasor measurement units (PMUs) and collected by Phasor Data Concentrators (PDCs) are sent to a control centre through communication channels. The state estimator (SE) will use these measurements and the available system model to estimate the system states and refine the measured data. Both the system model and the estimated system states are then input data to certain security or economic assessment programmes that are often used to investigate whether monitored conditions are satisfied. The monitored conditions can vary, depending on the design purpose of each controller; Examples are: voltage violation, dynamic security assessment, and contingency analysis. Once one of the conditions is violated, MPC based computation will be triggered. A final step is to implement the real time control computed to discard violations.

The steps of the MPC based coordinated voltage controller at a time instant include:

- Collect static values from the state estimator such as voltage magnitudes and angles, as well as dynamic variables such as generator angles, velocities, and real and reactive load recovery, depending on the purpose of use; collect target voltages from the security & economic programme, depending on the control scheme designed.
- Carry out sensitivity analysis to compute the sensitivities given the current state.
- Obtain the sensitivities which are required by the optimization problem formulation.
- Solve the optimization problem and implement only the first set of control actions.

12.3 MPC Based Voltage/Reactive Controller – an Example

Assume that a test system has an existing PI control unit, presented in Figure 12.4, which is to fulfil reactive power demand ($Q^{\text{PCC,ref}}$) requested by TSOs (e.g. to meet certain grid code requirements). For instance, the grid codes [27] and the European Network of Transmission System Operators for Electricity [28] define a set of common requirements at the interface between transmission and distribution systems (or wind farms in particular). By default, the PI controller would distribute total reactive power demand equally into available Var sources, resulting in losses not being minimized and potential voltage violation.

12.3.1 Control Scheme

The PI controller is beneficial in smoothly fulfilling the reactive power demand from external grids without expensive computation. The task is now to design a centralized MPC based controller which is capable of optimally distributing output of the PI controller into available Var sources and taking action on coordinating available voltage control variables to minimize the active power losses, subject to operational constraints described later.

The MPC based control scheme with this objective is sketched in Figure 12.5. The controller presented in this example focuses on slow control functionality, which is characterized by a slow response (e.g. 10 seconds to a few minutes) to adapt the DG response to changing quasi-steady-state requirements. In this example, there are two kinds of control variables:

$$\boldsymbol{u} = [\boldsymbol{q}^{\text{g}}, \boldsymbol{p}^{\text{g}}]^{\text{T}} \tag{12.9}$$

where ^{T} denotes array transposition, and $\boldsymbol{q}^{\text{g}}$, $\boldsymbol{p}^{\text{g}}$ are reactive power and active power injection of DGs, respectively.

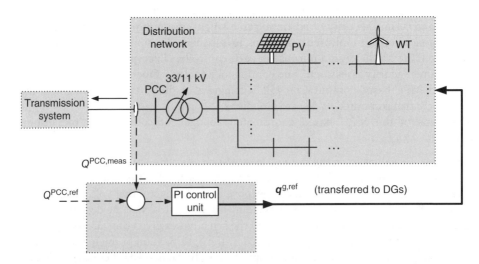

Figure 12.4 Existing centralized controller in the test system.

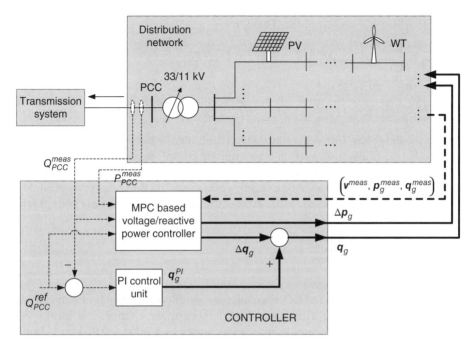

Figure 12.5 MPC based control scheme.

The control algorithm can be summarized as follows:

- Firstly, the MPC based controller collects necessary measurements such as (v^{meas}, $q^{g,meas}$, $p^{g,meas}$) where v are voltages at selected buses, and the superscript ^{meas} stands for measurements.
- Then, the controller, calculates output control actions such as changes of DGs reactive power set-points, Δq^g, and curtailment of active power injection of DGs, Δp^g, when needed. These control actions will be transmitted to corresponding components in the system to perform local control.

12.3.2 Overall Objective Function of the MPC Based Controller

Typically, the security & economic assessment programme in Figure 12.3 will define a target voltage for each bus in the network. These target voltages may follow a security or economic purpose, such as minimization of network losses. If some of the voltages fall in the undesirable region (predefined limits around the target voltages), the MPC controller will use the minimum control actions to bring these voltages within acceptable limits.

However, in order to investigate how the accuracy of the sensitivities impacts the performance of the controller, and to demonstrate that the MPC based controller can account for model inaccuracies (referring to inaccuracies in sensitivities), the MPC based controller in this example is set to incorporate the economic objective as well; hence the economic assessment programme is not needed in this example.

In the spirit of MPC, the objective is to compute the change of the control variables $\Delta u_{i|k}$ ($i = 1, \ldots, N_C$), shown in (12.10), in order to move actual voltages at time instant

k as quickly as possible to their optimal values:

$$\Delta \boldsymbol{u}_{i|k} = \left[\left[\Delta \boldsymbol{q}_{i|k}^g \right]^T, \left[\Delta \boldsymbol{p}_{i|k}^g \right]^T \right]^T \qquad \text{for } i = 1, \dots, N_C \tag{12.10}$$

with $\Delta \boldsymbol{u}_{i|k} = \boldsymbol{u}_{i|k} - \boldsymbol{u}_{i-1|k}$

The overall objective function of the MPC based controller is a function of multi-objectives corresponding to following control variables:

- *Reactive Power of DGs:* Different objective functions can be used by the system operators, beside traditional transmission losses minimization; for example, minimization of reactive power cost, minimization of deviations from contracted transactions, or minimization of the cost of adjusting reactive power control devices. In this example, losses minimization is selected, and the objective function (OF) at time instant k is therefore as follows:

$$OF_1(k) = \min \left(\left(\frac{\partial \boldsymbol{P}_{loss}}{\partial \boldsymbol{q}_g} \right)^T \Delta \boldsymbol{q}_g(k) \right) \tag{12.11}$$

where $\boldsymbol{P}_{loss}$ is a vector of the active power losses
- *Active Power Generation of DGs:* One economic operation objective is maximization of active power production from DGs. This is permanently satisfied at local level because DGs always try to capture maximum power by tracking actual conditions such as wind speed, or solar intensity. However, under certain operation conditions such as voltage violation, some curtailments of active power of DGs are requested to preserve normal operation. Consequently, minimization of active power curtailment becomes necessary in the network operation control:

$$OF_2(k) = \min \left(\sum_{j=1}^{N_g} \left| \Delta p_{gj}(k) \right| \right) \tag{12.12}$$

where N_g is the number of DGs

As mentioned earlier, the MPC based controller finds a sequence of control actions in N_C steps. Therefore, an overall objective is formulated as follows with incorporation of these objective functions in (12.11) and (12.12) under a relation described by a weighting matrix, $\boldsymbol{w}$, to penalize expensive control actions (curtailment of active power generation is defined as an expensive variable):

$$OF = \min \sum_{i=0}^{N_C-1} \left(\boldsymbol{w}^T \left[OF_{1,2}(k+i) \right]^T \right) \tag{12.13}$$

subject to the following constraints:
- Constraints of control variables

$$\boldsymbol{u}^{\min} \leq \boldsymbol{u}_{i|k} \leq \boldsymbol{u}^{\max}$$

$$\boldsymbol{u}_{i|k} = \boldsymbol{u}_{i-1|k} + \Delta \boldsymbol{u}_{i|k}$$

- Constraints of voltage

$$\boldsymbol{v}^{\min} \leq \boldsymbol{v}_{i|k} \leq \boldsymbol{v}^{\max}$$

$$\boldsymbol{v}_{i|k} = \boldsymbol{v}_{i-1|k} + \Delta \boldsymbol{v}_{i|k} = \boldsymbol{v}_{i-1|k} + \frac{\partial \boldsymbol{v}}{\partial \boldsymbol{u}} \Delta \boldsymbol{u}_{i|k}$$

– Constraints of reactive power exchange with the external grid

$$\left| \frac{\partial Q_{PCC}}{\partial u} \Delta u_{i|k} \right| \leq \varepsilon$$

$v_{i|k}$ is the set of predicted voltage magnitudes of monitored buses at time instant k and $\frac{\partial v}{\partial u}$ is the sensitivity matrix of monitored bus voltages with respect to the control variables. ε is a small tolerance coefficient.

Finally, it is worth mentioning that the MPC based controller requires three kinds of sensitivities to be provided: $\partial P_{loss}/\partial q_g$, $\partial V/\partial u$, and $\partial Q_{PCC}/\partial u$.

12.3.3   Implementation of the MPC Based Controller

In this example, the MPC based controller is triggered under predefined conditions that will be discussed later, to correct voltage profile and minimize total losses, as shown in Figure 12.6. As soon as it is triggered, the controller starts collecting available online measurements that are then used to estimate sensitivities. The process of determining control actions is performed afterwards incorporating the online measurements and sensitivities just estimated. The MATLAB nonlinear programming solver, fmincon, is used in this example to solve the objective function (12.13). Next, to avoid inaccurate measurements caused by transient phenomena after control action implementation, the new measurements will be collected after a time delay T_{delay}. N_{step} is a predefined value to limit the number of calculation iterations of the controller per triggering.

Figure 12.6 Flowchart of operational principle of MPC based controller.

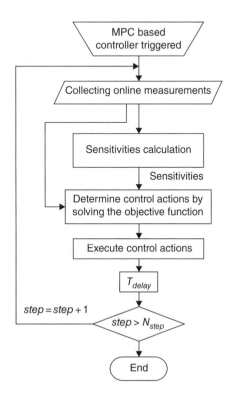

To relieve the computational burden, the controller is only triggered when necessary, under following conditions:

- *Voltage violation*
- *Considerable change of operation condition:* It is reasonable that when the networks change their operation condition significantly, optimal operational set-points of the networks need to be adjusted in response. The level of change could be specified based on all available measurements; this is, however, out of the scope of this example.

12.4 Test Results

12.4.1 Test System and Measurement Deployment

The test network in Figure 12.7 is taken from a UK generic distribution network (UKGDN), available in [26]. It comprises 75 load buses (except bus#1000 and bus#1100) and 22 distributed generator (DG) units (3 MW nominal capacity for each); it connects to the transmission system through a cable (with Thévenin reactance of 0.1 pu of 200 MVA short-circuit power used in this paper) and a 33/11 kV transformer equipped with OLTC, which is able to regulate voltage in the range of ± 10% of nominal

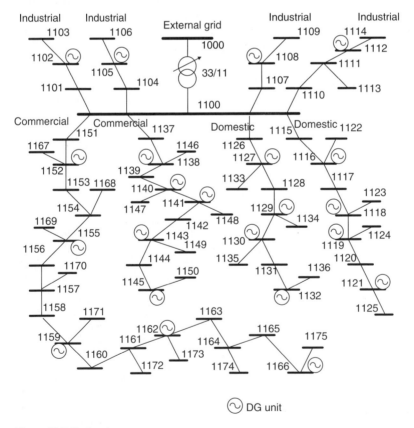

Figure 12.7 Test system.

Table 12.1 Parameters of the controller.

PI control unit		Corrective control unit	
K_I	K_P	T_{delay}	N_{step}
0.01	0.1	10 s	4

voltage at the distribution network side with 19 tap positions. The reactive power capacity of each DG is limited in the range of ± 2 MVAR.

It is assumed that measurement points are appropriately located in order to enable the state estimator to provide a good result.

12.4.2 Parameter Setup and Algorithm Selection for the Controller

The objective of the controller is to preserve the voltages in the distribution network within $\pm 5\%$ of their nominal values. Table 12.1 describes the necessary parameters that need to be set up for the controller operation.

As discussed before, N_{step} indicates the trade-off between controller performance and computational expense since a larger N_{step} translates to both higher performance and a higher computational burden. In this fashion, $N_{step} = 4$ is intuitively selected in this example. In addition, it is also assumed that the calculation time of the controller to give control actions is 2 s.

12.4.3 Results and Discussion

12.4.3.1 Loss Minimization Performance of the Controller

As discussed, the motivation of utilizing an MPC based controller is its capability to compensate for system model inaccuracies. To see how the inaccuracies would negatively affect the performance of the controller, the following scenarios are set:

- Scenario#1: At each sampling time, a standard optimal power flow calculation is carried out. This scenario potentially provides the best solution for the optimization problem and hence is considered as a base case for comparison.
- Scenario#2: Sensitivities are initially calculated by using the numerical approach, each obtained by a small perturbation to one control variable. They remain unchanged during the simulation time interval.
- Scenario#3: Sensitivities are frequently updated according to new operating points. Therefore, the accuracy level of the sensitivities is this scenario is much higher than the Scenario#2.

Figure 12.8 shows the response of the PI controller in fulfilling the reactive power demand requested by TSOs.

Figure 12.9 presents active power exchange with the transmission system. A higher exchange amount equates to lower losses incurred in the system. Alternatively, the exchange rate reflects the performance of the MPC based controller in tracking the reference values of monitored bus voltages, which are calculated for purpose of the losses minimization. Recall that power systems are nonlinear, and sensitivities are approximation values since they are estimated based on linearization of the power

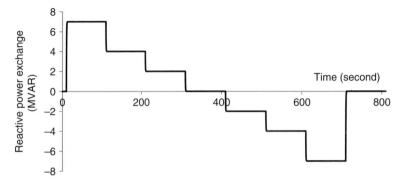

Figure 12.8 Correction of reactive power exchange.

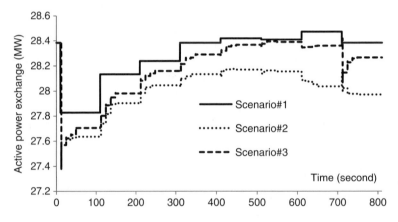

Figure 12.9 Active power exchange inversely proportional to reduction in losses.

system. Consequently, Scenario#1 constantly show the best performance because it uses a nonlinear optimization approach. Scenario#3 has higher accuracy of sensitivities than Scenario#2, so it yields a better performance.

The bus voltage #1166 would be the most vulnerable to power change of DGs empirically, so it is selected to display in Figure 12.10. It can be seen that even with the inaccuracy of sensitivities applied, the bus voltage always remain within acceptable limits. This is evidence to state that the MPC based controller is able to compensate for system model inaccuracies due to its closed-loop nature.

12.4.3.2 Voltage Correction Performance of the Controller

Initially all DGs operate optimally to keep reactive power exchange at zero. Afterwards at $t = 10$s and $t = 100$ s, a large amount of 20 MVAR export and then down to 5 MVAR, respectively, is requested by TSOs, and Figure 12.11 presents the response of the controller to fulfil this demand.

It can be seen from Figure 12.13 that during the time interval between $t = 10$s and $t = 100$ s when a large amount of reactive power is demanded, DGs closer to the substation try to fulfil reactive power exchange; meanwhile, other DGs at the end of several feeders, where voltages have reached their limits (±5% nominal voltage), absorb reactive

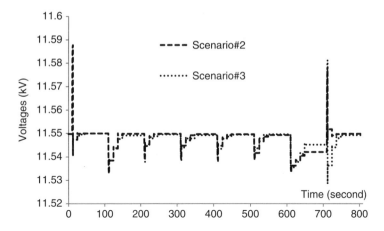

Figure 12.10 Voltage at bus#1166.

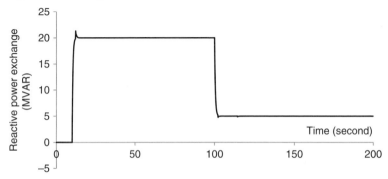

Figure 12.11 Correction of reactive power exchange.

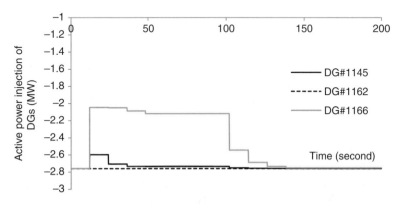

Figure 12.12 Active power injection of DGs.

power in order to keep voltages in limits. These DGs alone are, however, not able to fully correct voltages due to limits of their reactive power capability. Curtailment of active power generation of DGs is needed, as shown in Figure 12.12. As soon as the reactive power demand reduces to 5 MVAR at $t = 100$ s, the MPC based controller is aware

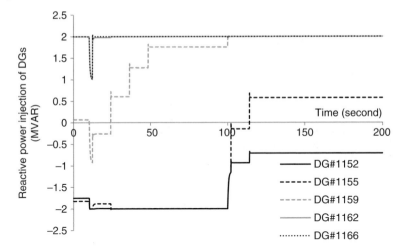

Figure 12.13 Reactive power injection of DGs.

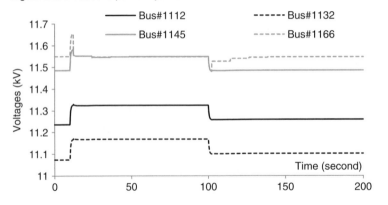

Figure 12.14 Voltage at several buses.

that active power curtailment should no longer be applied. However, to avoid excessive control actions caused by inaccurate model or measurement noises, the MPC based controller carefully follows sequence to gradually reduce the power curtailment, thereby always keeping the voltages in limits over the simulation time, as shown in Figure 12.14.

12.5 Conclusions

In this chapter, a centralized control scheme based on MPC has been presented for optimally coordinating available voltage control variables in the distribution network. By using the MPC technique, the controller yields advantages such as ease of integrating constraints, ability to discriminate expensive control actions, and compensation for system model inaccuracies. Future work should focuses on the controller's prediction skills, such as the ability to anticipate the effect of future actions of tap changers equipped in transformers. This may help avoiding temporary and unnecessary changes in the more expensive DG power generation.

References

1 L'Abbate, A., Fulli, G., Starr, F. *et al.* (2008) *Distributed Power Generation in Europe: Technical Issues for Further Integration*, Publications Office of the European Union.

2 US–Canada Power System Outage Task Force (2004) Final Report on the August 14, 2003 Blackout in the United States and Canada: Causes and Recommendations. Available: https://reports.energy.gov/

3 Union for the Coordination of Transmissions of Electricity (2009) *UCTE Operation Handbook*. Available: https://www.entsoe.eu/

4 Taylor, C. W. (1994) *Power system voltage stability*, McGraw-Hill, New York.

5 Van Cutsem, T., Vournas, C. (1998) *Voltage Stability of Electric Power Systems*, Kluwer, Boston.

6 Madureira, A.G., Pecas Lopes, J.A. (2009) Coordinated voltage support in distribution networks with distributed generation and microgrids. *Renewable Power Generation, IET*, **3**(4), 439–454.

7 Pham, H.V., Rueda, J.L., Erlich, I. (2014) Online optimal control of reactive sources in wind power plants. *IEEE Trans. Sustainable Energy*, **5**(2), 608–616.

8 Vargas, A., Samper, M.E. (2012) Real-time monitoring and economic dispatch of smart distribution grids: High performance algorithms for DMS applications. *IEEE Trans. Smart Grid*, **3**(2), 866–877.

9 Liew, S., Strbac, G. (2002) Maximising penetration of wind generation on existing distribution networks. *Pro. Inst. Elect. Eng., Gen., Transm., Distrib.*, **149**(3), 256–262.

10 Senjyu, T., Miyazato, Y. Yona, A., *et al.* (2008) Optimal distribution voltage control and coordination with distributed generation. *IEEE Trans. Power Delivery*, **23**(2), 1236–1242.

11 Gabash, A., Li, P. (2012) Active-reactive optimal power flow in distribution networks with embedded generation and battery storage. *IEEE Trans. Power Sytem*, **27**(4), 2026–2035.

12 Hajian, M., Glavic, M., Rosehart, W., *et al.* (2012) A chance constrained optimization approach for control of transmission voltages. *IEEE Trans. Power Syst.*, **27**(3), 1568–1576.

13 Chang, S.K., Albuyeh, F., Gilles, M., *et al.* (1990) Optimal real-time voltage control. *IEEE Trans. Power Syst.*, **5**(3), 750–758.

14 Vu, H., Pruvot, P. Launay, C., *et al.* (1996) An improved voltage control on large-scale power system. *IEEE Trans. Power Syst.*, **11**(3), 1295–1303.

15 Ranamuka, D., Agalgaonkar, A.P., Muttaqi, K.M. (2013) Online voltage control in distribution systems with multiple voltage regulating devices. *IEEE Trans. Sustainable Energy*, **5**(2), 617–628.

16 Dolan, M.J., Davidson, E.M., Kockar, I., *et al.* (2012) Distribution power flow management utilizing an online optimal power flow technique. *IEEE Trans. Power Syst.*, **27**(2), 790–799.

17 Zima, M., Andersson, G. (2003) *Stability assessment and emergency control method using trajectory sensitivities*. IEEE Power Tech, June 23–26, 2003, Bologna, Italy.

18 Larsson, M., Karlsson, D. (2003) Coordinated system protection scheme against voltage collapse using heuristic search and predictive control. *IEEE Trans. Power Syst.*, **18**(3), 1001–1006.

19 Beccuti, A., Demiray, T., Andersson, G., *et al.* (2010) A Lagrangian decomposition algorithm for optimal emergency voltage control," *IEEE Trans. Power Syst.*, **25**(4), 1769–1779.

20 Jin, L., Kumar, R., Elia, N. (2010) Model predictive control-based real-time power system protection schemes," *IEEE Trans. Power Syst.*, **25**(2), 988–998.

21 Glavic, M., Hajian, M., Rosehart, W., *et al.* (2011) Receding horizon multi-step optimization to correct nonviable or unstable transmission voltages. *IEEE Trans. Power Syst.*, **26**(3), 1641–1650.

22 Valverde, G., Van Cutsem, T. (2013) Model predictive control of voltages in active distribution networks. *IEEE Trans. Smart Grid*, **4**(4), 2152–2161.

23 Farina, M., Guagliardi, A., Mariani, F., *et al.* (2015) Model predictive control of voltage profiles in MV networks with distributed generation. *Control Engineering Practice, Elsevier*, **34**, 18–29.

24 Glavic, M., Van Cutsem, T. (2006) *Some reflections on model predictive control of transmission voltages.* Power Symposium, 38th North American, Sept. 17–19, 1991, Carbondale, USA.

25 Hug-Glanzmann, G. (2008) Coordinated power flow control to enhance steady-state security in power systems. Doctor of sciences thesis. Swiss Federal Institute of Technology Zurich.

26 United Kingdom Generic Distribution Network (UKGDS). [Online]. Available: http://sedg.ac.uk

27 Erlich, I., Winter, W., Dittrich, A. (2006) *Advanced grid requirements for the integration of wind turbines into the German transmission system.* Proc. 2006 IEEE Power Engineering Society General Meeting, Montreal, Canada.

28 European Network of Transmission System Operators for Electricity (2012) Network Code on Demand Connection. Available: https://www.entsoe.eu/

13

Multi-Agent based Approach for Intelligent Control of Reactive Power Injection in Transmission Systems

Hoan Van Pham[1,2] and Sultan Nasiruddin Ahmed[3]

[1] *Power Generation Corporation 2, Vietnam Electricity*
[2] *School of Engineering and Technology, Tra Vinh University, Vietnam*
[3] *FGH GmbH, Aachen, Germany*

13.1   Introduction

A power system is a highly dynamic system whose stability has to be ascertained in all cases; moreover, load centres expect a secure and reliable system. Voltage stability assessment is one of the major concerns in power system planning and secure operation as power grids span over several regions and sometimes even countries [1]. A direct link between the voltage and the reactive power makes it possible to control the voltage to desired values by control of the reactive power. The operator of the power system is responsible for controlling the transmission system voltage, which means having enough reactive power available to handle voltage violation conditions [2]. In normal conditions, the reactive power of the system not only dictates the voltage profile but also leads to losses. Loss minimization is an indispensable objective that must be considered in efficient power system operation [2–4]. Hence, to achieve certain global control objectives (e.g. N-1 secure operation, reactive power planning, minimization of losses, etc.) it is necessary to coordinate control actions among regional operators, but at the same time avoiding exposing local system data pertaining to regional infrastructure [5]. This chapter considers loss minimization as the objective function and maintaining voltage profile within safe limits as the constraint.

Optimization approaches based on power flow calculations often provide accurate results, but calculating nonlinear equations requires high computational time and resources, hence it is not suitable for real-time applications [6]. This chapter proposes an optimization approach in which the objective function is augmented to incorporate global optimization of a linearized large scale multi-agent power system using the Lagrangian decomposition algorithm [7, 8]. The aim is to maintain centralized coordination among agents via a master agent, leaving loss minimization as the only distributed optimization, which is analysed while protecting the local sensitive data. The efficiency of the local objective function stems from the use of active power loss sensitivity with respect to the control variables of the system. Control variables are defined as reactive power injection of generators and tap changer positions of transformers. The power loss sensitivities with respect to all control variables in the system are used in the first stages,

Dynamic Vulnerability Assessment and Intelligent Control for Sustainable Power Systems, First Edition.
Edited by José Luis Rueda-Torres and Francisco González-Longatt.

which are calculated from a linearized model of the system achieved through various control schemes and stored over regular intervals (viz. state estimation, PMU measurements, etc.) [9, 10]. Ultimately, this chapter has a centralized convex optimization problem which is solved using the aforementioned decomposition algorithm.

Lagrangian decomposition is a classical approach for solving constrained optimization problems. The augmented Lagrangian method adds an additional term to the unconstrained objective which mimics a Lagrange multiplier [11]. This method has been extensively used for solving numerous engineering problems, especially in the power systems field [12–14]. The advantage of this algorithm is the fact that local grid data does not have to be made globally available, which is often of crucial concern in actual power system operation, albeit degrading the computational performance since it involves iterating many times to reach graceful optimization. The optimization is carried out until the agents negotiate with the neighbouring areas on their inter-area variables (i.e. voltages at the interlinked buses).

The remainder of the chapter is structured as follows. Section 13.2 formulates the control problem exploiting power loss sensitivities. Section 13.3 proposes the augmented Lagrange formulation and its implementation. In Section 13.4 the proposed algorithm is implemented on a modified IEEE 30-bus system, and the performance analysis of the control scheme for various scenarios is studied. Section 13.5 concludes.

13.2 System Model and Problem Formulation

13.2.1 Power System Model

Governing any system comprises measuring system states and identifying an objective function which has to be achieved while concurrently tackling system constraints. A power system can be expressed in general differential-algebraic form given by (13.1)–(13.3):

$$\dot{x} = f(x, z) \tag{13.1}$$

$$g(x, z, u) = 0 \tag{13.2}$$

$$u^{\min} \leq u \leq u^{\max} \tag{13.3}$$

Here x corresponds to the dynamic state of generators and system loads, z corresponds to the algebraic voltages at buses, and u are the control variables (α tap changer set-points and q reactive power injection). Power system in real scenarios span several thousand kilometres, sometimes even countries and continents, mainly for utilization, efficiency, and management efficiency purposes.

If these regions are partitioned into N^a areas, each controlled by a local operator, this gives rise to an interconnected network of multiple operators with local supervisory control. Communication among these operators is imperative for an efficient and reliable system. Nevertheless, regional assets such as generation capacity, reactive power reserves, infrastructure details, and so on are to be protected from each other. To specify the interconnection among areas, Let v'_{ij} be the voltage magnitudes at the interconnection buses connected to area i of neighbouring areas j which is expected by area i; v_{ji} is the voltage magnitude at these buses expected by their own area j. Consensus among

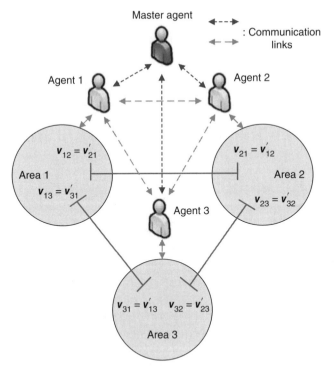

Figure 13.1 Partitioned power system managed by multi-agent system.

these areas implies the negotiation of the voltage profile at the coupling bus given as in (13.7). As shown in Figure 13.1, the regions are connected via the interconnecting buses or coupling buses with voltages v_{ij} (the solutions of the above equations) for N^a areas governing the whole system. The equations can then be written as:

$$\dot{x}_i = f_i(x_i, z_i) \tag{13.4}$$

$$g(x_i, z_i, u_i, v'_{ij}) = 0 \tag{13.5}$$

$$u_i^{min} \le u_i \le u_i^{max} \tag{13.6}$$

$$v'_{ij} = v_{ji} \tag{13.7}$$

where j is a subset (neighbouring area) of the power system (N^a areas) and $x_i, z_i, and\ u_i$ represent the state variables, dynamic states, and control variables of area i, respectively.

13.2.2 Optimal Reactive Control Problem Formulation

The purpose of an optimal reactive power dispatch is mainly to improve the voltage profile in the system and to minimize transmission losses. In this chapter, loss minimization is considered as the objective function J, while voltage profile is kept at values between 0.9 pu and 1.1 pu. The objective functions in (13.9) are introduced in a linearized form through sensitivities that are obtained by the linearization of the load flow equations around the nominal operating point as presented in the next section (Section 13.2.3). Changes of the control variables $\Delta w_i(k)$ at time-instant k are achieved by minimizing the objective functions (13.9):

$$\min J = \sum_{i=1}^{N^a} J_i \tag{13.8}$$

$$J_i = \left(\left(\frac{\partial P_i^{loss}}{\partial w_i} \right) \Delta w_i(k) \right) \tag{13.9}$$

$$w_i = \left[[u_i]^T, [v'_{ij}]^T \right]^T \tag{13.10}$$

subject to

$$v'_{ij} = v_{ji} \tag{13.11}$$

$$\Delta w_i^{min} \leq \Delta w_i(k) \leq \Delta w_i^{max} \tag{13.12}$$

$$\Delta v_i^{min} \leq \Delta v_i(k+1) \leq \Delta v_i^{max} \tag{13.13}$$

$$v_i^{min} \leq v_i(k+1) \leq v_i^{max} \tag{13.14}$$

$$v_i(k+1) = v_i(k) + \left(\frac{\partial v_i}{\partial w_i} \right)_D \Delta w_i(k) \tag{13.15}$$

where $\partial P_i^{loss}/\partial w_i$ and $\partial v_i/\partial w_i$ – that is, the loss sensitivities and voltage sensitivities of area i with respect to the control variables of area i and inter-area variables connected to area i, respectively – are calculated as in Section 13.2.3. v_i^{min} and v_i^{max} are the minimum and maximum acceptable voltages in area i, respectively. The constraints in the above equations are the result of linearization of the power system equations. They show the constraints of the voltage at the nodes, reactive power injections via devices, and transformer tap-positions. These parameters would be assessed and re-calibrated after every local optimization by the regional operators.

13.2.3 Multi-Agent Sensitivity Model

Sensitivities of a region are calculated based on sensitivity coefficients, the linearized factors arising from the nonlinear relationship among variables of the system. Load flow analysis is the backbone of sensitivity analysis, and through it the voltage magnitude and phase angle at each node and the complex power flowing in each transmission line can be obtained. The Inverse Jacobian matrix of the initial load flow analysis is manipulated to perform sensitivity analysis of the system, which in turn depends on the control parameters themselves; thus, the impact of change of magnitudes of control variables on control performance are significant and hence needs to be completely analysed for a wide range of changes. Since we have more than one control parameter, we need to develop an algorithm which studies the effect of concurrent effects of variation in multiple parameters. The calculation of the sensitivities in the multi-agent model has been split in two layers as seen in (13.16):

$$\frac{\partial P_i^{loss}}{\partial w_i} = \frac{\partial P_i^{loss}}{\partial v_i} \frac{\partial v_i}{\partial w_i} \tag{13.16}$$

13.2.3.1 Calculation of the First Layer

The objective is to minimize real power losses during the operation and control of a network. The real power loss P_i^{loss} of area i is represented by

$$P_i^{loss} = \sum_{k=1}^{N_{br}} G_k[v_h^2 + v_l^2 - 2v_h v_l \cos(\delta_h - \delta_l)] \tag{13.17}$$

where G_k is the conductance of line k which is connected between buses h and l in area i, and N_{br} is the number of branches of area i. In (13.17), the losses are represented by a nonlinear function of the bus voltages phase angles.

Then, the losses function is linearized as follows:

$$\frac{\partial P^{loss}}{\partial v_h} = G_k[2v_h - 2v_l \cos(\delta_h - \delta_l)] \tag{13.18}$$

$$\frac{\partial P^{loss}}{\partial v_l} = G_k[2v_l - 2v_h \cos(\delta_h - \delta_l)] \tag{13.19}$$

For every transmission line, the partial derivatives of P_i^{loss} with respect to the voltages at buses h and l are calculated. Partial derivatives pertaining to a certain bus are summed to form the power loss sensitivities with respect to all bus voltages in the system.

13.2.3.2 Calculation of the Second Layer

The vector of the control variables w_i is combination of three different vectors of the following variables: the inter-area variable v'_{ij}, reactive power injection of generators $q_{g,i}$, and tap ratio $\alpha_{tap,i}$. Therefore, the second layer was specified by computing the three sensitivities $\partial v_i/\partial v'_{ij}$, $\partial v_i/\partial q_{g,i}$ and $\partial v_i/\partial \alpha_{tap,i}$.

It is clear that $\partial v_i/\partial v'_{ij}$ is a unity vector and $\partial v_i/\partial q_{g,i}$ is the inversion of the Jacobian matrix calculated below:

Reactive power injection at bus k

$$q_k = v_k \sum_{m=1}^{N_{bus}} (G_{km} v_m \sin\theta_{km} - B_{km} v_m \cos\theta_{km}) \tag{13.20}$$

The Jacobian matrix is partly structured from partial derivatives of the reactive power injections as:

$$J_{km} = \frac{\partial q_k}{\partial v_m} = v_k(G_{km} \sin\theta_{km} - B_{km} \cos\theta_{km}) \tag{13.21}$$

$$J_{kk} = \frac{\partial q_k}{\partial v_k} = 2v_k(G_{kk} \sin\theta_{kk} - B_{kk} \cos\theta_{kk})$$

$$+ \sum_{m=1,m\neq k}^{N_{bus}} v_m(G_{km} \sin\theta_{km} - B_{km} \cos\theta_{km}) \tag{13.22}$$

In this chapter, changing the tap ratio of the transformer is equivalent to the injection of two reactive power increments into buses which are connected to the transformer terminals. Thus the sensitivities $v_i/\partial \alpha_{tap,i}$ are equivalent to two layers of sensitivities below:

$$\frac{\partial v_i}{\partial \alpha_{tap,i}} = \frac{\partial v_i}{\partial q_{tap,i}} \frac{\partial q_{tap,i}}{\partial \alpha_{tap,i}} \tag{13.23}$$

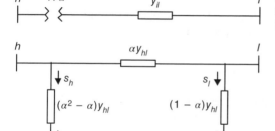

Figure 13.2 Model of tap changing transformer and its equivalent π circuit for the branch.

The sensitivities $\partial v_i/\partial q_{tap,i}$ is essentially a sub-matrix of the known $v_i/\partial q_{g,i}$. While $\partial v_i/\partial \alpha_{tap,i}$ is calculated from (13.35) and (13.36) for all branches equipped with tap changing transformers.

Transformer tap changing is more difficult to model since two buses are directly involved in the tap changing process. Let us consider a transformer connecting buses h and l with tap α, as shown in Figure 13.2. This branch can be represented by an equivalent π circuit.

The admittance of the branch is:

$$y_{hl} = g_{hl} + jb_{hl} \tag{13.24}$$

From Figure 13.2, the complex power injection to bus h is

$$s_h = p_h + jq_h = v_h i_h^* = v_h[v_h(\alpha^2 - \alpha)y_{hl}]^* \tag{13.25}$$

where * indicates the complex conjugate of the variable. So

$$s_h = v_h^2(\alpha^2 - \alpha)g_{hl} - jv_h^2(\alpha^2 - \alpha)b_{hl} \tag{13.26}$$

Similarly, the complex power injection to bus l is represented as

$$s_l = v_l^2(1 - \alpha)g_{hl} - jv_l^2(1 - \alpha)b_{hl} \tag{13.27}$$

From (13.26) and (13.27), the equations for q_h and q_l are

$$q_h = -v_h^2(\alpha^2 - \alpha)b_{hl} \tag{13.28}$$
$$q_l = -v_l^2(1 - \alpha)b_{hl} \tag{13.29}$$

If Δq_h is the increment of q_h with respect to voltage and tap position changes, then

$$\Delta q_h = \frac{\partial q_h}{\partial v_h}\Delta v_h + \frac{\partial q_h}{\partial \alpha}\Delta \alpha \tag{13.30}$$

However, for the power flow in Figure 13.2, we have

$$\Delta q_{th} = -\Delta q_h \tag{13.31}$$

So, differentiating (13.28) with respect to v_h and α, we have

$$\Delta q_{th} = 2b_{hl}v_h(\alpha^2 - \alpha)\Delta v_h + b_{hl}v_h^2(2\alpha - 1)\Delta \alpha \tag{13.32}$$

Similarly, differentiating (13.29) with respect to V_l and α, we have

$$\Delta q_{tl} = 2b_{hl}v_l(1 - \alpha)\Delta v_l - b_{hl}v_l^2\Delta \alpha \tag{13.33}$$

Moreover, (13.32) can be rewritten as follows

$$\Delta q_{th} = 2b_{hl}v_h\alpha(\alpha - 1)\Delta v_h + b_{hl}v_h^2(\alpha - 1)\Delta\alpha + b_{hl}v_h^2\alpha\Delta\alpha \qquad (13.34)$$

Since the value of α is close to unity, and Δv_h and $\Delta\alpha$ are small, we have:

$$\frac{\Delta q_{th}}{\Delta\alpha} = b_{hl}v_h^2\alpha \qquad (13.35)$$

Similarly, from (13.33):

$$\frac{\Delta q_{tl}}{\Delta\alpha} = -b_{hl}v_l^2 \qquad (13.36)$$

13.3 Multi-Agent Based Approach

13.3.1 Augmented Lagrange Formulation

It can be seen that the overall control problem (13.8)–(13.15) is not separable into sub-problems using the local variables of one agent i alone due to the interconnecting constraints (13.11). Therefore, a distributed algorithm based multi-agent system is introduced and presented in this section in order to achieve a global optimum of the whole system by separately solving local sub-problems.

In a multi-agent system, global optimization must be achieved while protecting regional data, hence we make use of the augmented Lagrangian method to integrate the interconnecting constraints of (13.11) into the global objective function (13.8) in the form of additional linear cost in terms of Lagrangian multipliers . The global objective now along with the losses and the inter-area constraints is as seen in (13.37):

$$L(\Lambda) = \sum_{i=1}^{N^a} \left(J_i + \sum_{j \in N_i} \left((\lambda_{ij})^{\mathrm{T}}(v'_{ij} - v_{ji}) + \frac{c}{2} \left\| v'_{ij} - v_{ji} \right\|_2^2 \right) \right) \qquad (13.37)$$

with the constraints being the ones from (13.12)–(13.15), and coefficient c is a positive scalar penalizing interconnecting constraint violations.

Then (13.37) can be decomposed into separate sub-problems so that they can be tackled and solved independently with $i = 1,..,N^a$ as:

$$\min_{w_i,v'_{ij}} J_i + \sum_{j \in N_i} \left(\left[(\lambda_{ij})^{\mathrm{T}} \ (-\lambda_{ji})^{\mathrm{T}} \right] \begin{bmatrix} v'_{ij} \\ v_{ij} \end{bmatrix} + \frac{c}{2} \left\| \begin{bmatrix} I & 0 \\ 0 & I \end{bmatrix} \begin{bmatrix} v'_{ji,prev} \\ v_{ji,prev} \end{bmatrix} - \begin{bmatrix} 0 & I \\ I & 0 \end{bmatrix} \begin{bmatrix} v'_{ij} \\ v_{ij} \end{bmatrix} \right\|_2^2 \right) \qquad (13.38)$$

where $v'_{ji.prev}$ and $v_{ji.prev}$ are v'_{ji} and v_{ji}, respectively, computed at the previous iteration for the other agents.

13.3.2 Implementation Algorithm

The implementation algorithm proposed in [11] is described in Figure 13.3. Firstly, each agent in turn updates its sensitivities and then minimizes its problem (13.38) to determine its optimal local and inter-area variables, while the variables of the other

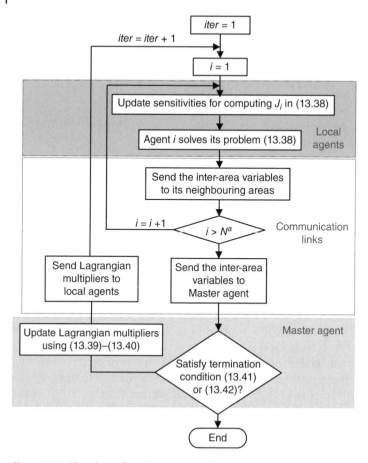

Figure 13.3 Flowchart of implementation algorithm.

agents stay fixed. After the last agent completes its optimization sub-problem, the inter-area variable is transmitted to the master agent where termination conditions such as agreement on the inter-area variables in (13.41) or maximum allowable number of iterations in (13.42) are checked. If the conditions are satisfied, the determined actions are implemented. Otherwise, the Lagrangian multipliers are updated following the strategy described by (13.39) and (13.40), and the whole process is then repeated in a further iteration:

$$\lambda_{ij,iter+1} = \lambda_{ij,iter} + c\left(v'_{ij,iter+1} - v_{ji,iter+1}\right) \tag{13.39}$$

$$\lambda_{ji,iter+1} = \lambda_{ji,iter} + c\left(v'_{ji,iter+1} - v_{ij,iter+1}\right) \tag{13.40}$$

$$v'_{ij,iter+1} - v_{ji,iter+1} < \varepsilon^v \tag{13.41}$$

$$iter \leq iter^{max} \tag{13.42}$$

with $\forall i, j \in N^a$

13.4   Case Studies and Simulation Results

13.4.1   Case Studies

Firstly, it is worth mentioning that the proposed approach can be extended to any number of interconnected areas with an arbitrary number of interconnection lines without conceptual modification to its structure.

The proposed method has been implemented on a modified IEEE 30-bus system taken from [15]. The test system is partitioned into three areas as shown in Figure 13.4 comprising ten transformers in total, equipped with tap-changers, each having two generators and several loads. There are 7 interconnection lines in total, corresponding to 14 inter-area variables.

13.4.2   Simulation Results

Note that the simulation results below are given with the initial control parameters set up as presented in Table 13.1. They are categorized into two groups: namely general parameters referring to the global control formulation, and those being relevant to the multi-agent based approach.

N_{step} indicates the trade-off between controller performance and computational expense since a larger N_{step} translates to both higher performance and a higher computational burden. In this fashion, $N_{step} = 4$ is intuitively selected in this example.

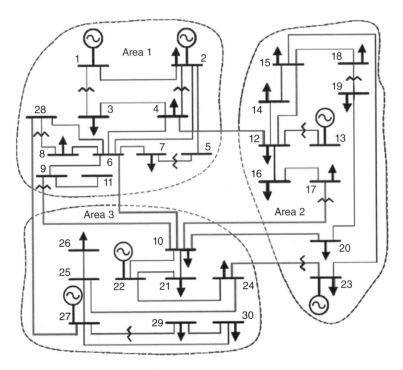

Figure 13.4 IEEE 30-bus modified system.

Table 13.1 Setup Parameters.

	General parameters				Parameters of multi-agent system		
ε_i^{loss} (MW)	$\|\Delta v_i^{max}\|$ & $\|\Delta v_i^{min}\|$ (p.u.)	$\|\Delta q_i^{max}\|$ & $\|\Delta q_i^{min}\|$ (MVAR)	$\|\Delta \alpha_i^{max}\|$ & $\|\Delta \alpha_i^{min}\|$ (p.u.)	ε^v (p.u.)	c	$iter^{max}$	
0.01	0.03	0.02	0.002	0.0005	9.5	300	

Table 13.2 Comparison on Power Loss Convergence.

Control circle No.	Multi-agent				Single-agent
	Area 1 (MW)	Area 2 (MW)	Area 3 (MW)	Total (MW)	Total (MW)
1	4.6341	1.6414	1.1029	7.3784	7.3784
2	4.3574	1.6158	1.0899	7.0631	6.679
3	4.1376	1.6066	1.0886	6.8327	6.2077
4	3.7285	1.5814	1.0581	6.368	5.83
5	3.4451	1.554	1.043	6.0421	5.5356
6	3.2059	1.5366	1.0328	5.7754	5.3163
7	3.0039	1.5293	1.0269	5.5601	5.1591
8	2.8541	1.5345	1.0204	5.409	5.0318
9	2.7493	1.5516	1.0136	5.3144	4.9593
10	2.6798	1.5637	1.0101	5.2537	4.9148
11	2.6479	1.5123	1.003	5.1633	4.8647
12	2.6202	1.5035	1.0016	5.1253	4.8545
13	2.5958	1.4718	1.008	5.0756	4.8055
14	2.5787	1.441	1.0135	5.0332	4.7998
15	2.5565	1.4132	1.0256	4.9953	
16	2.5408	1.3877	1.0342	4.9628	
17	2.5308	1.3587	1.0492	4.9387	
18	2.5283	1.3439	1.0705	4.9427	
19	2.5126	1.3258	1.0853	4.9237	

In addition, it is also assumed that the calculation time of the controller to give control actions is 2 s.

13.4.2.1 Performance Comparison Between Multi-Agent Based and Single-Agent Based System

It can be seen from Table 13.2 that the single-agent system provides a better convergence value of the losses compared to multi-agent. This likely stems from the fact that the higher values ε^v of the agreement on inter-area variables in (13.41) are, the worse the performance in terms of losses is. Moreover, the convergence speed of the single-agent system is faster as well.

In addition, the proposed control algorithm shows its capability in establishing cooperation between agents to achieve the global objective. This can be seen from Table 13.2, in that from control circle 12, the losses of area 3 show a trend pointing to an increase; in the meantime, losses of other areas decrease and the whole system loss decreases in response.

13.4.2.2 Impacts of General Parameters on the Proposed Control Scheme's Performance

In order to study the impacts of control parameters on performance, each simulation experiment below is carried out varying only one of the parameters while keeping the others unchanged. The parameters are set up at the values presented in Table 13.1.

Figure 13.5 depicts loss convergence with different change limits for selected voltages. As can be seen in the following, the narrower limits provide a better convergence value but with slower convergence speed within initial control circles that are far from the optimum operating state. The large limits often lead to fluctuation of convergence values, since the algorithm performance depends on sensitivities accuracy, which is inversely proportional to the limits; thus there is higher risk of triggering the termination condition (13.41).

Again, algorithm performance is adversely influenced by a larger change magnitude of control variables as presented in Figure 13.6 and Figure 13.7. This is clear because the sensitivities calculated are based on small changes of control variables.

13.4.2.3 Impacts of Multi-Agent Parameters on the Proposed Control Scheme's Performance

From Figure 13.8, it is likely that determining a correct value of the coefficient c in order to achieve the best performance will be difficult. Moreover, it should be highlighted that the coefficient c has a strong association with the control algorithm's performance.

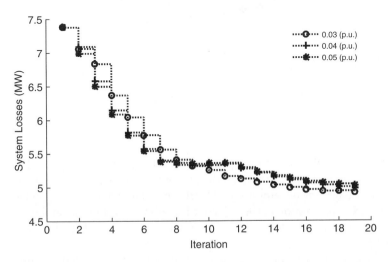

Figure 13.5 Loss convergence with different change limits of voltages $|\Delta v_i^{max}| = |\Delta v_i^{min}|$.

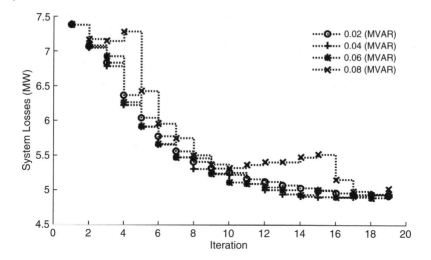

Figure 13.6 Loss convergence with different change limits of reactive power injection from generators.

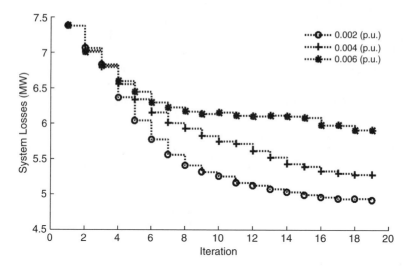

Figure 13.7 Loss convergence with different change limits of tap movement.

13.5 Conclusions

The need for coordination while protecting sensitive data in large interconnected grids for reliable operation of power system prompted the idea of a multi-agent based system. An effective augmented Lagrange decomposition algorithm was implemented and analysed by considering loss minimization as the objective function. This approach not only protects local data, but also gives a close performance to the conventional single-agent based system as seen in the simulation results.

On the other hand, improper selection of the general and multi-agent system parameters degrades the performance as shown in the control scheme's performance plots.

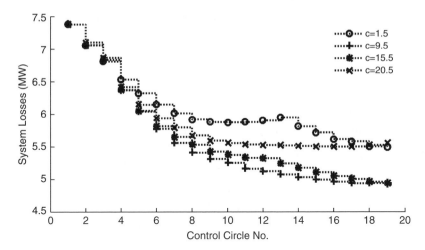

Figure 13.8 Loss convergence with different coefficient c.

Selection of these parameters for now was based on trial and error, but for future work, proper estimation of these parameters will play a vital role in the reduction of computation time and will also enrich the performance of the algorithm. In addition, further investigation is required to develop faster and more robust algorithms, and minimization of communication overheads should be included in the objective function.

The augmented Lagrange decomposition algorithm in this chapter has a disadvantage of no longer being separable across subsystems because of its dependence on the master agent. To achieve both separability and robustness for distributed optimization, we can instead use the alternating direction method of multipliers (ADMM) [16], which has been widely adopted in various industries due to its ease of decomposition and its convergence guarantees on a wide range of problems.

References

1 M. Larsson and D. Karlsson, "Coordinated system protection scheme against voltage collapse using heuristic search and predictive control," *IEEE Trans. Power Syst.*, vol. **18**, no. 3, pp. 1001–1006, Aug. 2003.

2 O. A. Mousavi and R. Cherkaoui, *Literature Survey on Fundamental Issues of Voltage and Reactive Power Control*. Ecole Polytech. Fédérale Lausanne Lausanne Switz., 2011.

3 A. M. Ramly, N. Aminudin, I. Musirin, D. Johari, and N. Hashim, "Reactive Power Planning for transmission loss minimization," in *Power Engineering and Optimization Conference (PEOCO), 2011 5th International*, 2011, pp. 116–120.

4 J. Zhu, K. Cheung, D. Hwang, and A. Sadjadpour, "Operation strategy for improving voltage profile and reducing system loss," *IEEE Trans. Power Deliv.*, vol. **25**, no. 1, pp. 390–397, Jan. 2010.

5 G. Glanzmann and G. Andersson, "Incorporation of n-1 security into optimal power flow for facts control," in *Power Systems Conference and Exposition, 2006. PSCE'06. 2006 IEEE PES*, 2006, pp. 683–688.

6 S. Boyd and L. Vandenberghe, *Convex Optimization*, Cambridge University Press, Cambridge, United Kingdom, 2004.

7 D. P. Bertsekas, *Constrained Optimization and Lagrange Multiplier Methods*, Academic Press, London, UK, 1982.

8 A. G. Beccuti, T. H. Demiray, G. Andersson, and M. Morari, "A Lagrangian decomposition algorithm for optimal emergency voltage control," *IEEE Trans. Power Syst.*, vol. **25**, no. 4, pp. 1769–1779, Nov. 2010.

9 Y. C. Chen, A. D. Dominguez-Garcia, and P. W. Sauer, "Online computation of power system linear sensitivity distribution factors," in *Bulk Power System Dynamics and Control-IX Optimization, Security and Control of the Emerging Power Grid (IREP), 2013 IREP Symposium*, 2013, pp. 1–4.

10 J. Peschon, D. S. Piercy, W. F. Tinney, and O. J. Tveit, "Sensitivity in power systems," *Power Appar. Syst. IEEE Trans. On*, no. 8, pp. 1687–1696, 1968.

11 R. R. Negenborn, B. De Schutter and J. Hellendoorn, "Multi-agent model predictive control for transportation networks: Series versus parallel schemes," *in Proc. Of the 12th IFAC Symposium on Information Control Problems in Manufacturing*, Saint-Etienne, France, 2006.

12 F. Zhuang and F. D. Galiana, "Towards a more rigorous and practical unit commitment by Lagrangian relaxation," *Power Syst. IEEE Trans. On*, vol. **3**, no. 2, pp. 763–773, 1988.

13 B. Kim and R. Baldick, "A comparison of distributed optimal power flow algorithms," *IEEE Trans. Power Syst.*, vol. **15**, no. 2, pp. 599–604, May 2000.

14 A. J. Conejo and J. A. Aguado, "Multi-area coordinated decentralized DC optimal power flow," *Power Syst. IEEE Trans. On*, vol. **13**, no. 4, pp. 1272–1278, 1998.

15 N. I. Deeb and S. M. Shahidehpour, "An efficient technique for reactive power dispatch using a revised linear programming approach," *Electr. Power Syst. Res.*, vol. **15**, no. 2, pp. 121–134, 1988.

16 S. Boyd, N. Parikh, E. Chu, B. Peleato, and J. Eckstein, "Distributed optimization and statistical learning via the alternating direction method of multipliers," *Foundations and Trends in Machine Learning*, vol. **3**, pp. 1–122, 2011.

14

Operation of Distribution Systems Within Secure Limits Using Real-Time Model Predictive Control

Hamid Soleimani Bidgoli[1], Gustavo Valverde[2], Petros Aristidou[3], Mevludin Glavic[1], and Thierry Van Cutsem[1]

[1] *Dept. of Electrical Engineering and Computer Science, University of Liege, Belgium*
[2] *School of Electrical Engineering, University of Costa Rica, Costa Rica*
[3] *School of Electronic and Electrical Engineering, University of Leeds, United Kingdom*

14.1   Introduction

The increasing penetration of distributed energy sources at distribution level is expected to create temporary over/under voltages and/or thermal overloads in Distribution Networks (DNs). Although these problems can be partly handled at the operational planning stage [1], the system may still undergo insecure operation conditions due to unforeseen events. In addition, reinforcing the network to deal with these temporary situations is not economically viable for the Distribution System Operator (DSO). Based on this, more flexible and coordinated control actions are required to cope with these operational challenges. To do so, the DSO should install measurement devices to monitor the system, and a control scheme to manage abnormal situations. The control of Dispersed Generation Unit (DGU) production is attractive, knowing that the violations take place only a fraction of the time [2].

The overall scope of this chapter is to introduce the reader to the problem of corrective control of voltages and congestion management of DNs based on Model Predictive Control (MPC) theory. The chapter presents the basic formulation of MPC and shows how it is adapted to predict and control the network condition in presence of a high penetration of DGUs. In addition, the primary objective is to present robust DN corrective control for temporary over/under voltages and thermal overloads as an alternative to network reinforcements. It is intended to provide comprehensive step-by-step formulations so that readers may reproduce or modify them based on their own needs. In addition, time domain simulations are included to demonstrate the effectiveness of the presented self-corrective controllers.

Corrective control can be realized according to different architectures: centralized [3–7], decentralized [8], distributed [5], and hierarchical [9]. The common assumption is that dispersed sources will modulate their production to support network controllability and flexibility. For example, a centralized controller based on sensitivity analysis is presented in [7]. The optimization problem minimizes the active power curtailment of the DGU production to regulate the voltages of the network. Using an Optimal Power

Dynamic Vulnerability Assessment and Intelligent Control for Sustainable Power Systems, First Edition.
Edited by José Luis Rueda-Torres and Francisco González-Longatt.
© 2018 John Wiley & Sons Ltd. Published 2018 by John Wiley & Sons Ltd.

Flow (OPF), Reference [5] discusses the impact of centralized and distributed voltage control schemes on potential penetration of DGUs.

In [8], a decentralized approach is proposed for the real-time control of both voltage and thermal constraints, using voltage and apparent power flow sensitivities to identify the most effective control actions. A hierarchical model based controller is suggested in [9] where the upper level coordinates different types of control devices relative to a set of prioritized objective functions.

The work in [10] presents a DN with a high penetration of wind power and photo-voltaic units both at the medium voltage and low voltage levels. Local voltage control is compared to coordinated voltage control, which acts on both active and reactive power of DGUs as well as the voltage set-point of the Load Tap Changer (LTC). Aiming to minimize network losses and keep voltages close to one per unit, a synergy between day-ahead (up to minute ahead) coordinated voltage control and real-time dynamic thermal rating is investigated in [11]. A 147-bus test DN planned for an actual geographical location is used to evaluate the proposed strategy.

Relatively few references deal with automatic, closed-loop control to smoothly steer a system back within security limits while compensating for model inaccuracies. MPC offers such a capability [12, 13] and there is a long history of industrial applications [14]. Reference [3] proposes a centralized voltage control scheme inspired by MPC. The problem was formulated as a receding-horizon multi-step optimization using a simple sensitivity model. This formulation is further extended in [4] to jointly manage voltage and thermal constraints.

Using a dynamic response model of the system, [15] suggests a multi-layer control structure for voltage regulation of DN. At the upper level, a static OPF computes the reference values of reactive power. The latter is communicated to the next layer, an MPC-based centralized controller, which handles the operational constraints. The authors in [16] present a dynamic control strategy in which fast and expensive sources, such as gas turbine generators, are used to modify the voltage and power balance of a distribution system during transients and which allows slower and cheaper generators to gradually take over after transients have died out.

Reference [17] proposes an MPC-based multi-step controller to correct voltages out of limits by applying optimal changes of the control variables (mainly active and reactive power of distributed generation and LTC voltage set-point) to smoothly drive the system from its current to the targeted operation region. The proposed controller is able to discriminate between cheap and expensive control actions and to select the appropriate set of control variables depending on the region of operation.

A key feature missing in many works is the capability of the controller to compensate for model inaccuracies and failures or delays in the control actions, which is achieved owing to the closed-loop nature of MPC.

Reference [18] designs a centralized, joint voltage and thermal control scheme, relying on appropriate measurement and communication infrastructures. DGUs are categorized as dispatchable and non-dispatchable, and three contexts of application are presented according to the nature of the DGUs and the aforementioned information exchanges. This extends the work of [4] by considering a new objective function. Namely, the deviations with respect to power schedules are minimized for the dispatchable DGUs, while the others operate as much as possible according to the maximum power tracking strategy. The formulation is such that corrections sent to DGUs vanish as soon as the operating constraints are no longer binding.

14.2 Basic MPC Principles

The name MPC stems from the idea of employing an explicit model of the controlled system to predict its future behavior over the next N_p steps. This prediction capability allows the solving of optimal control problems online, where the difference between the predicted output and the desired reference is minimized over a future horizon subject to constraints on the control inputs and outputs. If the prediction model is linear, then the optimization problem is quadratic if the objective is expressed through the ℓ_2-norm, or linear if expressed through the ℓ_1/ℓ_∞-norm [19].

The result of the optimization is applied using a receding horizon philosophy: At instant k, using the latest available measurements, the controller determines the optimal change of control variables from k to $k + N_c - 1$, in order to meet a target at the end of the prediction horizon, that is, at $k + N_p$. However, only the first component of the optimal command sequence ($\Delta u(k)$) is actually applied to the system. The remaining control inputs are discarded, and a new optimal control problem is solved at instant $k + 1$ with the new set of measurements that reflect the system response to the applied control actions at and before k. This idea is illustrated in Figure 14.1. As new measurements are collected from the plant at each instant k, the receding horizon mechanism provides the controller with the desired feedback characteristics.

As presented in Figure 14.1, the prediction horizon must be chosen such that it takes into account the expected effect of the computed control actions in the system. Based on this, the length of the prediction horizon should be at least equal to the length of the control horizon, that is, $N_p \geq N_c$. To decrease the computational burden, the lengths should be equal unless the controller is requested to consider changes happening beyond the control horizon.

14.3 Control Problem Formulation

The above principle is applied to voltage control and congestion management of DNs. The control variables are the active power (P_g) and reactive power (Q_g) of distributed generators and the voltage set-point of the LTC transformer (V_{tap}) at the bulk power supply point, grouped in the $m \times 1$ vector $u(k)$, at time k:

$$u(k) = [P_g^T(k), Q_g^T(k), V_{tap}(k)]^T \tag{14.1}$$

Figure 14.1 Prediction and control horizons.

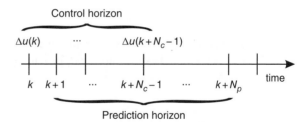

where ^{T} denotes vector transposition. MPC calculates the control variable changes $\Delta u(k) = u(k) - u(k-1)$ to bring back the monitored branch currents and bus voltages within permissible limits. The controlled variables are directly measured, and grouped in the $q \times 1$ vector $y(k|k)$.[1]

The MPC objective function may be set to minimize the sum of squared control variable changes or the sum of squared deviations between the controls and their references u_{ref}, as follows:

$$\min \underbrace{\sum_{i=0}^{N_c-1} ||\Delta u(k+i)||^2_{R_1}}_{\text{Objective 1}} + \underbrace{\sum_{i=0}^{N_c-1} ||u(k+i) - u_{ref}(k+i)||^2_{R_2} + ||\varepsilon||^2_S}_{\text{Objective 2}} \tag{14.2}$$

Note that both objectives have been included in (14.2). For the application of concern here, only one will be used depending on what is needed, as will be shown in the following sections. In matrix form, (14.2) can be written as:

$$\min \Delta u^T R_1 \Delta u + (u - u_{ref})^T R_2 (u - u_{ref}) + \varepsilon^T S \varepsilon \tag{14.3}$$

Here, Δu and u are $mN_c \times 1$ column vectors defined by:

$$\Delta u = \begin{bmatrix} \Delta u(k) \\ \Delta u(k+1) \\ \vdots \\ \Delta u(k+N_c-1) \end{bmatrix}, \quad u = \begin{bmatrix} u(k) \\ u(k+1) \\ \vdots \\ u(k+N_c-1) \end{bmatrix} \tag{14.4}$$

Note that u can be expressed in terms of Δu:

$$u = C_1 u(k-1) + C_2 \Delta u \tag{14.5}$$

with

$$C_1 = \begin{bmatrix} I \\ I \\ \vdots \\ I \end{bmatrix}, C_2 = \begin{bmatrix} I & 0 & \cdots & 0 \\ I & I & \cdots & 0 \\ \vdots & \vdots & \ddots & \vdots \\ I & I & \cdots & I \end{bmatrix} \tag{14.6}$$

I and 0 are the $m \times m$ identity and zero matrices, respectively.

Similarly, u_{ref} is the $mN_c \times 1$ vector of control variable references, which may vary in the control horizon, hence:

$$u_{ref} = \begin{bmatrix} u_{ref}(k) \\ u_{ref}(k+1) \\ \vdots \\ u_{ref}(k+N_c-1) \end{bmatrix} \tag{14.7}$$

In addition, ε is the vector of slack variables used to relax some inequality constraints in case of infeasibility. These variables are heavily penalized by the diagonal matrix S. Finally, R_1 and R_2 are $mN_c \times mN_c$ diagonal weighting matrices used to prioritize control actions.

---

1 $y(k|k)$ is read as prediction of y at time k given its state at time k

The above objective function is minimized subject to:

$$u^{MIN} \leq u \leq u^{MAX} \tag{14.8}$$

$$\Delta u^{MIN} \leq \Delta u \leq \Delta u^{MAX} \tag{14.9}$$

where u^{MIN}, u^{MAX}, Δu^{MIN}, and Δu^{MAX} are the lower and upper limits on control variables and on their rate of change, along the control horizon. As regards the predicted system evolution, the following equality constraint is also imposed, for $i = 1, 2, ..., N_p$:

$$y(k + i|k) = y(k + i - 1|k) + S_y \Delta u(k + i - 1) + S_\delta \delta \gamma(k + i) \tag{14.10}$$

where $y(k + i|k)$ is the predicted system output at time $k + i$ given the measurements at k, and S_y is the sensitivity matrix of those variables to control changes.[2].

The last term in (14.10) is included to account for the effect of a known disturbance δ in the predicted outputs. Hence, S_δ is the sensitivity of those variables to the known disturbance. In addition, γ is a binary variable equal to one for the instants when the known disturbance will occur, or zero otherwise.

The prediction is initialized with $y(k|k)$ set to the last received measurements. In compact form, (14.10) becomes

$$y = Fy(k|k) + \Phi_y \Delta u + \Phi_\delta \Gamma \delta \tag{14.11}$$

with,

$$y = \begin{bmatrix} y(k + 1|k) \\ y(k + 2|k) \\ \vdots \\ y(k + N_p|k) \end{bmatrix}, F = \begin{bmatrix} I \\ I \\ \vdots \\ I \end{bmatrix}, \Gamma = \begin{bmatrix} \gamma(k + 1) \\ \gamma(k + 2) \\ \vdots \\ \gamma(k + N_p) \end{bmatrix} \tag{14.12}$$

$$\Phi_y = \begin{bmatrix} S_y & 0 & \cdots & 0 \\ S_y & S_y & \cdots & 0 \\ \vdots & \vdots & \ddots & \vdots \\ S_y & S_y & \cdots & S_y \end{bmatrix}, \Phi_\delta = \begin{bmatrix} S_\delta & 0 & \cdots & 0 \\ S_\delta & S_\delta & \cdots & 0 \\ \vdots & \vdots & \ddots & \vdots \\ S_\delta & S_\delta & \cdots & S_\delta \end{bmatrix} \tag{14.13}$$

Note that Φ_y and Φ_δ are $qN_p \times mN_c$ and $qN_p \times N_p$ matrices, respectively.

Finally, the following inequality constraints are imposed to the predicted output:

$$-\varepsilon_1 \mathbf{1} + y^{LOW} \leq y \leq y^{UP} + \varepsilon_2 \mathbf{1} \tag{14.14}$$

ε_1 and ε_2 are the components of ε and $\mathbf{1}$ denotes a $qN_p \times 1$ unit vector. Additionally, y^{LOW} and y^{UP} are the lower and upper allowed output along the prediction horizon:

$$y^{LOW} = \begin{bmatrix} y^{low}(k + 1) \\ y^{low}(k + 2) \\ \vdots \\ y^{low}(k + N_p) \end{bmatrix}, y^{UP} = \begin{bmatrix} y^{up}(k + 1) \\ y^{up}(k + 2) \\ \vdots \\ y^{up}(k + N_p) \end{bmatrix} \tag{14.15}$$

where $y^{low}(k + i)$ and $y^{up}(k + i)$ are the minimum and maximum allowed output at time $k + i$.

---

2 This "instantaneous response" is a feature of the proposed model, but the general MPC applies to dynamic models.

In order to solve the above optimization problem, the equations have to be rearranged into a Quadratic Programming (QP) problem:

$$\min \frac{1}{2}\begin{bmatrix}\Delta u \\ \varepsilon\end{bmatrix}^T H \begin{bmatrix}\Delta u \\ \varepsilon\end{bmatrix} + f^T \begin{bmatrix}\Delta u \\ \varepsilon\end{bmatrix} \tag{14.16}$$

where

$$H = \begin{bmatrix} R_1 + C_2^T R_2 C_2 & 0 \\ 0 & S \end{bmatrix}, f = \begin{bmatrix} C_2^T R_2 (C_1 u(k-1) - u_{ref}) \\ 0 \end{bmatrix} \tag{14.17}$$

subject to:

$$\begin{bmatrix} \Delta u^{MIN} \\ 0 \end{bmatrix} \leq \begin{bmatrix} \Delta u \\ \varepsilon \end{bmatrix} \leq \begin{bmatrix} \Delta u^{MAX} \\ +\infty \end{bmatrix} \tag{14.18}$$

$$\begin{bmatrix} u^{MIN} - C_1 u(k-1) \\ y^{LOW} - Fy(k|k) - \Phi_\delta \Gamma \delta \\ -\infty \end{bmatrix} \leq A \begin{bmatrix} \Delta u \\ \varepsilon \end{bmatrix} \leq \begin{bmatrix} u^{MAX} - C_1 u(k-1) \\ +\infty \\ y^{UP} - Fy(k|k) - \Phi_\delta \Gamma \delta \end{bmatrix} \tag{14.19}$$

with

$$A = \begin{bmatrix} C_2 & 0_{mN_c \times 1} & 0_{mN_c \times 1} \\ \Phi_y & 1_{qN_p \times 1} & 0_{qN_p \times 1} \\ \Phi_y & 0_{qN_p \times 1} & -1_{qN_p \times 1} \end{bmatrix} \tag{14.20}$$

$0_{mN_c \times 1}$ is a $mN_c \times 1$ null matrix while $0_{qN_p \times 1}$ and $1_{qN_p \times 1}$ are $qN_p \times 1$ null and unit matrices.

In the present formulation, there are only two components in ε: the first one to relax the lower bound of all controlled variables and the second to relax the upper bound over the prediction horizon. In case of different types of controlled variables, such as bus voltages and branch currents, different slack variables are considered for them (i.e., two for bus voltage and one for branch current limits, see Section 14.5). In this case, matrix A is extended accordingly.

14.4   Voltage Correction With Minimum Control Effort

The first control goal analyzed in this chapter consists of maintaining the DN bus voltages within some predefined limits while minimizing the control deviations, as presented in Objective 1 in (14.2) (with $R_2 = 0$).

Initially, the operator will define a target voltage for each bus in the network. This target voltage may follow a security or economic purpose, such as network losses minimization. However, trying to reach the actual target values is impractical and likely infeasible. Alternatively, one can try to keep the network voltages within some limits around the target values. In what follows, these limits are referred to as *normal operation limits*, and the operation within these limits is the controller's main objective.

If any of the bus voltages violate the limits, the controller will use the minimum control actions to bring them within the acceptable limits. As the voltages are in the undesirable region but not in emergency, the controller will use the cheapest controls, since it is not economically justifiable to use expensive controls to maintain voltages in a narrow band of operation. For example, the requested band of operation for a monitored bus voltage may be set at [1.00 1.02] pu. Since the targeted operation might be the result of an OPF, the band might not be the same for all buses. Moreover, the range allowed for each controlled bus may depend on the importance of the customers connected to it or the cost associated with regulating the voltages within a narrow band of operation.

Conversely, if some bus voltages are operating in the unacceptable region, outside some *emergency limits*, the controller will use all available (both cheap and expensive) actions to return them to the undesirable region. The emergency limits can be the same for all buses. For instance, the operator can define that the network is under emergency conditions if any of the monitored voltages deviates from the range [0.94 1.06] pu. In practice, these limits are defined by the corresponding grid code.

Once the controller succeeds in bringing the voltages into the undesirable region, it will again use the cheapest control variables to reach the normal operation conditions.

Figure 14.2 summarizes the transitions between operation states after disturbances and the corresponding corrective actions. Note that there are cases where the correction of some bus voltages is infeasible with the available controls. Under these circumstances, the controller should, at least, apply the control actions that can bring the problematic voltages to a better operating point, even if it is outside the normal operational limits. The controller must do so until a feasible correction is found.

Note that when all voltages lie within the normal operational limits, no control actions are issued, that is, $\Delta u = 0$.

14.4.1 Inclusion of LTC Actions as Known Disturbances

A LTC is a slowly acting device that controls the distribution side voltage of the transformer by acting on its turn ratio. The LTC performs a tap change if the controlled voltage remains outside of a dead-band for longer than a predefined delay [20]. This delay is specified to avoid frequent and unnecessary tap changes that may reduce the LTC lifetime. If more than one step is required, the LTC will move by one step at a time with delay between successive moves.

The proposed controller leaves this local control unchanged but acts on the LTC set-point V_{tap} if appropriate. Therefore, tap changes will be triggered when changes in operating conditions make the controlled voltage V_{ctld} leave the dead-band, or when the controller requests a change of V_{tap} (and, hence, a shift of the dead-band) such that V_{ctld} falls outside the dead-band.

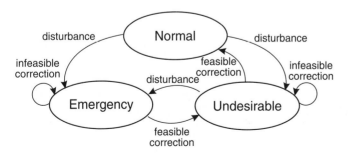

Figure 14.2 Operation states and corrective actions.

By assuming that a tap change produces ΔV_d variation in V_{ctld}, the number of tap changes can be roughly estimated by:

$$
N_{op} = \begin{cases} roundup[\frac{V_{ctld}-V_{tap}-d}{\Delta V_d}], & \text{if } V_{ctld} > V_{tap} + d \\[2mm] roundup[\frac{V_{tap}-d-V_{ctld}}{\Delta V_d}], & \text{if } V_{ctld} < V_{tap} - d \end{cases}
$$
(14.21)

where d is half the LTC dead-band and the function *roundup* provides the nearest upper integer. Note that ΔV_d is measurable after an LTC action occurs. This value is assumed constant and it is considered a known value for the controller.

From (14.21), the times of LTC actions can be estimated as:

$$
t_j^{act} = t_k + T_{f0} + T_f(j-1)
$$
(14.22)

for $j = 1, \ldots, N_{op}$. Here, t_k is the present time instant, T_{f0} is the time delay for the first step, and T_f is the time delay for subsequent tap steps. The controller can use this information to anticipate the future voltage changes due to the operation of the LTC. In order to do so, the controller must extend the prediction horizon N_p until the last predicted LTC control action is included. A general example of this is provided in Figure 14.3. At instant k, it is predicted from (14.21) and (14.22) that three tap changes will take place at t_1^{act}, t_2^{act}, and t_3^{act}, respectively, with t_2^{act} and t_3^{act} beyond the control horizon. In order to account for all these LTC actions, the controller extends the prediction horizon up to the smallest discrete time larger than t_3^{act}.

With the controller able to predict the effect of future LTC actions, it is possible to avoid premature and unnecessary output changes of DGUs since it can better decide whether or not the LTC actions are enough to correct the controlled voltages.

14.4.2 Problem Formulation

With the choice of objective function and control variables discussed above, the MPC-based controller has to solve the following QP problem:

$$
\min \Delta u^T R_1 \Delta u + \varepsilon^T S \varepsilon
$$
(14.23)

subject to:

$$
u^{MIN} \leq u \leq u^{MAX}
$$
(14.24)

$$
\Delta u^{MIN} \leq \Delta u \leq \Delta u^{MAX}
$$
(14.25)

$$
-\varepsilon_1 \mathbf{1} + V^{LOW} \leq V \leq V^{UP} + \varepsilon_2 \mathbf{1}
$$
(14.26)

Figure 14.3 Extension of prediction horizon to include predicted LTC actions.

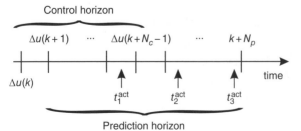

Control horizon

$\Delta u(k+1) \quad \cdots \quad \Delta u(k+N_c-1) \quad \cdots \quad k+N_p$

$\Delta u(k)$

t_1^{act} t_2^{act} t_3^{act}

time

Prediction horizon

V^{LOW} and V^{UP} are the lower and upper limits of the predicted voltages in V; the latter are computed by:

$$V = FV(k|k) + \Phi_V \Delta u + \Phi_d \Gamma \Delta V_d \tag{14.27}$$

where $V(k|k)$ is set to the last received bus voltage measurements, Φ_V contains the sensitivity matrix S_V of controlled bus voltages to control variables (see (14.13)), and Φ_d contains the sensitivities of controlled bus voltages to the LTC-controlled voltage (i.e., the last column of matrix S_V). Γ is an $N_p \times 1$ vector whose entries are 1 for the instants when the tap changes have been predicted, or 0 otherwise.

The sensitivity of bus voltages with respect to power injections can be obtained off-line from the transposed inverse of the power flow Jacobian matrix [2, 7]. The sensitivities of bus voltages with respect to the LTC-controlled voltage can be approximated by computing the ratio of variations of the monitored bus voltages to the LTC's controlled bus voltage due to a tap change. This information can be easily extracted from the solution of two power flow runs with a single tap position difference. The sensitivity matrix S_V can be updated infrequently as model errors will be compensated by the MPC scheme [17, 18].

The weight assigned to each control variable in R_1 should be related to the cost of the device to provide ancillary services. For example, the reactive power output of a DGU is considered cheaper than its active power output. The latter is actually considered an expensive control action and is not requested unless emergency conditions are encountered. If some of the voltages are outside the emergency limits, the controller will use all controls to correct them. Thus, active powers of DGUs are heavily weighted to minimize their use. Once the voltages reach the undesirable region, only the cheap control variables will be used.

The slack variables ε_1 and ε_2 in (14.26) are used to relax the voltage limits. The entries of the 2×2 diagonal matrix S should be given very high values.

The active power outputs of DGUs are constrained by their capacity. For example, the active power production of conventional synchronous machines is constrained by the turbine capacity. In renewable energy sources, where the production is driven by weather conditions, the corresponding variables of active power are upper bounded by the actual power extracted from the wind or the sun irradiance. This is, at any instant k, the controller cannot request more than the power that is being produced, but it can request active power reductions by partial curtailment. On the other hand, the reactive power output of renewable energy sources is considered fully controllable but subject to capacity limits.

Although the maximum reactive power production can be fixed for each DGU, it is desirable to update it with the actual terminal voltage and active power production [21], so that full advantage is taken from its capability. This information is used to update the limits on the control variables.

14.5   Correction of Voltages and Congestion Management with Minimum Deviation from References

The second formulation (Objective 2 in (14.2)) can accommodate various contexts of application, depending on the interactions and information transfers between entities acting on the DGUs, in accordance with regulatory policy. This leads to the definition of a number of operating modes, which are depicted in Figure 14.4.

Besides the monitoring of some bus voltages, active and reactive productions, and terminal voltages of DGUs, this controller variant also monitors the active and reactive power flows in critical (potentially congested) branches. Thus, the controller relies on a dedicated measurement and communication infrastructure but, as suggested in Figure 14.4, it could also rely on the results of a state estimator, for improved system monitoring.

Once the controller observes (or predicts) limit violations, it computes and sends active and/or reactive power corrections to the DGUs of concern. The latter are the differences between the reference and the computed controls, that is:

$$\Delta P_{cor}(k) = P_{ref}(k) - P_g(k) \tag{14.28}$$
$$\Delta Q_{cor}(k) = Q_{ref}(k) - Q_g(k) \tag{14.29}$$

Note that these corrections should stay at zero as long as no limit violation is observed (or predicted), and come back to zero as soon as operation is no longer constrained, as explained in what follows.

Furthermore, a distinction is made between *dispatchable* and *non-dispatchable* DGUs. The latter are typically wind turbine or photovoltaic units operated for Maximum Power Point Tracking (MPPT). In the absence of operating constraints, they are left to produce as much as can be obtained from the renewable energy source. The dispatchable units, on the other hand, have their production schedules P and Q, according to market opportunities or balancing needs, for instance.

14.5.1 Mode 1

This mode applies to non-dispatchable units. For MPPT purposes, at each time step k, the reference $P_{ref\ i}(k)$ of the i-th DGU should be set to the maximum power available on that unit. This information is likely to be available to the DGU MPPT controller, but is seldom transmitted outside. An alternative is to estimate that power from the measurements P_{meas}. Considering the short control horizon of concern here, a simple prediction is given by the "persistence" model:

$$P_{ref}(k + i) = P_{meas}(k) + \Delta P_{cor}(k - 1), \quad i = 0, ..., N_c \tag{14.30}$$

As long as no power correction is applied, the last term is zero and P_{meas} is used as a short-term prediction of the available power. When a correction is applied, the right-hand side in (14.30) keeps track of what was the available power before a correction started being applied. Using this value as reference allows resetting the DGUs under the desired MPPT mode as soon as system conditions improve.

A more accurate prediction can be used if data are available. That would result in the right-hand side of (14.30) varying with time $k + i$.

14.5.2 Mode 2

This mode applies to DGUs that are dispatchable but under the control of another actor than the DSO. Thus, the latter does not know the power schedule of the units of concern. In order to avoid interference with that non-DSO actor, the last measured power

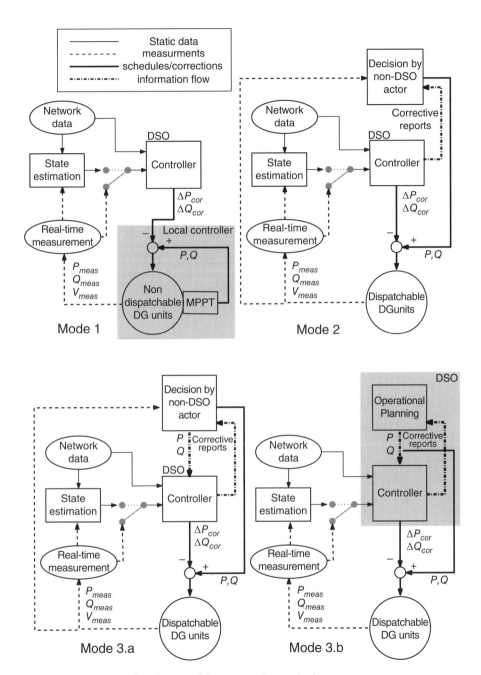

Figure 14.4 Contexts of application of the proposed control scheme.

productions are taken as reference values over the next N_c time steps:

$$P_{ref}(k + i) = P_{meas}(k), \quad i = 0, ..., N_c \tag{14.31}$$

On the other hand, if a control action has been applied by the DSO, to preserve network security, this action should not be counteracted by a subsequent non-DSO action in order to avoid conflict, leading for instance to oscillation. In other words, the DSO is assumed to "have the last word" in terms of corrective actions, since it is the entity responsible for network security.

In both Modes 1 and 2, the controller lacks information to anticipate the DGU power evolution. Hence, the corrections (14.28, 14.29) will be applied *ex post*, after the measurements have revealed the violation of a (voltage or current) constraint.

14.5.3   Mode 3

This mode relates to dispatchable DGUs whose power schedules are known by the controller, either because this information is transmitted by the non-DSO actors controlling the DGUs (see variant 3.a in Figure 14.4) or because the DSO is entitled to directly control the DGUs (see variant 3.b in Figure 14.4). The latter case may also correspond to schedules determined by DSO operational planning. Unlike in Mode 2, the schedule imposed to the units is known by the controller, which can anticipate a possible violation under the effect of the scheduled change and correct the productions *ex ante*. Although different from a regulatory viewpoint, Modes 3.a and 3.b are treated in the same way.

Figure 14.5 shows how the N_c future P_{ref} values are updated with the known schedule before being used as input for the controller. As long as the schedule does not change

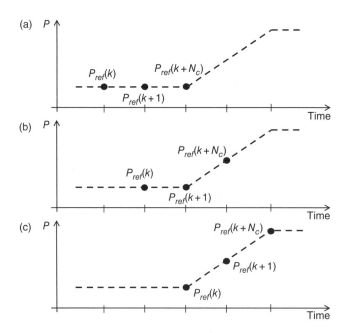

Figure 14.5 Mode 3: updating the P_{ref} values over three successive times.

within the N_c future time steps (see Figure 14.5a), $\boldsymbol{P}_{ref}$ remains unchanged; otherwise, the interpolated values are used.

In principle, the aforementioned choices also apply to the reference reactive powers $\boldsymbol{Q}_{ref}$. However, it is quite common to operate DGUs at unity power factor, to minimize internal losses, which amounts to setting $\boldsymbol{Q}_{ref}$ to zero, and corresponds to Mode 3.

To make system operation smoother and more secure, the identified limit violations and the corresponding corrections applied by the controller to DGUs should be communicated back to the non-DSO actors or the operational planners, as suggested by the dash-dotted arrows in Figure 14.4.

14.5.4 Problem Formulation

Here the objective is to minimize the sum of squared deviations, over the N_c future steps, between the controls and their references (i.e., Objective 2 in (14.2)). This leads to the following QP problem, with $\boldsymbol{R}_1 = 0$ in (14.3):

$$\min \ (\boldsymbol{u} - \boldsymbol{u}_{ref})^T \boldsymbol{R}_2 (\boldsymbol{u} - \boldsymbol{u}_{ref}) + \boldsymbol{\varepsilon}^T \boldsymbol{S}\boldsymbol{\varepsilon} \tag{14.32}$$

subject to:

$$\boldsymbol{u}^{MIN} \leq \boldsymbol{u} \leq \boldsymbol{u}^{MAX} \tag{14.33}$$

$$\Delta\boldsymbol{u}^{MIN} \leq \Delta\boldsymbol{u} \leq \Delta\boldsymbol{u}^{MAX} \tag{14.34}$$

$$-\varepsilon_1 \boldsymbol{1} + \boldsymbol{V}^{LOW} \leq \boldsymbol{V} \leq \boldsymbol{V}^{UP} + \varepsilon_2 \boldsymbol{1} \tag{14.35}$$

$$\boldsymbol{I} \leq \boldsymbol{I}^{UP} + \varepsilon_3 \boldsymbol{1} \tag{14.36}$$

$\boldsymbol{I}^{UP}$ is the upper limit of the predicted currents in $\boldsymbol{I}$, the latter are computed by:

$$\boldsymbol{I} = \boldsymbol{F}\boldsymbol{I}(k|k) + \boldsymbol{\Phi}_I \Delta\boldsymbol{u} \tag{14.37}$$

where $\boldsymbol{I}(k|k)$ is set to the last received branch current measurements, and $\boldsymbol{\Phi}_I$ contains the sensitivity matrix $\boldsymbol{S}_I$ of controlled branch currents to control variables.

The sensitivity matrix $\boldsymbol{S}_I$ should be updated more frequently, due to the higher variability of currents. The sensitivity of the branch current I_j with respect to the i-th DGU active (resp. reactive) power P_{gi} (resp. Q_{gi}) can be obtained as [4]:

$$\frac{\partial I_j}{\partial P_{gi}} \approx \frac{1}{V_k} \frac{P_j}{S_j} \frac{\partial P_j}{\partial P_{gi}} \approx \frac{P_j}{S_j} \tag{14.38}$$

$$\frac{\partial I_j}{\partial Q_{gi}} \approx \frac{1}{V_k} \frac{Q_j}{S_j} \frac{\partial Q_j}{\partial Q_{gi}} \approx \frac{Q_j}{S_j} \tag{14.39}$$

where P_j, Q_j, and S_j are respectively the active, reactive, and apparent power flows in the branch, and V_k is the voltage at the bus where the current is measured. The voltage is simply taken equal to 1 pu. The above approximations assume that P_j (resp. Q_j) does not change much when Q_{gi} (resp. P_{gi}) is varied, and the change of P_j (resp. Q_j) is equal to the change in P_{gi} (resp. Q_{gi}). Note that this approximation applies only if the j-th branch is on the path from the i-th DGU to the HV/MV tranformer; otherwise, a zero sensitivity is assumed since the branch current would not change with the DGU power change. Note also that using (14.38) and (14.39) requires having the branch equipped with active and reactive power flow measurements.

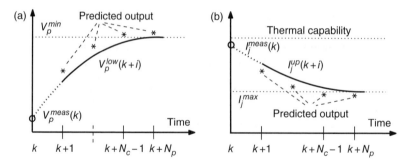

Figure 14.6 Progressive tightening of voltage and current bounds.

In this formulation, S_I should be updated at each discrete step while S_V may be kept constant at all operating points.

The diagonal weighting matrix R_2 allows prioritizing the controls, with lower values assigned to reactive than to active power deviations. Unlike the former formulation, DGU active power may be requested to change in undesirable conditions, not only in an emergency.

The new slack variable ε_3 is used to relax (14.36) in case of infeasibility; the entries of the 3×3 diagonal matrix S are given very high values.

To obtain a smooth system evolution, the bounds $V^{low}(k + i)$, $V^{up}(k + i)$ and $I^{up}(k + i)$ on the predicted voltages and currents are tightened progressively. An exponential evolution with time has been considered, as shown in Figure 14.6 for respectively a lower voltage (Figure 14.6a) and a current limit (Figure 14.6b). The circles indicate voltage or current values measured at time k, which fall outside the acceptable range. The limits imposed at the successive times $k + 1, \dots, k + N_p$ are shown with solid lines. They force the voltage or current of concern to enter the acceptable range at the end of the prediction horizon. Taking the lower voltage limit as an example, the variation is given by ($i = 1, \dots, N_p$):

$$V_p^{low}(k + i) = V_p^{min} - (V_p^{min} - V_p^{meas}(k))e^{-i/T_s}$$ (14.40)

where p is the bus of concern, T_s is a smoothing time constant and $V_p^{meas}(k)$ is the measurement received at time k. If it does not exceed the desired limit, that is, if $V_p^{meas}(k) \geq V_p^{min}$, the latter is used as constant bound in (14.35), at all future times, that is, $V_p^{low}(k + i) = V_p^{min}, \quad i = 1, \dots, N_p$.

Similar variations are considered for the upper voltage and the current limits. As regards the latter, the I^{max} value is set conservatively below the effective thermal capability monitored by the corresponding protection, as shown in Figure 14.6b.

14.6 Test System

The multi-step receding-horizon controllers presented above have been tested through simulations of a 75-bus, 11-kV radial DN hosting 22 DGUs; the line parameters are available in [22] and the network topology is shown in Figure 14.7.

The distribution system is connected to the external grid through a 33/11 kV transformer equipped with an LTC. The network topology consists of 8 feeders all directly

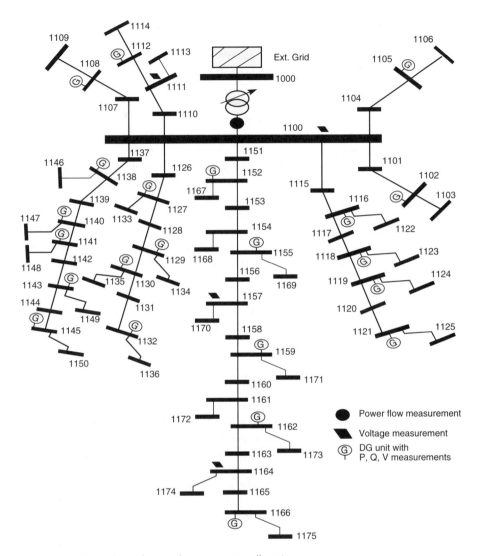

Figure 14.7 Network topology and measurement allocation.

connected to the main transformer, and they serve 38 loads modeled as constant current for active power and constant impedance for reactive power, and 15 more loads represented by equivalent induction motors.

In this test system, 13 out of the 22 DGUs are 3-MVA synchronous generators driven by hydro turbines with 2.55 MW of maximum capacity. The remaining nine DGUs are Doubly-Fed Induction Generators (DFIG) driven by wind turbines. Each DFIG is a one-machine equivalent of two 1.5 MW wind turbines operating in parallel. The nominal capacity of each DFIG is 3.33 MVA. The model of the wind turbine and its parameters were taken from [23].

The DFIGs operate in reactive power control mode. This is achieved by a PI controller that regulates the reactive power output according to the set-point value requested by the centralized controller. Compared to the synchronous machines, this reactive power control loop has a faster response.

The reactive power limits of DGUs are calculated, at any time k, given their actual operation conditions (P_g and V) and their nominal capacities [21]. Hence, any reactive power increase requested by the controller will not compromise the DGUs' active power output or violate the machine capacities.

It is assumed that the 22 DGUs are allocated with remote units that measure the active power, reactive power, and voltage magnitude at their terminals. These measurements are used by the centralized controller.

As regards load buses, the measurement configuration is such that no load is at a distance larger than two buses from a voltage monitored bus (see Figure 14.7). By following this rule, there are three load buses with monitored voltages. These voltage measurements along with the power output measurement of the DGUs make up a set of 71 measurements received by the controller.

It is not possible to ensure that non-monitored voltages will be within the desirable limits. However, by distributing the measurements all over the network, it is reasonable to expect that the voltages of non-monitored buses will be close to the voltages of the neighboring controlled buses.

The measurements are collected some time after the control actions are applied. This is to wait for the system response and to avoid making decisions based on measurements taken during transients.

Because of the fast sampling rate of modern monitoring units and their efficient communication links, it is assumed that the measurements are collected every 0.2s.

The measurements were simulated by adding white Gaussian noise restricted to $\pm 1\%$ for V measurements and $\pm 1\%$ of the respective DGU maximum power output for P_g and Q_g measurements. In order to filter out some of this noise, the controller uses the average of the 11 snapshots received over a time window of 2 s.

The controller sends corrections every 10 s. The prediction and control horizons are set to $N_p = N_c = 3$. This yields a good compromise between sufficient number of MPC steps and a short enough response time to correct violations.

It must be emphasized that the changes in operating point applied to the system, such as wind variations, load increases, and scheduled changes, have been made faster than in reality for a legible presentation of the results.

14.7 Simulation Results: Voltage Correction with Minimal Control Effort

In the first two scenarios (A and B), it is required that the monitored voltages remain within the [1.000 1.025] pu range while minimizing (14.23). In addition, the system is considered under emergency conditions when any bus reaches voltages outside the [0.940 1.060] pu interval.

The active and reactive power of DGUs are not allowed to change by more than 0.5 MW and 0.5 MVar respectively, while the LTC set-point is not allowed to change more than 0.01 pu per discrete time step.

In the objective function (14.23), identical costs have been assumed for all DGUs. Changes of the reactive power outputs cost the same as changes in LTC voltage set-point while the cost for active power changes is set 10 times higher. Moreover, the cost of using the slack variables is 1000 times higher than that of reactive power.

Since the variation of load powers with voltage are not well known in practice, the sensitivity matrix has been calculated by considering constant power models for all loads. In addition, the line parameters used to calculate the sensitivity matrix were corrupted by a random error whose mean value is zero and standard deviation is 10% of the actual line parameter. The objective is to demonstrate that the controller is robust and can compensate for these model inaccuracies.

14.7.1 Scenario A

The first scenario reported consists of the presence of over voltages in certain areas of the network caused by low power demands and high power production from DGUs. With the available set of measurements, it is found that the voltages in some areas are outside the normal operation constraints, but not enough to reach the emergency state. Here it is pointed out that, for legibility purposes, the following plots show the exact (noiseless) voltages and DGU power outputs, as opposed to the noisy and discrete measurements received and processed by the controller.

Figure 14.8 presents the obtained correction of the voltages starting at $t = 10$ s and changing every 10 s thereafter, under the effect of the controller adjusting the DGU reactive powers. Note that the controller yields an exponentially decreasing correction of the bus voltages until the latter reach the desired interval.

The selected curves in Figure 14.8, correspond to the most representative voltages in the network. The only load bus voltage reported is v1159 whereas the remaining voltages correspond to DGU buses.

Although most of the DGUs reduced their reactive power production, the ones connected to buses 1102 and 1108 were requested to increase (see Figure 14.9). This is because the controller anticipated that the voltages near bus 1108 would violate the lower limit of 1 pu. Hence, the reactive powers of these machines increase to maintain those voltages within limits.

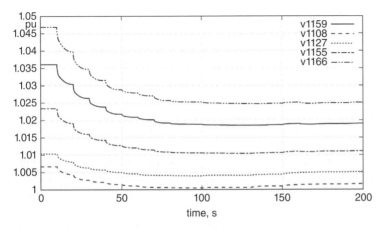

Figure 14.8 Scenario A: Voltage correction.

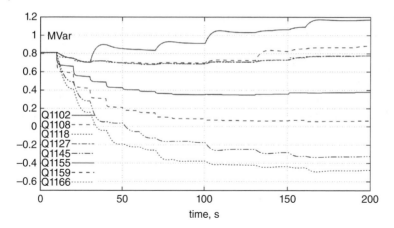

Figure 14.9 Scenario A: Reactive power output of the DGUs.

From Figure 14.9 it is seen that the maximum change of reactive power output occurs in the DGU at bus 1166. This machine is requested to absorb reactive power until the most problematic voltage v1166 is corrected.

As soon as all the monitored voltages are brought back inside the limits, the controller does not request further changes in $\Delta \boldsymbol{u}$. This condition is met at $t = 190$ s. These results confirm that even in the presence of inaccurate sensitivity values, the controller is able to correct the voltages.

14.7.2 Scenario B

The second scenario reported consists of an external voltage drop that affects the voltages at all buses. This disturbance was simulated by a negative step change of 0.08 pu of the Thévenin equivalent voltage, at the primary side of the main transformer. For this case, the LTC is coordinated along with the DGUs to correct the voltage drop in the DN. The purpose of this test is to demonstrate the prediction capability of the proposed controller and how this avoids unnecessary changes of alternative controls.

Figure 14.10 presents the voltage correction at a sample of monitored buses. The problem is corrected at about 90 s after the operation of the LTC and a few changes of the DGU power outputs. Just after the disturbance at $t = 1$ s, the difference between V_{ctld} and its set-point is enough to trigger eight tap movements (see (14.21)). The first tap operation occurs at $t = 21$ s and the subsequent ones occur in steps of 10 s until $t = 91$ s. After this time, no more LTC actions are required.

The tiny voltage corrections seen mainly at bus 1166 at $t = 130$ s and $t = 160$ s were triggered by the noise affecting voltage measurements that indicated small violations of the monitored voltages.

Table 14.1 details the future tap movements as anticipated by the controller. Equations 14.21 and 14.22 are used for this purpose. It is found that the LTC will operate seven times. The controller makes a rough estimate of the LTC time actions. For example, it wrongly anticipated that the first LTC action would occur at $t = 30$ s, while it actually occurred at $t = 21$ s. This example confirms that there is no need to

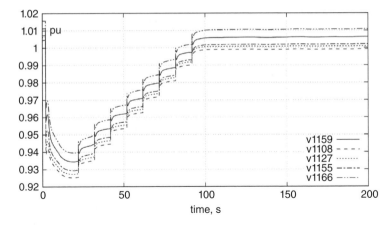

Figure 14.10 Scenario B: Voltage correction.

Table 14.1 Controller Anticipation of the LTC Actions

$t[s]$	N_{op}	N_p	t_1	t_2	t_3	t_4	t_5	t_6	t_7
10	7	9	30	40	50	60	70	80	90
20	7	8	30	40	50	60	70	80	90
30	7	7	30	40	50	60	70	80	90
40	6	6	40	50	60	70	80	90	–
50	5	5	50	60	70	80	90	–	–
60	3	3	60	70	80	–	–	–	–
70	3	3	70	80	90	–	–	–	–
80	2	3	80	90	–	–	–	–	–
90	1	3	90	–	–	–	–	–	–

know the exact moment at which the LTC will act. In fact, any mistake in prediction will be corrected and updated when new measurements are available. The important aspect is that the controller can anticipate future actions and use this information to avoid more expensive control actions.

In order to capture all the anticipated future events due to LTC actions, the controller extended its prediction horizon length. For example, at $t = 10$ s, the controller anticipated that the last control action would occur at $t = 90$ s. Hence, it automatically increases N_p from three to nine steps. For $t > 50$ s, N_p is reset to three.

Figure 14.11 presents the coordinated reactive power output of the DGUs. Since the controller is able to predict the effect of the LTC's future actions in the controlled voltages, it did not require significant changes of Q_g. A similar behavior occurs for changes in active power outputs during emergency conditions. One could argue that there is a cost associated with many tap movements due to wear of equipment. This could be included in the cost function to reduce the number of tap changes by making use of other available control variables.

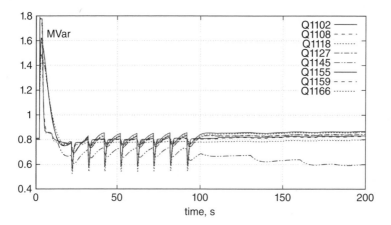

Figure 14.11 Scenario B: Reactive power output of the DGUs.

14.8 Simulation Results: Voltage and/or Congestion Corrections with Minimum Deviation from Reference

The last three scenarios reported in this section evaluate the control scheme presented in Section 14.5. The matrices R_2 and S are diagonal with entries equal to 1 for reactive powers, 25 for active powers, 500 for the slack variables ε_1 and ε_2, and 5000 for ε_3. In addition, only the power output of DGUs are considered control variables.

14.8.1 Scenario C: Mode 1

All 22 DGUs are assumed to be driven by wind turbines, operated for MPPT. Thus the control of all DGUs is in Mode 1 (see Figure 14.4). Initially, the dispersed generation exceeds the load, and the DN is injecting active power into the external grid. At the same time, the DGUs are operating at unity power factor, and the DN draws reactive power from the external grid.

A 10% increase in wind speed takes place from $t = 20$ to $t = 80$ s, as shown in Figure 14.12. This results in an increase of the active power flow in the transformer, as shown in Figure 14.13. At $t = 60$ s, the thermal limit of the latter, shown with heavy line in Figure 14.13, is exceeded. This is detected by the controller through a violation of the constraint (14.36) at $t = 70$ s.

The controller corrects this congestion problem by acting first on the DGU reactive powers, which have higher priority through the weighting factors. Figure 14.14 shows that the controller makes some DGUs produce reactive power, to decrease the import (and, hence, the current) through the transformer. The latter effect can be seen in Figure 14.13. However, the correction of DGU reactive powers alone cannot alleviate the overload, and from $t = 80$ s on, the controller curtails the active power of wind turbines as shown by Figure 14.12. The overload is fully corrected at $t = 120$ s.

To illustrate the ability of the proposed control scheme to steer the DGUs back to MPPT, the system operating conditions are relieved by simulating a 4.1 MW load increase starting at $t = 170$ s. The corresponding decrease of the active power flow in the transformer can be observed in Figure 14.13. This leaves some space to restore part

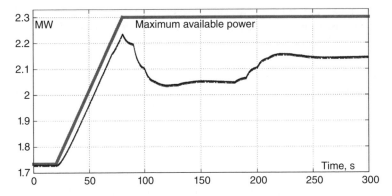

Figure 14.12 Scenario C: Active power produced by DGUs.

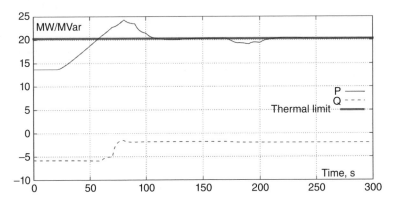

Figure 14.13 Scenario C: Power flows in the transformer.

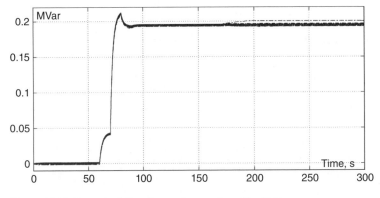

Figure 14.14 Scenario C: Reactive power produced by DGUs.

of the curtailed DGU active powers. As expected, the controller increases the DGU productions until the transformer current again reaches its limit, at around $t = 210$ s. Figure 14.12 shows that, indeed, the active productions get closer to the maximum power available from wind.

The unpredicted thermal limit violation caused by the initial wind increase was corrected *ex post*. It is interesting to note that, on the contrary, when taking advantage of the load relief, the controller steers the system in such a way that it does not exceed the thermal limit.

14.8.2 Scenario D: Modes 1 and 2 Combined

It is now assumed that nine units, connected to buses 1118, 1119, 1129, 1132, 1138, 1141, 1155, 1159, and 1162 (see Figure 14.7), are driven by wind turbines and are non-dispatchable. They are thus operated in Mode 1. However, since the wind speed is assumed constant in this scenario, the productions of those units remain constant.

The remaining 13 DGUs use synchronous generators and are dispatchable. They are operated in Mode 2. An increase of their active power by an actor other than the DSO, thus not known by the controller, takes place from $t = 100$ to $t = 140$ s, as shown in Figure 14.15. The schedule leaves the reactive powers unchanged. Since the initial network voltages are close to the admissible upper limit, shown by the heavy line in Figure 14.16, the system experiences high voltage problems.

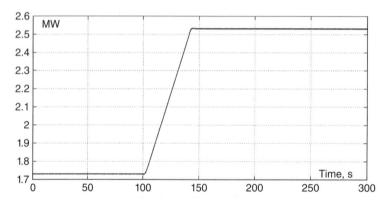

Figure 14.15 Scenario D: Active power produced by dispatchable units.

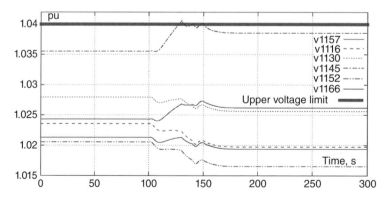

Figure 14.16 Scenario D: Bus voltages.

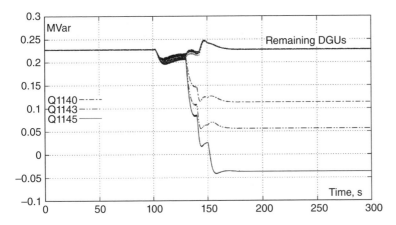

Figure 14.17 Scenario D: Reactive power produced by dispatchable units.

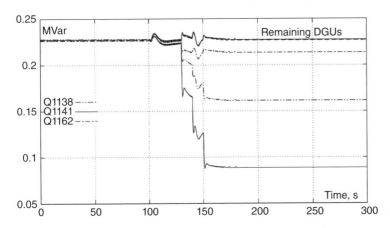

Figure 14.18 Scenario D: Reactive power produced by non-dispatchable units.

The controller does not send corrections until $t = 130$ s, when the voltage at bus 1145 exceeds the limit. Over the 40 seconds that follow this limit violation, the controller adjusts the reactive powers of both dispatchable and non-dispatchable units, as shown by Figures 14.17 and 14.18. It is easily seen that different corrections are applied to different DGUs, depending on their locations in the system. It is also seen from Figure 14.16 that the voltage at bus 1145 crosses the limit several times, followed by reactive power adjustments. These *ex post* corrections were to be expected since, in this example, the DGUs are either in Mode 1 or in Mode 2.

14.8.3 Scenario E: Modes 1 and 3 Combined

In this last scenario, some DGUs are non dispatchable and operated in Mode 1, while the dispatchable ones are operated in Mode 3, with their schedules known by the controller. The latter may come, for instance, from operational planning decisions.

Two successive changes of DGU active powers are considered: (i) an unforeseen wind speed change from $t = 20$ to $t = 70$ s increasing the production of the non-dispatchable units, and (ii) a power increase of the dispatchable units scheduled to take place from

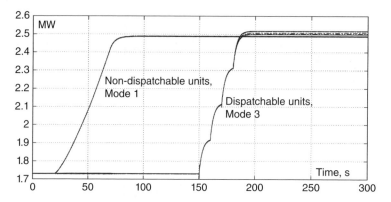

Figure 14.19 Scenario E: Active power produced by various units.

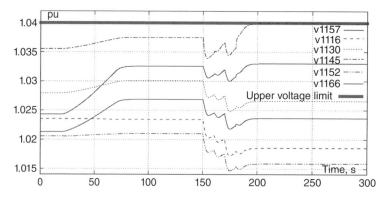

Figure 14.20 Scenario E: Bus voltages.

$t = 150$ to $t = 190$ s. The corresponding active power generations are shown in Figure 14.19.

Figure 14.20 shows the resulting evolution of a few bus voltages. The increase in wind power makes them approach their limit, shown by the heavy line. Without a corrective action, the subsequent scheduled change would cause a limit violation. However, the latter change is anticipated by the controller, through the P_{ref} values updated as shown in Figure 14.5. Therefore, the controller anticipatively adjusts the DGU reactive powers, as seen in Figures 14.21 and 14.22, and no voltage exceeds the limit while all the active power changes are accommodated. The controller anticipative behavior is clearly seen in Figure 14.20, where the voltage decrease resulting from the reactive power adjustment counteracts the increase due to active power increase, leading the highest voltage to land on the upper limit.

14.9 Conclusion

This chapter has presented a scheme for corrective control of voltages and congestion management of DNs based on MPC. It is demonstrated how temporary abnormal conditions can be overcome with a correct selection of DGU power outputs.

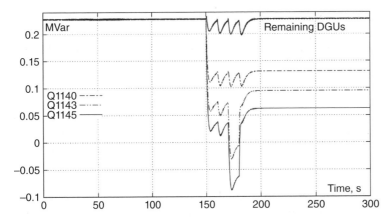

Figure 14.21 Scenario E: Reactive power produced by dispatchable units.

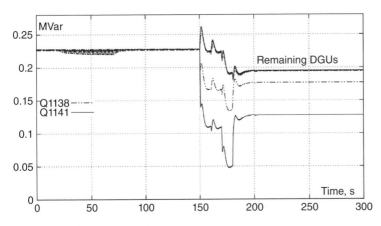

Figure 14.22 Scenario E: Reactive power produced by non-dispatchable units.

The main features of the MPC-based scheme are recalled hereafter:

- Bus voltages and branch currents are controlled such that they remain within an acceptable range of operation. Hence, the controllers do not act unless these limits are violated or the reference values of productions have been changed.
- The controllers discriminate between cheap and expensive control actions and can select the appropriate set of control variables depending on the regions of operation.
- Being based on multiple time step optimization, these controllers are able to smoothly drive the system from its current to the targeted operation region.
- Due to the closed-loop nature of MPC, the control schemes can compensate for model inaccuracies and failure or delays of the control actions.
- Lastly, owing to their anticipation capabilities, these controllers also take into consideration the requested actions that will be applied in the future. This is a common situation when the LTC of the transformer is requested to operate at time k but acts later due to standard control delays. Accounting for future control actions avoids the premature and maybe unnecessary dispatch of other control actions.

Thanks to the repeated computations, MPC offers some inherent fault-tolerance capability (particularly with respect to modeling errors and control failure). Illustrative examples pertaining to voltage control can be found in [3, 13]. The fault-tolerance can be further increased if the MPC is supported by an intelligent fault identification and detection scheme. With this support, even re-configurable control can be achieved [12].

Improved modeling could deal with the dynamic response of the controlled devices. The DGUs considered in this chapter relate to power-electronics interfaces and, hence, were assumed to react faster than the MPC sampling period of—typically—10 seconds. This justifies using a static representation, through sensitivity matrices, to predict the future system evolution. Dynamic models would be required for slower responding devices or if the MPC sampling period was decreased. It has also been suggested that artificial neural networks could be a viable option to further enhance MPC performance [12].

References

1 Q. Gemine, E. Karangelos, D.E. and Cornelusse, B. (2013) Active network management: planning under uncertainty for exploiting load modulation, in *IREP Symposium-Bulk Power System Dynamics and Control -IX, Rethymnon, Greece.*

2 Borghetti, A., Bosetti, M., Grillo, S., Massucco, S., Nucci, C., Paolone, M., and Silvestro, F. (2010) Short-term scheduling and control of active distribution systems with high penetration of renewable resources. *IEEE Systems Journal*, **4** (3), 313–322.

3 Valverde, G. and Van Cutsem, T. (2013) Control of dispersed generation to regulate distribution and support transmission voltages, in *Proceedings of IEEE PES 2013 PowerTech conference.*

4 Soleimani Bidgoli, H., Glavic, M., and Van Cutsem, T. (2014) Model predictive control of congestion and voltage problems in active distribution networks, in *Proc. CIRED conference, paper No 0108.*

5 Vovos, P.N., Kiprakis, A.E., Wallace, A.R., and Harrison, G.P. (2007) Centralized and distributed voltage control: Impact on distributed generation penetration. *IEEE Transactions on Power Systems*, **22** (1), 476–483.

6 Dolan, M., Davidson, E., Kockar, I., Ault, G., and McArthur, S. (2012) Distribution power flow management utilizing an online optimal power flow technique. *IEEE Transactions on Power Systems*, **27** (2), 790–799.

7 Zhou, Q. and Bialek, J. (2007) Generation curtailment to manage voltage constraints in distribution networks. *IET Generation Transmission and Distribution*, **1** (3), 492–498.

8 Sansawatt, T., Ochoa, L.F., and Harrison, G.P. (2012) Smart decentralized control of DG for voltage and thermal constraint management. *IEEE Transactions on Power Systems*, **27** (3), 1637–1645.

9 Hambrick, J. and Broadwater, R.P. (2011) Configurable, hierarchical, model-based control of electrical distribution circuits. *IEEE Transactions on Power Systems*, **26** (3), 1072–1079.

10 Leisse, I., Samuelsson, O., and Svensson, J. (2013) Coordinated voltage control in medium and low voltage distribution networks with wind power and photovoltaics, in *IEEE PES PowerTech Conference*, Grenoble, p. 6.

11 Degefa, M., Lehtonen, M., Millar, R., Alahäivälä, A., and Saarijärvi, E. (2015) Optimal voltage control strategies for day-ahead active distribution network operation. *Electric Power Systems Research*, **127**, 41–52.

12 Maciejowski, J.M. (2002) *Predictive Control With Constraints*, Prentice-Hall.

13 Glavic, M., Hajian, M., Rosehart, W., and Van Cutsem, T. (2011) Receding-horizon multi-step optimization to correct nonviable or unstable transmission voltages. *IEEE Transactions on Power Systems*, **26** (3), 1641–1650.

14 Qin, S. and Badgwell, T.a. (2003) A survey of industrial model predictive control technology. *Control Engineering Practice*, **11** (7), 733–764.

15 Farina, M., Guagliardi, A., Mariani, F., Sandroni, C., and Scattolini, R. (2015) Model predictive control of voltage profiles in MV networks with distributed generation. *Control Engineering Practice*, **34**, 18–29.

16 Falahi, M., Lotfifard, S., Ehsani, M., and Butler-Purry, K. (2013) Dynamic model predictive-based energy management of DG integrated distribution systems. *IEEE Transactions on Power Delivery*, **28** (4), 2217–2227.

17 Valverde, G. and Van Cutsem, T. (2013) Model predictive control of voltages in active distribution networks. *IEEE Transactions on Smart Grid*, **4** (4), 2152–2161.

18 Soleimani Bidgoli, H., Glavic, M., and Van Cutsem, T. (2016) Receding-horizon control of distributed generation to correct voltage or thermal violations and track desired schedules, in *Proc. 19th Power Systems Computation Conference (PSCC)*, Genoa, Paper. No. 114, p. 7.

19 Bemporad, A. and Morari, M. (1999) Robust model predictive control: A survey. *Robustness in identification and control*, **245**, 207–226.

20 Van Cutsem, T. and Vournas, C. (1998) *Voltage Stability of Electric Power Systems*, Springer.

21 Valverde, G. and Orozco, J. (2014) Reactive power limits in distributed generators from generic capability curves, in *Proc. IEEE PES 2014 General Meeting*, Washington DC., p. 5.

22 Centre for Sustainable Electricity and Distributed Generation, *United Kingdom Generic Distribution Network*, available online: https://github.com/sedg/ukgds.

23 G. Tsourakis, B.N. and Vournas, C. (2009) Effect of wind parks with doubly fed asynchronous generators on small-signal stability. *Electrical Power System Research*, **79** (1), 190–200.

15

Enhancement of Transmission System Voltage Stability through Local Control of Distribution Networks

Gustavo Valverde[1], Petros Aristidou[2], and Thierry Van Cutsem[3]

[1] *School of Electrical Engineering, University of Costa Rica, Costa Rica*
[2] *School of Electronic and Electrical Engineering, University of Leeds, United Kingdom*
[3] *Dept. of Electrical Engineering and Computer Science, University of Liege, Belgium*

15.1   Introduction

The most noticeable developments foreseen in power systems involve Distribution Networks (DNs). Future DNs are expected to host a large percentage of renewable energy sources [1], and small Dispersed Generation Units (DGUs) at distribution level are expected to supply a growing percentage of demand [2]. This proliferation of DGUs and the advances in infocommunications are key drivers of the transformations seen today in DNs.

The increased penetration of DGUs has given rise to new operational problems in DNs, such as over-voltages and thermal overloads at times of high DG production and low load. In response to these problems, several DN control schemes have been proposed to dispatch the DGUs and ensure secure network operation [3]. For instance, in [4], two coordinated voltage control algorithms are proposed. The first is based on simple control rules in which the Load Tap Changer (LTC) of the main transformer is used to control voltages. If this action is unable to restore all network voltages within limits, the resource with the highest effect on the most problematic bus is selected. The second algorithm is based on optimization—more precisely, a mixed integer nonlinear programming problem is used to minimize costs of network losses and generation curtailment, subject to voltage limits, active and reactive power limits of Distributed Energy Resources (DER), and branch power flow limits. The discrete control variables contain transformer tap position and switched capacitors while the continuous variables are the set points of real and reactive power or terminal voltages of DER.

The reactive power coordination scheme presented in [5] minimizes the number of transformer tap operations while satisfying voltage limits of the feeder, branch flow constraints, and transformer capacity constraints. The work in [6] presents a voltage control scheme to coordinate the operation of multiple regulating devices and DGUs in Medium Voltage (MV) distribution systems under variability of system topology and DGU availability.

In [7], the authors propose a Model Predictive Control (MPC) approach to regulate DN voltages using DGUs coordinated with the main transformer's LTC. Due to

Dynamic Vulnerability Assessment and Intelligent Control for Sustainable Power Systems, First Edition.
Edited by José Luis Rueda-Torres and Francisco González-Longatt.

the closed-loop nature of MPC, the proposed control scheme can account for model inaccuracies and failure or delays of the control actions. The authors showed that anticipation of future LTC actions provide better and more efficient control decisions. This work was later extended in [8] to incorporate congestion management of DN lines. Moreover, a variant MPC formulation is presented in [9]. Here, the objective is to minimize the deviations of DGU active and reactive powers from reference values. DGU powers are restored to their desired schedule as soon as operating conditions allow doing so. Three modes of operation of the proposed controller are presented, involving dispatchable units as well as DGUs operated to track maximum power output, as detailed in Chapter 14.

These advanced control schemes rely on modern communication infrastructure and are currently at the heart of Smart Grid developments. They offer an attractive alternative to expensive network reinforcement, as long as stressed operating conditions prevail for a limited duration only. In addition to alleviating DN problems, they provide the possibility to actively support the transmission grid, if the legislative framework allows it and the controllers are designed to do so.

This chapter demonstrates how such advanced DN control schemes, and in particular Volt-Var Control (VVC), can affect long-term voltage stability of Transmission Networks (TNs). VVC refers to the process of managing voltage levels and reactive power throughout a distribution grid. It involves controlling devices that inject or consume reactive power to regulate the voltage profile, collaborating with existing equipment that directly control DN voltages.

In traditional DNs, the primary components for controlling voltage are LTCs, voltage regulators, and capacitor banks. The first two refer to variable ratio transformers connecting the DN to the TN or placed within the DN, and adjusted to raise or lower voltage as required. Capacitor banks are connected through mechanical breakers to the distribution grid, and when activated they inject reactive power for voltage boosting and/or power factor correction.

Modern VVCs, such as the ones referred to this chapter, rely on an advanced communication infrastructure to coordinate DGUs to inject/consume reactive power to mitigate dangerously low or high voltage conditions. These controllers usually incorporate the actions of traditional devices, such as LTCs or the activation of capacitor banks.

Beyond maintaining a stable voltage profile, VVCs have other potential benefits. For example, by acting on reactive power flows in the DN, they can alleviate thermal overloads [8], correct the power factor at the connection point with the TN [10], or minimize network losses. Moreover, using DGUs located close to the loads, VVCs offer additional flexibility to perform load voltage reduction. This well-known technique exploits the load sensitivity to voltage to decrease the active and reactive power demand [11–13]. Although not as effective as (under-voltage) load shedding [14], it may be applied as a first line of defense [15] or to reduce peak demands and energy consumption [16, 17].

Finally, if the VVCs are extended to allow the control of flexible loads, storage units, or the active power output of DGUs, then great flexibility is offered to manage the DN and support the TN in case of emergencies.

The remainder of this chapter is organized as follows. Section 15.2 provides a brief review of long-term voltage instability mechanisms and some conventional solutions used today. Then, Section 15.3 analyzes the impact that VVCs can have on long-term voltage stability of TNs as well as the mechanisms available to enhance stability. In Sections 15.4 and 15.5, a combined Transmission and Distribution (T&D) test-system is used to provide an example of this behavior and some long-term instability counter-measures. Finally, conclusions are drawn in Section 15.6.

15.2 Long-Term Voltage Stability

Long-term voltage instability results from the inability of the combined transmission and generation system to deliver the power requested by loads at least at some nodes [11, 18, 19]. It is a slow phenomenon with a timescale of up to several minutes. This type of instability usually occurs when the system is highly stressed (high demand of active and reactive power) and is initiated by component outages (line, transformer, generator, etc.). Some of the driving forces behind long-term voltage instability are the lack of fast reactive power resources, the action of OverExcitation Limiters (OELs) enforcing gen-erator reactive power limits, and load power restoration. The last of these can be due to the LTCs trying to recover DN voltages, and therefore the load powers, or even due to the self-restoration of loads.

Long-term voltage instability is caused by the non-existence of a post-disturbance equilibrium or the lack of attraction towards the stable post-disturbance equilibrium point [11, 19], for example due to delayed corrective actions. To analyze this behavior, let us consider a DN feeding loads, as shown in Figure 15.1. The voltage on the TN and DN sides of the transformer are V and V_d, respectively. The complex power enter-ing the transformer on the TN side is $P + jQ$, and leaving the transformer on the DN side $P_d + jQ_d$. The loads are sensitive to voltage and hence the net load power can be considered a function of V_d.

The evolution of this long term voltage instability can be explained with the help of Figure 15.2a, showing the loadability curve in the (P, Q) space of load powers [11, 20]. In this space, there exists a feasible region bounded by Σ_1 in which the combined genera-tion and transmission system model has (at least) one equilibrium point. For load powers outside this region, the (algebraic) equations characterizing that system in steady state have no solution [11]. The example system initially operates at the long term equilibrium point A with V_d equal to the LTC setpoint value V_d^o and a power consumption of $S(0^-)$.

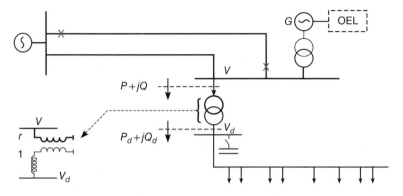

Figure 15.1 Simple T&D system with DN controlled by LTC.

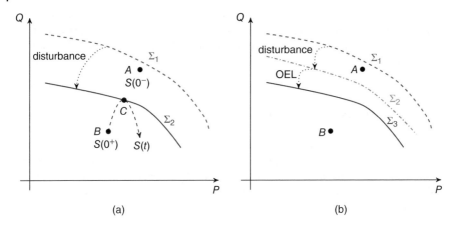

Figure 15.2 Loadability curves (a) Disturbance and restoration (b) Disturbance followed by OEL action.

Let us consider an incident in the transmission system that abruptly causes the loadability surface to shrink, so that point A is left outside the post-disturbance feasible region Σ_2. This means that the original load demand cannot be met any more. If the system is short-term voltage stable, then it will settle at the short-term equilibrium B, inside the post-disturbance feasible region Σ_2. The load power consumption at this point ($S(0^+)$) is lower, due to the voltage drop in both V and V_d caused by the disturbance.

Starting from point B, the load restoration mechanisms try to bring the power back to point A. A common load restoration mechanism is due to LTC actions adjusting the transformer ratio r in successive steps to restore V_d (close) to V_d^o. If these tap changes were successful, and assuming a negligible LTC deadband, the power on the distribution side would be restored to the pre-disturbance value $S(0^-)$. Using the transformer model in the lower left of Figure 15.1, it is easily shown that the transmission power $P + jQ$ would also be restored to its pre-disturbance value.

However, this restoration is not possible as point A lies now outside the feasible region Σ_2. At some point C, the trajectory of the restored load powers $S(t)$ (shown in Figure 15.2a) will reach the loadability curve and turn backwards. Thus, the LTC will fail to restore the distribution voltage to V_d^o and, hence, the power to its pre-disturbance value. These unsuccessful attempts make the transmission voltage V fall below an acceptable value.

Real-life systems are obviously more complex, with a load power space of much higher dimension. Furthermore, the limits on generator reactive power enforced by OELs, for instance, contribute to further shrinking the feasible region. For the simple system of Figure 15.1, the feasible region shrinks under the effects of the line outage followed by the reduction of the field current of generator G located next to the load. This sequence is depicted in Figure 15.2b.

15.2.1 Countermeasures

The main objectives of instability countermeasures are to restore a long-term equilibrium (fast enough so to be in the region of attraction) and stop system degradation. Traditionally, there are several defense mechanisms proposed and applied to achieve

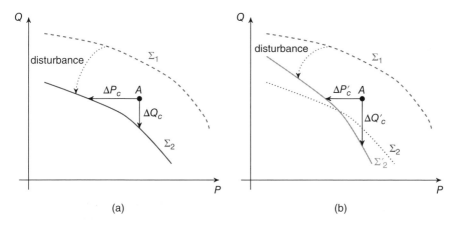

Figure 15.3 Restoration of equilibrium point with corrective actions (a) Curve more sensitive to reactive power (b) Curve more sensitive to active power.

these goals, such as reactive compensation switching/boosting, LTC emergency controls, generation rescheduling, and load reduction.

If we assume a load restoration to the pre-disturbance operating point A, the corrective actions need to establish a new operating point inside the feasible region Σ_2. From Figure 15.3a, it can be seen that a feasible point can be established by decreasing the load active or reactive power consumption by ΔP_c or ΔQ_c respectively, or through a combined reduction.

Some countermeasures act only in one direction and others in both. For example, reactive compensation reduces the net reactive power consumption; increasing active power generation of local units decreases the net active power consumption seen from the TN; while depressing the DN voltages or shedding some loads decreases both active and reactive power consumption. It can be seen in Figure 15.3a that along each direction there is a different distance to the loadability surface Σ_2. Thus, a combination of corrective actions is necessary to achieve the optimal effect.

Moreover, depending on the characteristics of the system after the disturbance, one of the two directions can be more effective than the other. This behavior is shown in Figure 15.3b, where for the feasible region defined by Σ'_2 decreasing the active power consumption is more effective than its reactive counterpart.

In selection of corrective actions, the impact on consumers should be taken into consideration. Reactive power compensation through shunt switching and generator boosting has no direct adverse effect on customers. Increasing active power generation of local units does not impact customers either, but depends on the available reserves and usually comes at a higher cost. Depressing the DN voltages to exploit Conservation Voltage Reduction (CVR) can be an effective countermeasure but is harmful for consumers if the DN voltages remain uncontrollably low, which may affect or disconnect some appliances such as electronic devices. Finally, when the corrective action to be taken is load shedding, this severely impacts consumers and the amount of load to be shed should be the minimum possible.

In addition to defining the type and extent of corrective actions needed, the timing is also of great importance. If the control action is delayed too much then the system

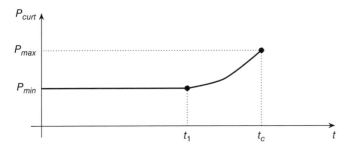

Figure 15.4 Evolution with time of the minimum load curtailment needed to stabilize the system [11].

may exit the region of attraction rendering the countermeasure useless. Hence, countermeasures with no adverse effect on consumers should be employed as soon as possible while the remainder should be delayed as needed to make sure that they are absolutely necessary. It should be kept in mind though, that the more a countermeasure is delayed, the larger a corrective action could be needed to save the system. For example, Figure 15.4 shows the minimum load reduction needed to save the system as a function of the time to take this action, counted since the disturbance occurrence. Up to a specific time instant, this amount is constant, while after that a larger amount of load needs to be curtailed to save the system. Finally, there is a no-return point in time after which the system cannot be saved.

In this section some basic notions of long-term voltage stability have been introduced, necessary to understanding and analyzing the impact of VVCs at distribution and transmission level. Readers interested in further information concerning this subject can refer to textbooks such as [11, 18].

15.3 Impact of Volt-VAR Control on Long-Term Voltage Stability

In this section we describe how VVC can affect long-term voltage stability. First, we show how incorrectly designed VVCs can precipitate voltage instability. Then, we explain how VVCs can be employed in emergency situations to support the TN.

Let us consider the same simple system of Figure 15.1, though now the DN is equipped with a centralized VVC acting on the DGUs, the reactive power compensation devices, the LTC setpoint, and so on. This updated system is sketched in Figure 15.5. While a centralized controller is assumed in this chapter, most of the results derived below can be applied to distributed or local controllers as well.

As discussed in the Introduction, the primary purpose of VVC is to dispatch the available distributed resources to keep the voltages within an acceptable range and alleviate any security problems within the DNs. For example, in case of depressed voltages inside the DN, the VVC will try to support and restore the voltages, usually using reactive power injection from the DGUs or other devices. Similarly, in case of thermal overloads, the VVC will dispatch the DGUs in such way as to alleviate the problem [9], using reactive or even in some instances active power modulation (with dispatchable DGUs, flexible loads, etc.).

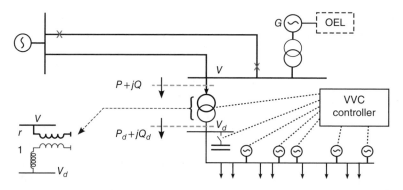

Figure 15.5 Simple system with DN controlled by LTC and DGUs.

Most of the VVC algorithms, either local, distributed, or centralized, assume a strong TN. That is, in their design and testing, stiff transmission voltages are considered. Unfortunately, this hypothesis is no longer valid in degraded operating conditions when interactions between transmission and distribution systems become critical. In such situations, VVCs unaware of the system weakening and operating under the strong TN assumption could end up precipitating the developing voltage instability. Moreover, due to the high speed of action of power-electronic based interfaces (present in many DGUs), the instability could occur much faster than with conventional machines; thus, it becomes harder to detect and apply countermeasures [15].

Let us assume again an incident in the transmission system that causes the loadability curve to shrink, so that point A is left outside the post-disturbance feasible region Σ_2, as shown in Figure 15.6a. The system settles at the short-term equilibrium B, inside the post-disturbance feasible region, with a lower load power consumption due to the voltage drop caused by the disturbance. Activated by the depressed voltages inside the DN, the VVC will try to restore the voltages to the pre-disturbance values (V_d^o) and,

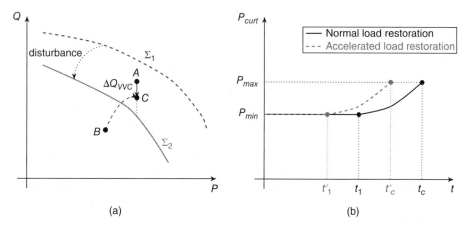

Figure 15.6 Impact of VVCs on long-term voltage stability (a) Disturbance and voltage restoration with VVC (b) Minimum load reduction needed with time after disturbance with effect of VVC.

assuming a very accurate voltage control with negligible dead-band, will consequently be restoring the load consumption to the pre-disturbance value (point A).

The VVC can achieve this voltage restoration by dispatching the DGUs to inject more reactive power. If a total of ΔQ_{VVC} is injected into the DN to support the voltages, then a new equilibrium point C is targeted, as shown in Figure 15.6a. Therefore, two mechanisms are working in parallel, the load-voltage restoration increasing the active and reactive power consumption and the VVC increasing the reactive power injection. If the targeted operating point C lies within the feasible region defined by Σ_2 (and VVC acts fast enough to have attraction), then the system will be stabilized by the VVC. Otherwise, this point cannot be reached, and the system will require further corrective actions or it will collapse. Note that in real systems, the situation is more complex as along with the reactive power injection, the system network losses are also modified, thus a positive or negative change in power demand can be observed as well.

Furthermore, starting from point B, the VVC accelerates the voltage (and hence the load power) restoration procedure adding to the effect of LTCs. This acceleration is due to the much faster action times of DGUs compared to LTC devices. As explained in the previous section, the timing of applying the necessary corrective actions is important in order to stay within the region of attraction of the feasible point. Thus, by accelerating the voltage restoration procedure, the available time to apply the countermeasures becomes shorter as well, as shown in Figure 15.6b. This behavior can invalidate existing long-term voltage instability countermeasures set in place by transmission system operators.

15.3.1 Countermeasures

As described in Section 15.2.1, any countermeasure should try to quickly restore a long-term equilibrium and stop system degradation. In addition to the traditional countermeasures described previously, VVCs can be used to restore an equilibrium point by decreasing the net active (resp. reactive) power consumption by ΔP_c (resp. ΔQ_c), or through a combination of actions. Due to the flexibility of DGUs, DNs equipped with a VVC have higher controllability and, if appropriately designed, they can support the TN in stressed operating conditions [10, 21]. Moreover, VVCs offer a faster way to implement some countermeasures thus avoiding further degradation or loss of attraction due to delays.

As a first action, the reactive power injection of DGUs and other reactive compensation devices controlled by VVC can be used to directly act on ΔQ_c. This is usually a natural response by the VVC when depressed voltages are detected. However, in emergency situations, the objective of strictly controlling the DN voltage profile can be relaxed in favor of supporting the TN voltages through coordinated reactive power compensation [7, 10].

In addition, the VVC can control the active power set-points of DERs to provide ΔP_c, if these control actions are available in the system. Such devices could be DGUs with the capability of increasing their production, energy storage systems, or even flexible loads that can decrease their consumption. To the same end, some load shedding can be selected as a last option if other methods fail.

Finally, VVCs can be used to implement CVR and exploit load voltage reduction providing both ΔP_c and ΔQ_c [15]. Traditionally, load voltage reduction is actuated through a reduction of the voltage set-points of LTCs controlling TN-DN transformers, and acting on a single distribution bus. However, VVCs can be used to complement LTC voltage set-point reduction with coordinated control of DGUs. As the DGUs are located closer to the loads and have faster reaction time than LTCs, they can be used to depress the voltages more effectively and quickly. Furthermore, while an LTC only monitors a single distribution bus, VVC can monitor the voltages inside the entire DN and ensure that a minimum emergency voltage profile is maintained throughout the network.

15.4 Test System Description

In this section, the combined T&D system used to demonstrate the above analysis is detailed (Subsection 15.4.1), along with the VVC used for the control of the DNs (Subsection 15.4.2), and some emergency detection methods employed to activate the support of VVCs to the TN (Subsection 15.4.3).

15.4.1 Test System

To show the notions and techniques analyzed above, a large-scale system is used involving a TN including multiple DNs with dispersed generation. The TN part re-uses the Nordic test system set up by the IEEE PES Task Force on Test Systems for Voltage Stability and Security Analysis [22]. Its one-line diagram is shown in Figure 15.7. It includes 68 buses and 20 synchronous machines modeled with their excitation systems, voltage regulators, power system stabilizers, speed governors, and turbines.

The original TN model included 22 aggregate, voltage-dependent loads behind explicitly modeled TN-DN transformers. Six of them were replaced by 40 detailed distribution systems, behind new TN-DN transformers of smaller powers. The six buses were chosen in the Central area because it is the most impacted by voltage instability.

Each DN is a replica of the same medium-voltage distribution system, whose one-line diagram is shown in Figure 15.8. It consists of eight 11-kV feeders all directly connected to the TN-DN transformer, involving 76 buses and 75 branches. The various DNs were scaled to match the original (aggregate) load powers, while respecting the nominal values of the TN-DN transformers and other DN equipment.

The transformer connecting each DN to the TN is equipped with an LTC controlling its distribution side voltage. To avoid artificial synchronization of transformers, the delays on tap changes were randomized around their original values. Thus, the first tap change takes place 28–32 s after leaving the voltage dead-band and the subsequent changes occur every 8–12 s.

Each DN serves 38 voltage sensitive loads and 15 represented by equivalent induction motors. The voltage sensitive loads are modeled as:

$$P = P_0 \left(\frac{V}{V_0} \right)^\alpha \quad Q = Q_0 \left(\frac{V}{V_0} \right)^\beta \tag{15.1}$$

with $\alpha = 1.0$ (constant current) and $\beta = 2.0$ (constant impedance), respectively. V_0 is set to the initial voltage at the bus of concern, while P_0 and Q_0 are the corresponding initial active and reactive powers. The equivalent motors are representative of small industrial

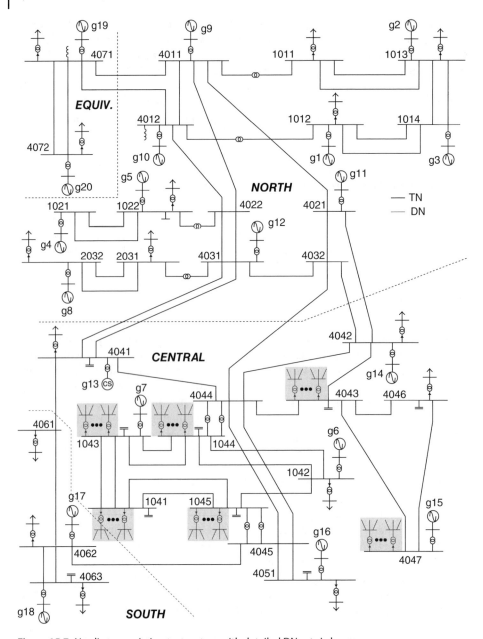

Figure 15.7 Nordic transmission test system with detailed DNs at six buses.

motors, with constant torque and a third-order model (the differential states being rotor speed and flux linkages) [18].

Moreover, it includes 22 DGUs, of which 13 are 3-MVA synchronous generators and the remaining are 3.3-MVA Doubly Fed Induction Generators (DFIGs) [23]. The reactive power of each 3-MVA synchronous generator is adjusted to the value received from the VVC using a local proportional-integral control loop. The latter acts on the set-point

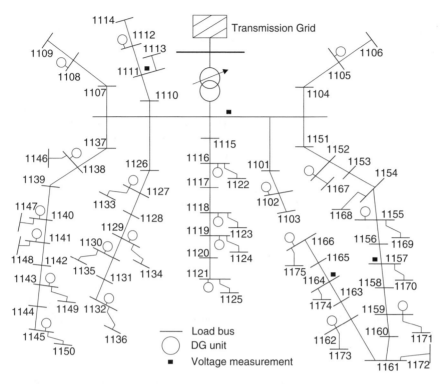

Figure 15.8 Topology of each of the 40 distribution networks.

of a voltage regulator (responding to faster changes). Similarly, the DFIGs operate in reactive power control mode (see [23]), to meet the power requested by the VVC.

The combined transmission and distribution model includes 3108 buses, 20 large and 520 small synchronous generators, 600 motors, 360 DFIGs, 2136 voltage-dependent loads, and 56 LTC-equipped transformers.

15.4.2 VVC Algorithm

Each of the 40 DNs is equipped with a separate centralized VVC using the MPC principle. This controller is summarized below and detailed in Chapter 14.

The control variables are the active and reactive power set-points of the DGUs:

$$u(k) = [P_g(k)^T \ Q_g(k)^T]^T \tag{15.2}$$

where ^{T} denotes array transposition and k is the discrete time.

At time k, the controller uses a sensitivity-based model to predict the behavior of the system over a future interval with N_p discrete steps (for $i = 1, \ldots, N_p$):

$$V(k + i|k) = V(k + i - 1|k) + \frac{\partial V}{\partial u} \Delta u(k + i - 1) \tag{15.3}$$

where $\Delta u(k) = u(k) - u(k - 1)$ is the vector of control changes and $V(k + i|k)$ the one of predicted bus voltages at time $k + i$ given the measurements at time k, $V(k|k)$ is set to the measured values received at time k. Matrix $\frac{\partial V}{\partial u}$ contains the sensitivities of voltages to control variables.

The sequence of control changes optimizes a multi-time step objective under constraints. The following quadratic objective is considered at time k:

$$\min_{\Delta u, \varepsilon} \sum_{i=0}^{N_c-1} \|\Delta u(k+i)\|_R^2 + \|\varepsilon\|_S^2 \tag{15.4}$$

where the squared control changes aim at distributing the effort more evenly over the DGUs and the time steps. R is a weight matrix used to force control priorities. For example, higher costs are assigned to variations of P_g compared to Q_g. The slack variables $\varepsilon = [\varepsilon_1 \ \varepsilon_2]^T$ are used when some of the constraints, detailed hereafter, make the optimization problem infeasible. These variables are heavily penalized using the weight matrix S, to keep them at zero when the problem is feasible.

The minimization is subjected to the following inequalities:

$$u^{min} \leq u(k+i) \leq u^{max} \quad i = 0, 1, \ldots, N_c - 1 \tag{15.5a}$$

$$\Delta u^{min} \leq \Delta u(k+i) \leq \Delta u^{max} \quad i = 0, 1, \ldots, N_c - 1 \tag{15.5b}$$

$$-\varepsilon_1 \mathbf{1} + V^{min} \leq V(k+N_p|k) \leq V^{max} + \varepsilon_2 \mathbf{1} \tag{15.5c}$$

The constraints (15.5a) are associated with the permitted range of control variables. The active powers, updated from real-time measurements, are already at their upper limits. The reactive power limits are updated with the voltage and active power as described in [24], based on the information in [25] for the DFIGs. The constraints (15.5b) are associated with the rate of change of control variables. In inequalities (15.5c), 1 denotes a vector of ones, while V^{min} (resp. V^{max}) is the desired lower (resp. upper) limit on the voltages. It must be emphasized that voltages are forced to reintegrate their limits at the prediction horizon, that is, at time step $k + N_p$ only.

In principle, the above control scheme is designed to control DN voltages only. The relation of the VVC to TN voltage stability is established by the voltage limits V^{min} and V^{max} set in (15.5c). The selection of these limits can make a difference between reaching a new long term equilibrium point or not, as will be shown in the following sections.

Measurements involve DGU active/reactive powers and terminal voltages, as well as voltages at three load buses. The latter were selected so that no load is more than two branches away from a voltage monitored bus. The measurement noise was simulated by adding a Gaussian random variable limited to ± 0.01 pu for voltages, and $\pm 1\%$ of the DGU maximum active and reactive powers.

15.4.3 Emergency Detection

As discussed in the previous sections, the main objective of VVCs is to keep voltages within some secure limits and to optimize DN operation. To be able to support the TN in emergency situations, some detection mechanisms need to be in place. One such mechanism is LIVES (standing for Local Identification of Voltage Emergency Situations), which was originally proposed in [26]. This mechanism detects the unsuccessful attempt of LTCs to restore their DN voltages, as sketched in Figure 15.2a. To achieve this, LIVES monitors each LTC independently, thus providing a distributed—and more robust—emergency detection mechanism.

Another detection approach is to monitor the reactive reserve of the large, TN-connected generators. It is quite common to have generator reactive power measurements in the SCADA system of TN control centers. A comparison with capability curves can flag a generator that has switched under limit. This information can make up an alternative alarm signal to be send to VVCs, combining simplicity with anticipation capability. This simple emergency detection was used in the simulation reported hereafter.

15.5 Case Studies and Simulation Results

In this section, the case studies used to support the analysis presented earlier in this chapter and their simulation results are shown. Table 15.1 offers an overview of these case studies. They are divided into three groups, namely: stable cases, unstable cases, and unstable cases with the use of emergency corrective actions.

As the timing of controls is critical, long-term dynamic simulations were performed to simulate the system responses to large disturbances in the TN. The simulations were performed with the RAMSES software developed at the University of Liège [27].

15.5.1 Results in Stable Scenarios

For these cases, a TN disturbance is considered that is not threatening to the system stability. Thus, the disturbance causes the loadability curve to shrink but the pre-disturbance point A is still within the feasible region, as shown in Figure 15.9a. The disturbance considered is the outage of line 4061-4062 (see Figure 15.7).

15.5.1.1 Case A1

In this first scenario, the VVCs are not in service. The DGUs operate with constant reactive powers and do not take part in voltage control. This leaves only the traditional voltage control by LTCs. Hence, the DNs are considered passive.

The long-term evolution of the system until it returns to steady state is shown in Figures 15.10 and 15.11. It is driven by the LTCs, in response to the voltage drops

Table 15.1 Overview of simulated scenarios.

Case	VVC	Stable	Corrective Action	Contingency	Section
A1	no	yes	-	outage ofline 4061-4062	15.5.1
A2	yes		-		
B1	no	no	-	5-cycle short-circuit near bus 4032, cleared by opening line 4032-4042	15.5.2
B2	yes		-		
C1	no	no	CVR with LTC reversal		
C2	yes	yes	CVR with VVC set-point change		15.5.3
C3			Active power decrease with flexible loads		

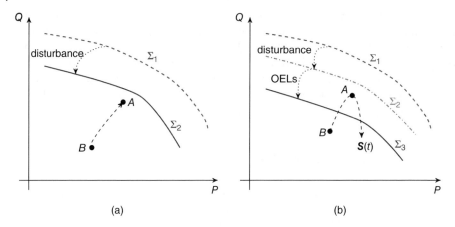

Figure 15.9 Simplified loadability curves of considered case studies (a) Case study A (b) Case study B and C.

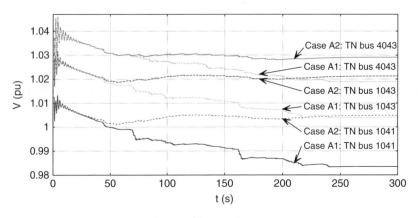

Figure 15.10 Cases A1 & A2: voltages at three TN buses.

initiated by the line tripping. There are some 112 tap changes in all 40 DNs. Figure 15.10 shows the TN voltage evolution at three representative buses of the Central area. The voltage at bus 1041 is the most impacted but remains above 0.985 pu. All DN voltages are successfully restored in their dead-bands by the LTCs, which corresponds to a stable evolution [11]. For instance, Figure 15.11 shows the voltage evolution at two DN buses: 01a, controlled by an LTC with a [1.02 1.03] pu dead-band, and 01a-1171, located further away in the same DN.

15.5.1.2 Case A2
The same disturbance as in Case A1 is considered, but the VVCs are now active. For simplicity, all components of V^{min} have been set to 0.98 pu and those of V^{max} to 1.03 pu. This interval encompasses all LTC deadbands, so that there is no conflict between an LTC and the VVC.

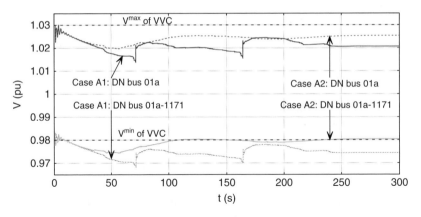

Figure 15.11 Cases A1 & A2: voltages at two DN buses.

The corresponding TN and DN voltage evolutions can be found in Figures 15.10 and 15.11, for easier comparison. With respect to Case A1, a steady state is reached at almost the same time, while the TN voltages are slightly higher. The voltages at DN buses not directly controlled by LTCs (such as 01a-1171 in Figure 15.11) are restored above V^{min} by the VVC.

It is worth mentioning that the number of tap changes has decreased from 112 to 35, showing that the sharing of the control effort by active DNs reduces the wear of LTCs.

The DN buses such as 01a-1171 in Figure 15.11 have their voltages increased by the additional reactive power produced by the coordinated DGUs. For instance, in Case A2, the DG participations decrease the net reactive power load seen by the TN by almost 90 Mvar (see Figure 15.12), which contributes to increasing TN voltages. Note, however, that a hidden effect of this load voltage increase is an increase of active and reactive powers of the voltage sensitive loads. While very moderate in this scenario, this effect will be predominant in the next ones.

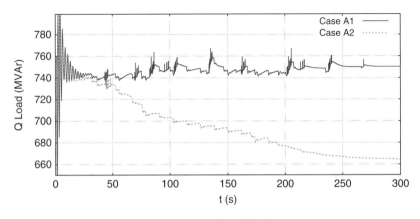

Figure 15.12 Cases A1 & A2: reactive power from TN to DNs.

15.5.2 Results in Unstable Scenarios

For these cases, a critical TN disturbance is considered leading to long-term voltage instability. The disturbance considered is a 5-cycle short-circuit near bus 4032, cleared by opening line 4032-4042 (see Figure 15.7). First, the disturbance causes the loadability curve to shrink but the pre-disturbance point *A* is still within the feasible region. Nevertheless, a series of generator OEL actions in the TN further shrink the feasible region leading to a system collapse. This sequence is depicted in Figure 15.9b.

15.5.2.1 Case B1

Similarly to Case A1, in this scenario the DN voltages are only controlled by the LTCs and the VVCs are inactive. The voltage evolutions at two TN buses are shown in Figure 15.13. The sustained voltage sag caused by the LTCs and OELs ends with a system collapse due to the loss of synchronism of machine g6 at $t \simeq 225$ s. The maximum power that can be provided to loads is severely decreased by the initial line outage as well as cascading field current reductions due to seven OELs acting before $t \simeq 147$ s. At the same time, the LTCs unsuccessfully attempt to restore DN voltages, which is impossible since load powers cannot be restored at their pre-disturbance values. The static aspects of this mechanism were analyzed in Section 15.2.

15.5.2.2 Case B2

In this scenario, in addition to LTCs, the VVCs are also activated. The resulting voltage evolutions can be found in Figures 15.13 and 15.14, while Figure 15.15 shows the total active and reactive powers transferred from the TN to the DNs.

Since the DGUs have their reactive power productions increased to correct DN voltages, the net reactive power load seen by the TN decreases progressively, passing from 750 to 200 Mvar in 150 s. This results in TN voltages dropping significantly less than in Case B1, until $t \simeq 150$ s (see Figure 15.13). In principle, this reactive support could prevent voltage instability (see Figure 15.6a), but it is not sufficient in this case, where five OELs are activated successively. The limitations of generator reactive powers are identified in Figure 15.16. At $t \simeq 165$ s, soon after the fifth OEL is activated, the system collapses with all generators in the Central and South areas, except g14, going out of step with the rest of the system.

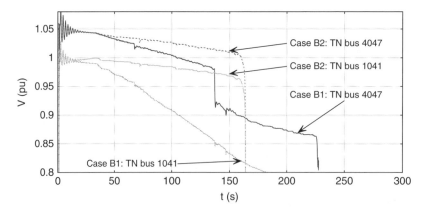

Figure 15.13 Cases B1 & B2: voltages at two TN buses.

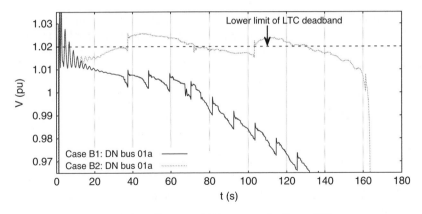

Figure 15.14 Cases B1 & B2: voltage at a DN bus controlled by an LTC.

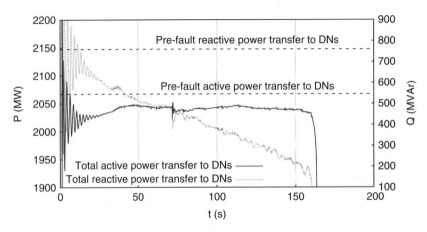

Figure 15.15 Case B2: total active and reactive power transfer from TN to DNs.

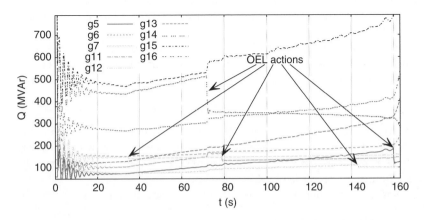

Figure 15.16 Case B2: reactive power produced by TN-connected generators.

This severe outcome is partly explained by the fact that the VVCs promptly restore voltages, and hence, load powers. Consequently, when generators connected to the TN stop controlling their voltages, under the effect of OELs, they are facing a higher load power compared to Case B1. The restoration of DN voltages is illustrated by the dotted curve in Figure 15.14. The voltage is boosted by the controller in significantly less time than with a traditional LTC, with the result that the LTC makes two tap changes only. The resulting fast restoration of load active power is shown in Figure 15.15, where the small, 25-MW non restored power is caused by the (intentional) voltage dead-bands of the controllers.

The overall effect of voltage control at DN level is to precipitate system collapse, while making the latter less predictable from the mere observation of voltages. From a system operation viewpoint, this case is worse than Case B1. The static aspects of this behavior were also explained in Section 15.2.

15.5.3 Results with Emergency Support From Distribution

In this section, Case B is revisited assuming some remedial actions to secure system operation. This requires a timely detection of the emergency condition, which was discussed in Section 15.4.3.

The remedial action considered in the sequel is a decrease of DN voltages exploiting the sensitivity of load power to voltage. This effect strongly depends on the type of loads [11, 18, 28]. As long as a decrease in the voltage magnitude leads to decreased load consumption (in active and/or reactive power), the mechanism is beneficial to system stability. However, some loads (for instance induction motors and electronically controlled loads) have the tendency to rapidly restore their consumed power after a voltage decrease. Moreover, induction motors could eventually exhibit increased consumption when approaching their stalling point [18].

The voltage decrease ΔV should be as large as possible while ensuring a minimum acceptable voltage at all DN nodes [28]. When implemented through LTC voltage decrease alone, this reduction is applied at only one point of the DN. This has led system operators to conservatively set ΔV to 0.05 pu [11, 18]. On the other hand, acting on multiple DGUs offers better controllability by adjusting DN voltages at multiple points. Therefore, more points of the DN have their voltages decreased while still remaining above the acceptable limit.

Moreover, as the LTC controllers involve mechanical components, their speed of action is limited and can be insufficient to counteract instability. Acting on DGUs offers a significantly faster control of DN voltages.

15.5.3.1 Case C1

This case involves the classical LTC voltage set-point reduction without any contribution by the VVCs. Thus, this scheme is to be compared to Case B1, from which it differs after $t \simeq 75$ s, when an alarm is received from the generator limitation. At that time, the LTC voltage set-points are decreased by $\Delta V = 0.05$ pu.

Figure 15.17 shows the voltage evolution at two TN buses. It can be seen that the amount of load decrease is not enough to stabilize the system. Further tests showed that $\Delta V = 0.08$ pu would be needed. However, in that case, several DN bus voltages are unacceptable, lower than 0.90 pu.

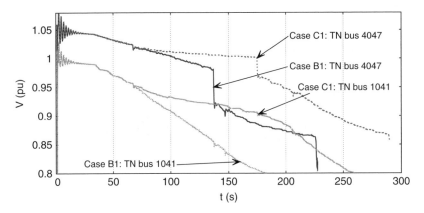

Figure 15.17 Cases B1 & C1: voltages at two TN buses.

15.5.3.2 Case C2

Case C2 is to be compared with Case B2, from which it differs after $t \simeq 75$ s, when the generator limitation alarm is received. At that time, the above mentioned ΔV corrections are applied. While in the unstable Case C1, the LTCs alone were unable to restore DN voltages, in this case, they contribute, together with VVCs, to depressing the DN voltages.

Voltage evolutions at TN buses are shown in Figure 15.18, together with those of the uncontrolled Case B1, for comparison purposes. The TN voltages are smoothly stabilized. Figure 15.19 shows the corresponding voltage evolutions in one DN, together with the changing V^{min} and V^{max} limits. It is seen that the DN voltages are promptly brought within the new desired range.

The voltage reduction causes the decrease of active and reactive power transfer from the TN to DNs, as shown in Figure 15.20. As explained in Section 15.2, the reactive power transferred between the TN and DNs varies with the reactive power produced by the DGUs, the reactive power consumed by the loads, and the network losses. First, the DGUs are directed by the VVCs to reduce and maintain the DN voltages within the emergency limits. To achieve this, they decrease their reactive power productions. Then,

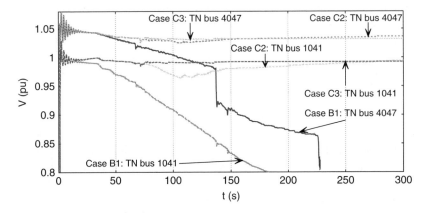

Figure 15.18 Cases B1, C2 & C3: voltages at two TN buses.

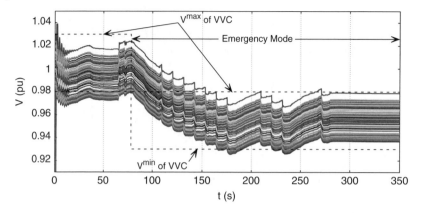

Figure 15.19 Case C2: voltages at various DN buses of the same feeder.

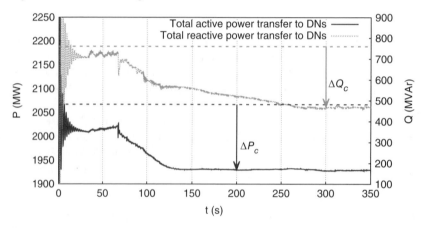

Figure 15.20 Case C2: total active & reactive power transfer from TN to DNs.

the reduced DN voltages lead to decreased reactive power consumption by the loads and decreased losses. In the system studied, the benefit brought by the load decrease outweighs the negative impact of the reduced reactive support from DGUs. Thus, the restored power point is moved as shown in Figure 15.3b, entering the feasible region.

The evolution of reactive powers of TN-connected generators for Case C2 is shown in Figure 15.21. It can be seen that no further generator limitation takes place after $t = 75$ s, in contrast to Case B2 where the system eventually collapses after the successive field current limitations of generators g5 and g6, as shown in Figure 15.16. Furthermore, the non-limited generators are relieved, as shown by Figure 15.21.

Effect of time delay in applying the corrective actions To assess the robustness of the scheme to delays in emergency detection signals, three cases were considered, each with an additional delay of 30 s added to the original alarm time of 75 s. With delays up to 60 s (i.e., with an emergency alarm up to $t = 145$ s), the system can be saved. This is made possible by the fast response of DGUs steered by the VVC. As explained in Section 15.2, there exists a no-return point after which the corrective actions are unable to restore stability as the system has exited the region of attraction, as confirmed by the curves in Figure 15.22.

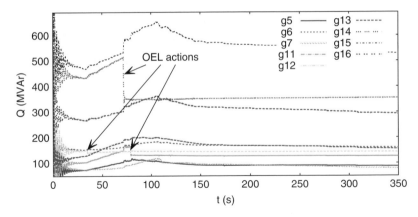

Figure 15.21 Case C2: reactive power produced by TN-connected generators.

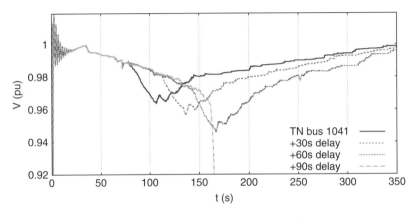

Figure 15.22 Case C2: Effect of time delay on emergency signal.

Sensitivity of CVR-based countermeasure to load types The emergency technique presented in this Chapter exploits the reduction of load consumption when voltages are decreased. Thus, the sensitivity of loads towards voltage variations is critical to its success. This sensitivity is taken into account through the parameters α and β, used in the load model (15.1).

These parameters are selected according to the mixture of loads present in the system. An indicative list can be found in [11, 18]. When the induction motors are modeled explicitly, such as in the test system used, the values $(\alpha, \beta) = (1.0, 2.0)$ are frequently used.

These load model parameters determine the load power decrease for a given voltage reduction. Figure 15.23 shows the voltage at a TN bus in Case C3 ($\Delta V = -0.05$ pu), with two different sets of parameters. The first pair $(\alpha, \beta) = (0.8, 1.0)$ is an extreme case, where the loads have very low sensitivity to voltage; while the second is the one used in the simulations previously shown. It can be seen that, even though in both cases the system is saved, with more sensitive loads the voltage recovers a little faster to a higher value.

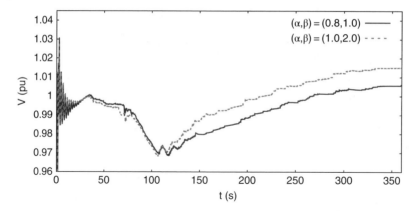

Figure 15.23 Case C2: voltage at a TN bus with two different load model parameter pairs; voltage reduction $\Delta V = -0.05$ pu.

In addition, a parametric study was performed to check the effectiveness of the method as a function of the load model parameters and the emergency voltage reduction (ΔV). Case C2 was considered. The simulation was repeated for several values of ΔV ($-0.05, \ldots, -0.01$ pu) and using a wide range of α ($0.8, \ldots, 2.0$) and β ($1.0, \ldots, 3.0$) parameter pairs. For $\Delta V = -0.05$ pu and $\Delta V = -0.04$ pu, the system is stabilized for all (α, β) pairs. For smaller voltage reduction, Figure 15.24 shows the values for which the system can be stabilized.

According to the plot, in this test system, active power reduction is more effective than its reactive power counterpart. For less sensitive active power loads (lower α values), voltage instability completely depends on α irrespective of the β value. As active power load is more sensitivity to voltage, β starts to have some more importance. Moreover, it is confirmed that, as loads get more voltage sensitive (higher α and β values), a smaller ΔV is sufficient to stabilize the system.

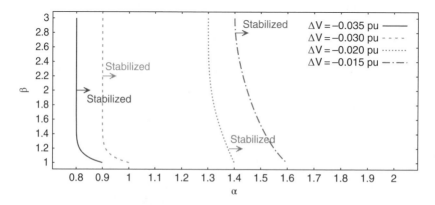

Figure 15.24 Case C2: regions of successful stabilization in the (α,β) space, for various voltage reduction values ΔV.

15.5.3.3 Case C3

In the last case, we consider the availability of flexible loads inside the DNs. More specifically, in case of emergency, all DNs can defer up to 5% of their original load consumption in increments of 1%. This case can be compared to Case B2.

The VVC controller uses the DGU reactive power to keep the DN voltages within the secure limits as defined in Section 15.4.2. At $t \simeq 75$ s, when the generator limitation alarm is received, the VVCs take a snapshot of their corresponding TN bus voltage and try to stop the degradation by keeping the voltages at that level. To this purpose they decrease their flexible load consumption by 1% at every control action taking place every ≈ 10 s.

The TN voltage evolutions at two buses are shown in Figure 15.18, together with those of the uncontrolled Case B1 and the stabilized using CVR Case C2. The TN voltages are smoothly stabilized. Figure 15.25 shows the corresponding voltage evolutions in one DN. It is seen that the DN voltages are brought within the desired ranges.

The combined corrective actions introduced by the DG reactive power injection and the flexible loads is shown in Figure 15.26. Of course, this power reduction of the flexible loads needs to be restored at a later stage when the system is out of the emergency state

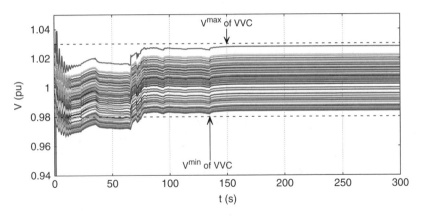

Figure 15.25 Case C3: voltages at various DN buses of the same feeder.

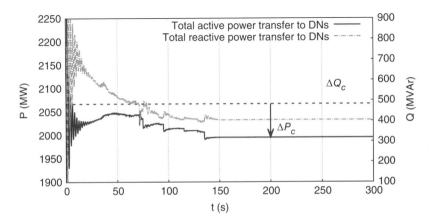

Figure 15.26 Case C3: total active & reactive power transfer from TN to DNs.

(e.g., when the tripped line is reconnected or new generators are put online). However, this procedure is not shown in this chapter.

15.6 Conclusion

This chapter demonstrated how advanced DN control schemes, and in particular Volt-Var Control (VVC), can affect the long-term voltage stability of Transmission Networks (TNs).

Specifically, the chapter has reported on simulations of a large-scale test system including one TN and multiple DNs. The latter are assumed to be active by controlling the reactive power of DGUs, with the aim of keeping DN voltages within a desired range. The DNs are controlled independently of each other, and of the TN. In this study, a specific MPC-based scheme was considered. However, the results are representative of future smart grids in which DGUs will contribute to regulating voltages at other buses than the one controlled by the LTC.

In normal operating conditions, the case study shows some benefits of VVC after a disturbance affecting the TN: (i) by increasing the DGU reactive powers to support DN voltages, the power factor at the TN/DN connection point is improved; (ii) the number of LTC steps is decreased.

On the contrary, in long-term voltage instability scenarios:

- VVC is impacted by the growing weakness of the TN system, which behaves very differently from what is assumed in the VVC scheme, set up for "normal" conditions. The main discrepancy stems from LTCs inability to restore DN voltages as expected.
- VVC control may, in turn, impact system dynamics. This is the case when DGUs are controlled concurrently (and not in complement) to LTCs to restore the DN voltages, and hence the load powers. This fast load power restoration may precipitate system collapse, making the situation more severe than with classical LTC control.

On the other hand, VVC can contribute to corrective actions against voltage instability by decreasing the DN voltages, thereby reducing the power consumed by loads, through the reduction of reactive power production of DGUs. The above action can be triggered:

- either locally, observing the unsuccessful attempt of the LTC to restore its DN voltage, when the VVC is designed to complement LTC operation;
- or centrally, from an alarm signal issued at TN level, when the VVC is set to operate concurrently with the LTC.

References

1 Bayod-Rújula, A.A. (2009) Future development of the electricity systems with distributed generation. *Energy*, **34** (3), 377–383, doi: 10.1016/j.energy.2008.12.008.
2 Lopes, J.P., Hatziargyriou, N.,Mutale, J., Djapic, P., and Jenkins, N. (2007) Integrating distributed generation into electric power systems: A review of drivers, challenges and opportunities. *Electric Power Systems Research*, **77** (9), 1189–1203, doi: 10.1016/j.epsr.2006.08.016.

3 Joos, G., Ooi, B., McGillis, D., Galiana, F., and Marceau, R. (2000) The potential of distributed generation to provide ancillary services, in *2000 Power Engineering Society Summer Meeting (Cat. No.00CH37134)*, vol. 3, IEEE, vol. 3, pp. 1762–1767, doi: 10.1109/PESS.2000.868792.

4 Kulmala, A., Repo, S., and Järventausta, P. (2014) Coordinated voltage control in distribution networks including several distributed energy resources. *IEEE Transactions on Smart Grid*, **5** (4), 2010–2020, doi: 10.1109/TSG.2014.2297971.

5 Agalgaonkar, Y.P., Pal, B.C., and Jabr, R.A. (2014) Distribution voltage control considering the impact of pv generation on tap changers and autonomous regulators. *IEEE Transactions on Power Systems*, **29** (1), 182–192, doi: 10.1109/TPWRS.2013.2279721.

6 Ranamuka, D., Agalgaonkar, A.P., and Muttaqi, K.M. (2016) Online coordinated voltage control in distribution systems subjected to structural changes and dg availability. *IEEE Transactions on Smart Grid*, **7** (2), 580–591, doi: 10.1109/TSG.2015.2497339.

7 Valverde, G. and Van Cutsem, T. (2013) Model predictive control of voltages in active distribution networks. *IEEE Transactions on Smart Grid*, **4** (4), 2152–2161, doi: 10.1109/TSG.2013.2246199.

8 Soleimani Bidgoli, H., Glavic, M., and Van Cutsem, T. (2014) Model predictive control of congestion and voltage problems in active distribution networks, in *Proc. CIRED conference, paper No 0108*.

9 Soleimani Bidgoli, H., Glavic, M., and Van Cutsem, T. (2016) Receding-horizon control of distributed generation to correct voltage or thermal violations and track desired schedules, in *Proc. 19th Power Systems Computation Conference (PSCC)*, Genoa, Paper. No. 114, p. 7.

10 Valverde, G. and Van Cutsem, T. (2013) Control of dispersed generation to regulate distribution and support transmission voltages, in *Proceedings of IEEE PES 2013 PowerTech conference*, doi: 10.1109/PTC.2013.6652119.

11 Van Cutsem, T. and Vournas, C. (1998) *Voltage Stability of Electric Power Systems*, Springer.

12 Vournas, C. and Karystianos, M. (2004) Load tap changers in emergency and preventive voltage stability control. *IEEE Transactions on Power Systems*, **19** (1), 492–498, doi: 10.1109/TPWRS.2003.818728.

13 Vournas, C., Metsiou, A., Kotlida, M., Nikolaidis, V., and Karystianos, M. (2004) Comparison and combination of emergency control methods for voltage stability, in *2004 IEEE Power Engineering Society General Meeting*, vol. 2, IEEE, vol. 2, pp. 1799–1804, doi: 10.1109/PES.2004.1373189.

14 Hiskens, I.A. and Gong, B. (2006) Voltage stability enhancement via model predictive control of load. *Intelligent Automation & Soft Computing*, **12** (1), 117–124, doi: 10.1080/10798587.2006.10642920.

15 Aristidou, P., Valverde, G., and Van Cutsem, T. (2015) Contribution of distribution network control to voltage stability: A case study. *IEEE Transactions on Smart Grid*, pp. 1–1, doi: 10.1109/TSG.2015.2474815.

16 Crider, C. and Hauser, M. (1990) Real time t&d applications at virginia power. *IEEE Computer Applications in Power*, **3** (3), 25–29, doi: 10.1109/67.56579.

17 Wang, Z. and Wang, J. (2014) Review on implementation and assessment of conservation voltage reduction. *IEEE Transactions on Power Systems*, **29** (3), 1306–1315, doi: 10.1109/TPWRS.2013.2288518.

18 Taylor, C.W. (1994) *Power System Voltage Stability*, Mc Graw Hill.

19 Löf, P.A., Hill, D.J., Arnborg, S., and Andersson, G. (1993) On the analysis of long-term voltage stability. *International Journal of Electrical Power & Energy Systems*, **15** (4), 229–237, doi: 10.1016/0142-0615(93)90022-F.

20 Dobson, I. and Lu, L. (1993) New methods for computing a closest saddle node bifurcation and worst case load power margin for voltage collapse. *IEEE Transactions on Power Systems*, **8** (3), 905–913, doi: 10.1109/59.260912.

21 ENTSO-E (2012) Network code on demand connection, *Tech. Rep.* URL https://www.entsoe.eu.

22 IEEE PES Task Force (2015) Test system for voltage sability analysis and security assessment, *Tech. Rep.*, PES-TR19.

23 Ellis, A., Kazachkov, Y., Muljadi, E., Pourbeik, P., and Sanchez-Gasca, J. (2011) Description and technical specifications for generic WTG models: A status report, in *Proc. IEEE PES 2011 Power Systems Conference and Exposition (PSCE)*, doi: 10.1109/PSCE.2011.5772473.

24 Valverde, G. and Orozco, J. (2014) Reactive power limits in distributed generators from generic capability curves, in *Proc. IEEE PES 2014 General Meeting*.

25 Engelhardt, S., Erlich, I., Feltes, C., Kretschmann, J., and Shewarega, F. (2011) Reactive power capability of wind turbines based on doubly fed induction generators. *IEEE Transactions on Energy Conversion*, **26** (1), 364–372, doi: 10.1109/TEC.2010.2081365.

26 Vournas, C. and Van Cutsem, T. (2008) Local identification of voltage emergency situations. *IEEE Transactions on Power Systems*, **23** (3), 1239–1248, doi: 10.1109/TPWRS.2008.926425.

27 Aristidou, P., Fabozzi, D., and Van Cutsem, T. (2014) Dynamic simulation of large-scale power systems using a parallel Schur-complement-based decomposition method. *IEEE Transactions on Parallel and Distributed Systems*, **25** (10), 2561–2570, doi: 10.1109/TPDS.2013.252.

28 Wang, Z. and Wang, J. (2014) Review on implementation and assessment of conservation voltage reduction. *IEEE Transactions on Power Systems*, **29** (3), 1306–1315, doi: 10.1109/TPWRS.2013.2288518.

16

Electric Power Network Splitting Considering Frequency Dynamics and Transmission Overloading Constraints

Nelson Granda[1] and Delia G. Colomé[2]

[1] *Facultad de Ingeniería Eléctrica y Electrónica, Escuela Politécnica Nacional, Quito, Ecuador*
[2] *Instituto de Energía Eléctrica, Universidad Nacional de San Juan - CONICET, San Juan, Argentina*

16.1 Introduction

Electrical Power System (EPS) security refers to the ability of the system to overcome imminent disturbances without loss of load. Security not only includes stability but also integrity and evaluation of EPS operation considering under/over voltages, under/over frequency, and overloads. Security control aims at decision making considering different time horizons and detail levels in order to prevent catastrophic contingencies and blackouts.

Controlled power system islanding is considered as the last emergency control action to halt failure propagation, avoiding uncontrolled power system separation and preventing wide area blackouts. In order to design a *controlled islanding scheme (CIS)*, it is necessary to define electrical areas with adequate generation/demand balance and transmission facilities that can be opened. A global survey shows that controlled islanding schemes represent around 7% of protection schemes installed worldwide to safeguard EPS integrity [1]. A carefully planned and implemented controlled islanding scheme, besides preventing blackouts, allows: satisfying power demand for some users, maintaining operating generation facilities within safe limits, and assuring stable operation of the newly formed island through control actions that guarantee suitable variations of frequency, voltage, and current variables. In addition, power system restoration becomes more efficient when controlled islanding is applied first [2].

Several methodologies for controlled power system islanding proposed in the literature develop or use different algorithms and techniques to solve the problem; however, most of them can be synthesized in a two stage scheme: (i) Vulnerability Assessment and (ii) Islanding process, as shown in Figure 16.1.

16.1.1 Stage One: Vulnerability Assessment

The goal is to assess the risk level presented by an EPS during a specific static or dynamic operating condition regarding the occurrence of cascading events [3]. Additionally, vulnerability assessment determines the type and timing of control actions to be applied in order to avoid a system collapse. It must answer the following questions: When should

Dynamic Vulnerability Assessment and Intelligent Control for Sustainable Power Systems, First Edition.
Edited by José Luis Rueda-Torres and Francisco González-Longatt.

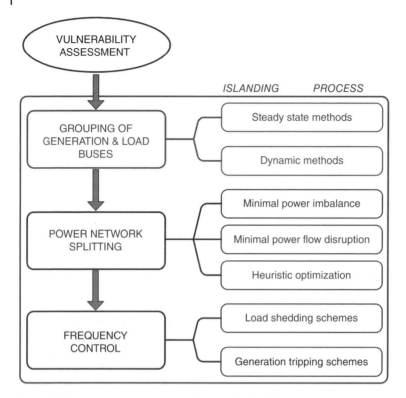

Figure 16.1 Synthesis of proposed methodologies for CIS.

the islanding process start? What are the variables to be monitored in order to define the onset of the islanding process? The analysis must consider possible contingencies and changes in operating conditions, establishing indicators to assess the vulnerability of the EPS [4].

16.1.2 Stage Two: Islanding Process

When vulnerability assessment points out that the power system is on the verge of collapse, the islanding process is started:

- *Grouping of generation and load buses:* This involves determining the groups of generation and load buses that will shape the islands. Proposed approaches to grouping generation buses divide into: (i) Steady state methods, mainly based on power flow equations [5–7] and (ii) Dynamic analysis methods, which consider the dynamic characteristics of generators and loads as well as network topology [8–12].

  Slow coherency has been the most used off-line approach to determine the *Groups of Coherent Generators (GCG)* in an EPS. However, system operation conditions change continuously during real-time: generation-load pattern, network topology, type, length, and place of faults can alter dynamic generator coherency. WAMS measurements enable dynamic behavior of EPS to be observed and might be used to assess generators' coherency during real-time operation. Due to the great amount of data delivered by WAMS, data mining and computational intelligence techniques are natural routes to solving the problem [13].

- *Power network splitting:* In this phase, transmission lines to be opened so as to form the islands are defined. Electric network partitioning is a complex optimization problem due to the unavoidable combinatorial explosion of the search space. Moreover, it is necessary to guarantee both speed and accuracy in determining the final islanding strategy. Proposed methods devised to solve the problem can be classified according to the objective function: (i) minimal power imbalance—this means minimizing the imbalance between generated and consumed active power into each island [5, 8, 10, 14–16]; (ii) minimal power flow disruption—the objective function minimizes the change of power flows at each island after system separation [17–20]; and (iii) heuristic optimization, which allows formulating complex objective functions with several variables to be minimized [21, 22]. Constrained spectral [19, 20] and multilevel recursive [15, 17] partitioning algorithms are the most promising candidates for the task.

- *Frequency control:* It is imperative to ensure integrity and survival of each island after system separation by controlling frequency, voltage, and current variables. The main proposed approaches focus on frequency control in each island after islanding through: (i) *Under-frequency load shedding schemes (UFLS)*, and (ii) *Generation tripping schemes (GTS)*. After islanding, it is virtually impossible to get islands with exact generation-load matching; for this reason UFLS is usually applied to balance active power and control excessive frequency excursions. In the case of islands with a generation surplus, only reference [8] proposes a methodology to control overfrequency by generation tripping; therefore, a GTS able to define the amount and place of generation to be tripped must be defined.

Assessment and control of overloaded transmission facilities after system islanding have not been adequately addressed in approaches proposed to date. Generally, it is assumed that there are sufficient resources to alleviate overloads; furthermore, changes in power injections aimed to control overloads may produce changes in active power balance and consequently in island frequency.

The present work introduces a new methodology to solve the power network splitting problem considering dynamic frequency behavior and overloading limits of transmission facilities. A network splitting mechanism based on the shortest path concept is devised to define the island boundaries; it takes into account active power imbalance constraints inside each island. Power imbalance constraint limits are determined based on maximum and minimum frequency deviations obtained by means of a reduced frequency response model. Load shedding and generation tripping schemes are defined in cases where the active power imbalance constraint is binding. Finally, an approach based on shift factors is used to determine the additional amount of load to be shed in order to alleviate possible overloads. The flowchart of the proposed methodology is shown in Figure 16.2.

In this chapter, it is assumed that vulnerability assessment and real-time generator coherency information is available and has been obtained by approaches [23, 24], respectively. This information is used as input to the proposed network splitting mechanism, introduced in Section 16.2. The methodology to determine limits for the active power imbalance constraint is presented in Section 16.3. After islands and tie-lines have been

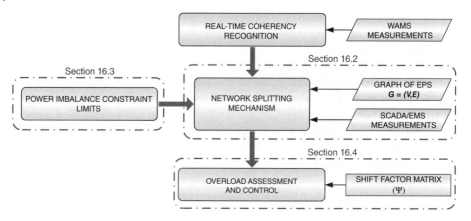

Figure 16.2 Flowchart of the proposed methodology.

defined, the procedure applied to assess and alleviate overloaded transmission links is presented in Section 16.4. A study case is shown in Section 16.5, where the proposed methodology is applied to avoid system collapse due to cascade tripping. Finally, the main conclusions and recommendations for future work are given in Section 16.6.

16.2 Network Splitting Mechanism

The proposed network splitting mechanism belongs to the minimal power imbalance approaches and determines the composition of each island and the transmission lines that should be opened in order to feature the islands. Firstly, a simple weighted undirected graph of the EPS is obtained off-line and it is continuously reduced and refreshed in real-time with measured active power injections and network topology. Later, through a graph partitioning procedure, generation and load buses are assigned to each GCG—information delivered by the real-time coherency recognition stage—in such a way that the active power imbalance constraint is enforced. During this process, the amount of load or generation that must be shed in order to enforce the active power imbalance constraint into each island is calculated. Finally, tie-lines are determined through a search procedure.

Dynamic frequency behavior is indirectly considered through the power imbalance constraint. Allowed limits for frequency variation, imposed by the user or country technical codes, are translated into active power imbalance by means of analytical expressions obtained from the reduced frequency response model of each island [25]. The flowchart of the proposed network splitting mechanism is shown in Figure 16.3.

Most of the calculations associated with the proposed approach are done in a real-time environment and fed with measurements data gathered by WAMS/PMU and SCADA/EMS: voltage and current phasors, frequency, generated and consumed active power, power flow in transmission facilities, and breaker status. Off-line computations are related to the procurement and identification of models to be employed in real-time, mainly a graph representation of the power network and a reduced model of the power-frequency control system of each generator.

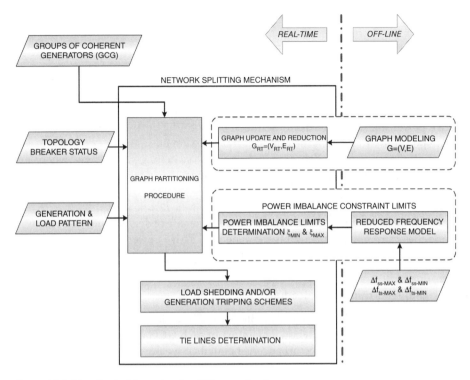

Figure 16.3 Flowchart of the network splitting mechanism.

16.2.1   Graph Modeling, Update, and Reduction

The procedure starts in an off-line environment, modeling the EPS through a simple weighted undirected graph **G**, as (16.1):

$$
\begin{aligned}
\mathbf{G} &= (\mathbf{V_G}, \mathbf{E_G}) \\
\mathbf{V_G} &= \{v, w(v)\} \\
\mathbf{E_G} &= \{u, v, w(e)\}
\end{aligned}
\tag{16.1}
$$

where $\mathbf{V_G}$ is the vertices set with its weights *w(v)* that represents power system buses and net active power injection on each bus, respectively; $\mathbf{E_G}$ is the edges set with its weights *w(e)* that represents power system transmission links and their distance measures as defined in (16.2), respectively. In transmission networks, line resistance is usually small and can be neglected.

$$
w(e) = |Z_{u-v}| = \sqrt{R_{u-v}^2 + X_{u-v}^2} \approx X_{u-v}
\tag{16.2}
$$

Later, in a real-time environment, the graph **G** is continuously refreshed with generation-load patterns and breaker status data. The updated graph $\mathbf{G_{RT}}$ is conveniently reduced following the guidelines in [26]:

- *Equivalence of parallel lines:* This is applied when two or more transmission links are connected between two buses *u* and *v*; the equivalent impedance is calculated by (16.3):

$$\frac{1}{Z_{EQ}} = \frac{1}{Z_1 + Z_2 + \cdots + Z_n} \approx \frac{1}{X_1 + X_2 + \cdots + X_n} \tag{16.3}$$

- *Removal of degree-one nodes:* This is the case of radial connected buses, where the interconnection link between buses u, and v is removed as well as bus v. In order to maintain the active power balance after graph reduction, it is necessary to move the active power injection P_{INJ} from the radial bus v to the upstream bus u adding the active power losses P_{LOSS} in the removed transmission link.
- *Removal of degree-two nodes:* This is the case of switching substations, where net active power injection is zero. The equivalent impedance is calculated by (16.4):

$$Z_{EQ} = Z_1 + Z_2 \approx X_1 + X_2 \tag{16.4}$$

- *Removal of step-up transformers:* This is a particular case of degree-one node removal, where the active power injection is the generated active power. The aforementioned reduction rules are illustrated in Figure 16.4.

16.2.2 Graph Partitioning Procedure

In this stage, the updated and reduced graph $\mathbf{G_{RT}}$ will be divided into a number of islands equal to the number of GCGs. First, generators of each GCG are sorted in descending order according to their pre-disturbance generated active power P_G, and all buses are

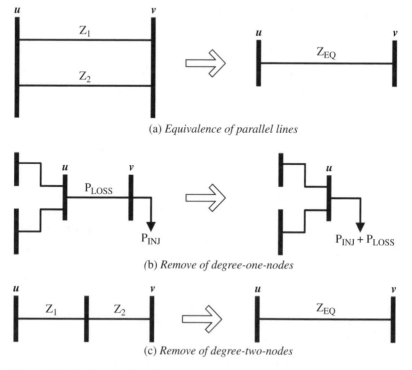

(a) *Equivalence of parallel lines*

(b) *Remove of degree-one-nodes*

(c) *Remove of degree-two-nodes*

Figure 16.4 Graph reduction rules.

labeled as unvisited. Through a recursive allocation procedure, load buses are appended to the closest GCG. Starting with the first GCG, the generator with the largest generated active power P_G is chosen as the pivot generator and labeled as visited. Then the closest load bus to the pivot generator is calculated considering the impedance magnitude as edge weight. Finding the closest load bus implies a minimization problem, given by (16.5), and Dijkstra's shortest path algorithm is used because of its calculation speed [27]:

$$\min_{k=1,\dots,N_L} [D_{jk}(B_{Gj}, B_{Lk})] \tag{16.5}$$

where $\mathbf{D_{jk}}$ denotes a vector with the shortest paths between pivot generation bus $\mathbf{B_{Gj}}$ and load buses $\mathbf{B_{Lk}}$, and N_L is the number of non-visited load buses. When the impedance magnitude (16.2) is used as edge weight, the graph partitioning procedure assigns only the closest electrical buses into each island. The closest load bus to the pivot generator is added to the GCG and labeled as visited.

Each time a new load bus is added to a GCG, it is imperative to make sure that the active power imbalance constraint (16.6) is not binding. This means that the active power imbalance on each island—total generated active power minus total assigned load demand—remains within limits ξ_{max}, ξ_{min}. Cases where the active power imbalance constraint is binding are solved by means of UFLS and GTS, as explained in Section 16.2.3.

$$\xi_{min} < \sum_{i=1}^{NG_{GCG}} P_{Gi} - \sum_{k=1}^{NL_{GCG}} P_{Lk} < \xi_{max} \tag{16.6}$$

Limit values (ξ_{max}, ξ_{min}) determine the frequency behavior on each island: wider limits imply greater frequency variations on each island, and less load to shed when tripping tie-lines to create the islands.

The recursive allocation procedure concludes when all generators have been checked as visited, whereupon a connectivity check is performed in order to find isolated load buses, which will be added to the closest GCG as explained in the next paragraph. The flowchart of the graph partitioning procedure is presented in Figure 16.5.

16.2.3 Load Shedding/Generation Tripping Schemes

When a disturbance produces the tripping of generation or load facilities, the active power balance is broken, so that some loads cannot be served by the online generators and are not visited during the recursive allocation procedure. These loads are attached to the closest GCG according to (16.7), where $\mathbf{D_{kj}}$ denotes a vector with the shortest paths between pivot load bus $\mathbf{B_{Lk}}$ and generation buses $\mathbf{B_{Gj}}$, and N_G is the number of generation buses.

$$\min_{j=1,\dots,N_G} [D_{kj}(B_{Lk}, B_{Gj})] \tag{16.7}$$

The active power excess provided by these buses must be shed. In the case of a generation surplus, those generators that should be tripped can also be identified. The amount of active power generation or load to be tripped is defined by the active power imbalance constraint limits and has no relation with traditional UFLS or GTS.

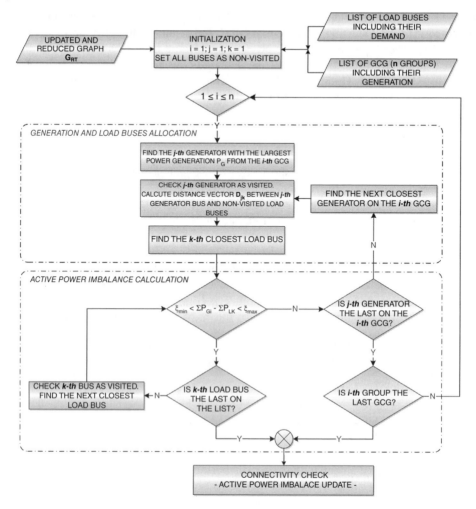

Figure 16.5 Graph partitioning procedure.

16.2.4 Tie-Lines Determination

After all load buses have been attached to a specific GCG, it is possible to determine the tie-lines that need to be opened in order to create the islands. A search procedure is used to look for transmission lines whose endpoints belong to a different GCG, to become tie-lines.

16.3 Power Imbalance Constraint Limits

Maximum (ξ_{max})and minimum(ξ_{min}) values for the active power imbalance constraint on each island are determined by means of analytical expressions derived from the reduced frequency response model, which essentially represents the average response of all generators in the island following a generation/load imbalance [28].

Inter-machine oscillations during a disturbance are ignored by the model; however, results are still valid since only coherent generators are put together into each island.

Several approaches have been proposed to calculate frequency behavior after a disturbance that produces active power imbalances in the EPS; the most popular are:

- Analytical expressions and nomograms [29–31], which are easy to use and fast to calculate, but the results are approximate.
- Time-domain simulation of the complete EPS model [32–34], including full speed regulator-prime mover models, boiler dynamics, turbine conduit dynamics, AGC, and so on. These approaches provide the exact frequency response with computational costs that can be high depending on the size and complexity of EPS under analysis.
- Equivalent models of the power-frequency control system [25, 28, 35]—usually known as frequency response models—whose accuracy is similar to the simulation of the complete EPS model but with reasonable computational costs. Depending on the generator's technology, complex transfer functions might be used to model the turbine-governor set of each generator [36], obtaining the so called "complete frequency response model." In this work, the frequency response model of every generator is reduced by using only a first-order approximation of the speed regulator-prime mover set, which allows using a "reduced frequency response model" of each island and formulating analytical expressions to calculate the maximum transient frequency deviation.

16.3.1 Reduced Frequency Response Model

The dynamic frequency response of an isolated EPS following a generation-load imbalance $\Delta P_O(s)$ can be calculated through a reduced frequency response model with: equivalent system inertia H_{EQ}, generator's permanent droop R_j, and system load damping D. All dynamic models of speed regulators and prime movers are represented with an equivalent first-order model. The dynamic frequency response of the reduced model in the Laplace domain is given by (16.8):

$$\Delta\omega(s) = -\Delta P_o(s) \cdot \frac{g_1(s)}{g_2(s)}$$

$$g_1(s) = \prod_{i=1}^{N}(1 + s\,T_i) \tag{16.8}$$

$$g_2(s) = (2\,s\,H_{EQ} + D)\prod_{i=1}^{N}(1 + s\,T_i) + \sum_{j=1}^{N}\left[\frac{Km_j}{R_j}(1 + s\,F_j\,T_j)\prod_{i=1,i\neq j}^{N}(1 + s\,T_i)\right]$$

where Km_j, F_i, and T_i are parameters of the equivalent first-order model of speed regulator-prime mover of each generator [25]. These parameters are obtained by fitting the response of the equivalent first-order model to the response of the detailed dynamic model using a non-linear least-squares search method [37]. The time response of the reduced frequency response model is given by (16.9):

$$\Delta\omega(t) = \Delta P_o \cdot \sum_{i=1}^{N+1}\frac{A_i}{p_i}(1 - e^{p_i\,t}) \cdot U(t) \tag{16.9}$$

where $\mathbf{A_i}$ and $\mathbf{p_i}$ denote poles and zeros of the transfer function (16.8), respectively. The time instant $\mathbf{t_{min}}$ when the maximum transient frequency deviation occurs is determined by taking the derivative of (16.9) and solving the resulting equation (16.10):

$$\frac{d\Delta\omega(t)}{dt} = -\Delta P_o \cdot \sum_{i=1}^{N+1} A_i \cdot e^{p_i \cdot t_{min}} \cdot U(t) = 0 \tag{16.10}$$

The maximum transient frequency deviation $\mathbf{\Delta f_{MAX}}$ is calculated by replacing $\mathbf{t_{min}}$ into (16.9). This model has been extended in order to include not only thermal but also hydraulic generators. The steady state frequency deviation $\mathbf{\Delta f_{ss}}$ after primary frequency regulation is calculated using (16.11), where $\mathbf{R_{EQ}}$ is the equivalent system droop of each island and is determined considering only those generators participating in primary frequency control.

$$f_{ss} = f_{nom} + \Delta f_{ss}$$
$$\Delta f_{ss} = \frac{R_{EQ}}{1 + R_{EQ} \cdot D} \cdot \Delta P_o \tag{16.11}$$

The application domain of each model is shown in Figure 16.6, where thin continuous lines correspond to frequency on each bus given by the time-domain simulation of the full EPS model. It can be seen that before 4 s, the time evolution of the complete frequency response model and the reduced model are identical; however after this moment they drift apart. The aforementioned models' response fits well with the frequency evolution obtained by time-domain simulation. The maximum transient frequency deviation obtained by solving (16.8), (16.9), and (16.10) is represented with a diamond. Finally, the steady state frequency deviation obtained with (16.11) agrees well with the steady state frequency obtained by time-domain simulation.

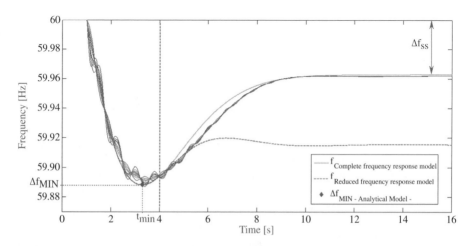

Figure 16.6 Frequency response model application domain.

16.3.2 Power Imbalance Constraint Limits Determination

Using (16.9)–(16.11) for the maximum transient and steady state frequency deviation, the power imbalance constraint limits (ξ_{max}, ξ_{min}) are calculated. As input to the procedure, values of the maximum allowed steady state frequency deviation ($\Delta f_{ss\text{-}MAX}$) and maximum allowed transient frequency deviation ($\Delta f_{ts\text{-}MAX}$) are needed. Usually, these values are defined by country regulations, for example, ENTSO-E standards dictate that, for Continental Europe, the maximum steady state frequency deviation shall be ±200 mHz and the maximum instantaneous frequency deviation shall be ±800 mHz [38]. The procedure to determine active power constraint limits ξ_{max} and ξ_{min} is shown in Figure 16.7.

First, the active power imbalance (ΔP_{MAX}) that produced the maximum allowed steady state frequency deviation ($\Delta f_{ss\text{-}MAX}$) is calculated using (16.11). The obtained ΔP_{MAX} value is then introduced in (16.9) and (16.10) in order to determine the maximum expected transient frequency deviation (Δf_{MAX}) on each island. If Δf_{MAX} is less than $\Delta f_{ts\text{-}MAX}$, this implies that a smaller active power imbalance will produce transient and steady state frequency deviations less than those defined by the user, and consequently ΔP_{MAX} becomes the upper limit ξ_{max} for the active power imbalance constraint. On the other hand, if Δf_{MAX} is greater than $\Delta f_{ts\text{-}MAX}$ then (16.9) is used to calculate a new ΔP_{MAX} value, thereby ensuring that frequency deviations remain under

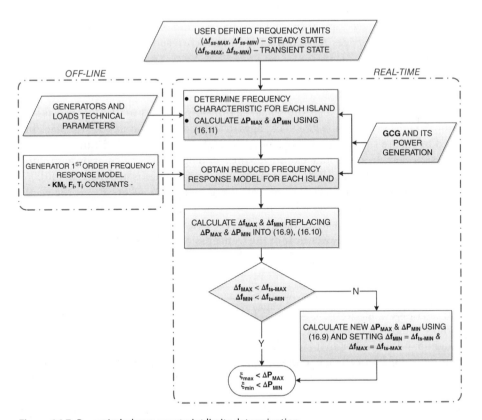

Figure 16.7 Power imbalance constraint limits determination.

user defined limits. In this case, new $\mathbf{\Delta P_{MAX}}$ value becomes the upper limit ξ_{max} for the active power imbalance constraint.

The whole procedure is repeated for determining the lower limit ξ_{min} for the active power imbalance constraint, where minimum allowed steady state frequency deviation ($\mathbf{\Delta f_{ss\text{-}MIN}}$) and minimum allowed transient frequency deviation values ($\mathbf{\Delta f_{ts\text{-}MIN}}$) are used.

16.4 Overload Assessment and Control

So far, generation, load, and topology of each island have been determined. A DC power flow is used to assess overloads on each island and a procedure based on shift factors is employed to determine the amount of load to be shed in order to alleviate the expected overloads. EPS facilities have different overloading levels depending on their constructive characteristics. The proposed scheme aims to control transmission link overloads that must be relieved in short time periods, which is the case for devices based on power electronics such as FACTS, HVDC stations, and power converters associated with solar and wind generation.

The procedure described below is applied to each island and starts with calculation of shift factor matrix $\mathbf{\Psi}$ through (16.12), where $\mathbf{B}$ is the diagonal branch susceptance matrix, $\mathbf{B}^-$ is the reduced nodal susceptance matrix, and $\mathbf{A}^-$ is the reduced incidence matrix. In reduced matrices $\mathbf{A}^-$ and $\mathbf{B}^-$, the column corresponding to the system slack bus is filled with zeros.

$$\mathbf{\Psi} = \mathbf{B} \cdot [\mathbf{A}^- \cdot \mathbf{B}^-, 0] \tag{16.12}$$

The shift factor matrix $\mathbf{\Psi}$, also known as the *transmission sensitivity matrix*, gives the variations in flow for each line due to changes in the nodal injections, with the slack bus assumed to ensure the real power balance. The shift factor matrix is a function of the transmission element susceptances and network topology [39]. A calculation alternative is to estimate matrix $\mathbf{\Psi}$ from PMU measurements in real-time, as proposed in [40].

After this, active power flows $\mathbf{P_{u\text{-}v}}$ are calculated by (16.13), where $\mathbf{P_{inj}}$ is the vector of active power injections on each bus. In the case of overloaded transmission links, the surplus power flow $\mathbf{\Delta P_{u\text{-}v}}$ is determined using (16.14), where $\mathbf{P_{u\text{-}v}}$ and $\mathbf{P_{u\text{-}v}}^{MAX}$ are the calculated and maximum power flow of the transmission link, respectively.

$$\mathbf{P}_{u-v} = \mathbf{\Psi} * \mathbf{P}_{inj} \tag{16.13}$$

$$\mathbf{\Delta P}_{u-v} = \mathbf{P}_{u-v} - \mathbf{P}_{u-v}^{MAX} \tag{16.14}$$

A set of Nb load buses participating in load shedding must be defined according to their entry values in $\mathbf{\Psi}$; those buses with higher shift factors are candidate buses for load shedding. Remember that element ψ^i_{u-v} of matrix $\mathbf{\Psi}$ represents the sensitivity of the flow on line $u\text{-}v$ to a change in the power injection at bus i; therefore, shedding load in bus i with the highest element ψ^i_{u-v} will produce the biggest change in flow of line $u\text{-}v$.

The surplus power flow $\mathbf{\Delta P}_{u-v}$ is distributed among the candidate buses Nb through factor FD_i calculated by (16.15), that weights individual bus demand with respect to total demand that will be shed. Finally, the real amount of load ($PL_i{}^{Lshd}$) to be shed on bus i is obtained by (16.16):

$$FD_i = \frac{PL_i}{\displaystyle\sum_{i=1}^{Nb} PL_i} \tag{16.15}$$

$$PL_i^{Lshd} = \frac{FD_i \cdot \Delta P_{u-v}}{\psi_{u-v}^i} \tag{16.16}$$

Reactive power to be shed ($QL_i{}^{Lshd}$) on bus i is determined considering that the power factor remains constant, so that (16.17) holds:

$$QL_i^{Lshd} = QL_i \cdot \frac{\Delta P_{u-v}}{\displaystyle\sum_{i=1}^{Nb} PL_i \cdot \psi_{u-v}^i} \tag{16.17}$$

The use of DC power flow involves knowing in advance the shift factor matrix $\mathbf{\Psi}$. Calculating $\mathbf{\Psi}$ implies computing matrices $\mathbf{B}$, $\mathbf{B}^-$, and $\mathbf{A}^-$ as shown in (16.12); the computational cost of $\mathbf{\Psi}$ can be important when considering a large EPS. One way to speed up the calculation of $\mathbf{\Psi}$ associated with each island is to determine periodically (on line) the matrix $\mathbf{\Psi}$ of the entire EPS before separation into islands. Then, in real-time, the matrix $\mathbf{\Psi}$ associated with each island is calculated through elementary matrix operations on the entire system matrix $\mathbf{\Psi}$, as proposed in [41], accelerating the DC power flow calculation process. Overload assessment and control based on DC power flow delivers good results; however, when high reactive power flows in the network, results may be conservative.

16.5 Test Results

The IEEE 10-machine 39-bus test system (New England system) is used to simulate a cascade tripping of transmission lines with the subsequent loss of stability and system collapse. Through the application of the proposed methodology, the system collapse is averted, keeping the main electrical variables within normal operating limits. Generator 10 is modeled as a hydraulic unit in order to show the performance of the reduced frequency response model to predict transient frequency deviations. A traditional UFLS scheme, whose first step trips at 59.4 Hz, is also modeled. The test system is shown in Figure 16.8.

16.5.1 Power System Collapse

The power system is feeding a total load of P = 5478.3 MW, Q = 1178.6 Mvar, and under these circumstances the trip of line 14-15 at 0.2 s, due to a three-phase fault, causes the overload and subsequent tripping of line 03-04 at 10 s. The cascade tripping produces two electrical areas tied by a radial link. Vulnerability assessment, implemented as proposed in [23], determines that the areas lose synchronism at 12.18 s with subsequent system collapse. The difference between the *Center of Inertia (COI)* angle of each area

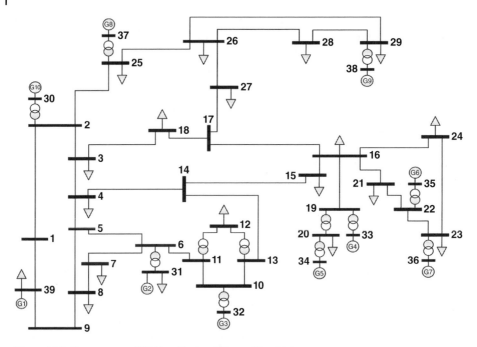

Figure 16.8 Test system—IEEE New England 10-machine 39-bus system.

and the COI angle of the entire system is used as stability index: when the index value exceeds 180 degrees, it is said that the area is unstable.

Time evolution of generator rotor angles is shown in Figure 16.9 with thin continuous lines, while the thick continuous line represents system COI angle evolution (δ_{COI}). Dotted and dashed lines depict the COI-referred rotor angles of areas 1 ($\delta_{Area1\text{-}COI}$) and

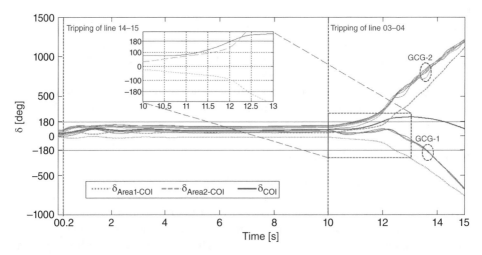

Figure 16.9 Collapse case—COI-referred and generator rotor angles [deg].

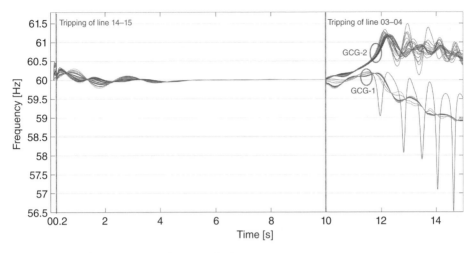

Figure 16.10 Collapse case—Bus frequency [Hz].

2 ($\delta_{\text{Area2-COI}}$), respectively. It can be seen that the COI-referred rotor-angle corresponding to area 2 reaches 180 degrees difference first. It can be observed that after tripping of line 03-04 at 10 s, the system separates into two electrical areas. The time evolution of frequency in all buses is depicted in Figure 16.10, where a similar separation into two areas is also observed.

16.5.2 Application of Proposed Methodology

After the trip of line 03-04, the real-time coherency recognition procedure identifies GCG-01 = {G1,G2,G3} and GCG-02 = {G4,G5,G6,G7,G8, G9}. This grouping is consistent with the network topology, since after tripping of line 03-04, generators {G1,G2,G3} are connected with the remaining system through a weak radial interconnection (line 01-02 and line 01-39).

Input values for the tolerance limits determination procedure are: $\Delta\mathbf{f}_{\text{ss-MAX}} = 60.4\,\text{Hz}$, $\Delta\mathbf{f}_{\text{ss-MIN}} = 59.6\,\text{Hz}$, $\Delta\mathbf{f}_{\text{ts-MAX}} = 61.0\,\text{Hz}$ and $\Delta\mathbf{f}_{\text{ts-MIN}} = 59.4\,\text{Hz}$. It is noteworthy that maximum and minimum transient frequency limits corresponds to over-speed protection of generators and UFLS settings, respectively. Applying the procedure described in Section 16.3 to Island 1, it is found that the power imbalance $\Delta\mathbf{P}_{\text{MIN}}$ obtained when considering $\Delta\mathbf{f}_{\text{ss-MIN}} = 59.6\,\text{Hz}$ produces an expected transient frequency deviation greater than the minimum allowed ($\Delta f_{\text{MIN}} > \Delta\mathbf{f}_{\text{ts-MIN}}$). This means that after EPS splitting, frequency in Island 1 will drop beyond 59.4 Hz, triggering the first step of the traditional UFLS. Therefore, a new value for $\Delta\mathbf{P}_{\text{MIN}}$ is calculated by setting $\Delta f_{\text{MIN}} = \Delta\mathbf{f}_{\text{ts-MIN}} = 59.4\,\text{Hz}$, and the result becomes the lower limit for the power imbalance constraint in Island 1. In Island 2, maximum and minimum expected transient frequency deviations are within the specified limits.

The frequency evolution on each island is presented in Figure 16.11; time-domain simulation results of the complete frequency response model are represented with solid lines, and maximum and minimum transient frequency values obtained by solving (16.8)–(16.10) are represented by diamonds. The agreement between the two models is remarkable: frequency error is less than 0.1% and time error is less than 1.5 %.

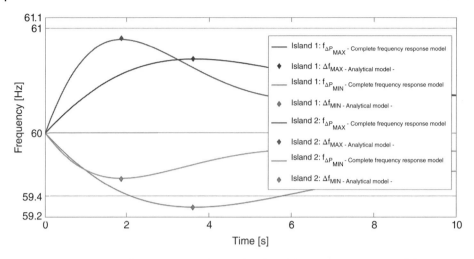

Figure 16.11 Frequency deviation—Complete and analytical frequency response models [Hz].

The proposed procedure to determine power imbalance constraint limits allows the coordination between traditional UFLS and CIS. User can define minimum transient frequency deviation after islanding in such a way that traditional UFLS does not act or triggers only up to a certain step.

After finishing the recursive allocation procedure (Figure 16.5), power imbalance into each island lies between calculated limits; however, the connectivity check determines that load bus 4 is isolated and cannot be added to Island 2 because there is no possible electrical connection between them. Therefore, load bus 4 is added to Island 1 with a consequent change in the power imbalance. The load shedding/generation trip procedure determines that 785.2 MW of load must be tripped in Island 1 so that frequency stays within the user specified limits. Island 2 is generation rich and 490 MW of generated power in bus 37 must be tripped to avoid triggering over-speed protections. Finally, the tie-line determination procedure establishes that line 01-02 will be tripped. The aforementioned loads, generator, and tie-line are tripped together at 11.5 s, before the system loses synchronism.

Once the islands have been determined, the overload assessment procedure shows that line 05-06 is overloaded. Following the procedure stated in Section 15.4, an additional 83 MW of load are shed in buses 4 and 8 in order to alleviate the overload. A side effect of this additional load shedding is that minimum transient and final steady state frequencies increase.

Results obtained with the proposed methodology are summarized in Table 16.1. In the first part, maximum and minimum values for transient as well as steady state frequency deviation are shown. In the second part, the active power imbalance evolution through each stage of the network splitting mechanism is shown. Stages of active power imbalance updating are: Graph Partitioning Procedure (GPP), Connectivity Check (CC), and Load Shedding/Generation Trip Schemes (LS/GT) for frequency and overload control. Finally, in the third part, control actions are detailed, including additional load shedding at buses 4 and 8 for overload control.

Table 16.1 Results of the proposed methodology.

ISLAND 1			ISLAND 2		
ANALYTICAL FREQUENCY RESPONSE MODEL					
$\Delta P_{MAX} =$	430.2	[MW]	$\Delta P_{MAX} =$	736.6	[MW]
$\Delta f_{ss\text{-}MAX} =$	0.40	[Hz]	$\Delta f_{ss\text{-}MAX} =$	0.40	[Hz]
$\Delta f_{ts\text{-}MAX} =$	0.71	[Hz]	$\Delta f_{ts\text{-}MAX} =$	0.90	[Hz]
[a] $\Delta P_{MIN} =$	−228.1	[MW]	$\Delta P_{MIN} =$	−355.1	[MW]
$\Delta f_{ss\text{-}MIN} =$	−0.21	[Hz]	$\Delta f_{ss\text{-}MIN} =$	−0.40	[Hz]
$\Delta f_{ts\text{-}MIN} =$	−0.38	[Hz]	$\Delta f_{ts\text{-}MIN} =$	−0.44	[Hz]
ACTIVE POWER IMBALANCE EVOLUTION					
GPP	−127.2	[MW]	GPP	277.2	[MW]
CC	−1013.3	[MW]	CC	1163.3	[MW]
LS/GT	−228.1	[MW]	LS/GT	673.3	[MW]

CONTROL MEASURES				
Load Shedding Scheme			Generation Tripping Scheme	
Bus	Load		Generator	Generated Power
4	301.4 + 48	[MW]	37	490	[MW]
7	32	[MW]		
8	164.3 + 35	[MW]		
12	3.5	[MW]		
31	4.6	[MW]		
39	279.5	[MW]		
Total	868.3	[MW]	Total	490	[MW]

a) Minimum transient frequency is set to 59,4 Hz, according to specified limits
and ΔP_{MIN} is calculated thereafter.

The dynamic performance of the main system variables after controlled system islanding is included in Table 16.2. These values were obtained by time-domain simulation of
the full power system model. Maximum and minimum transient frequency deviation
after islanding are shown in the first part of the table, while the steady state magnitudes
(at 40 s of simulation) of frequency and voltages are presented in the second part.

Table 16.2 Dynamic performance of main system variables.

	ISLAND 1			ISLAND 2	
f_{MIN} [Hz]	59.53	[Hz]	f_{MIN} [Hz]	60.1	[Hz]
f_{MAX} [Hz]	60.21	[Hz]	f_{MAX} [Hz]	60.79	[Hz]
f_{40s} [Hz]	59.72	[Hz]	f_{40s} [Hz]	60.26	[Hz]
$V_{MIN, 40s}$ [pu]	0.991	[pu]	$V_{MIN, 40s}$ [pu]	1.000	[pu]
$V_{MAX, 40s}$ [pu]	1.055	[pu]	$V_{MAX, 40s}$ [pu]	1.048	[pu]

The frequency in Island 1 after system separation is above 59.4 Hz, thus avoiding activation of the traditional UFLS scheme. This is an important advantage considering that an unexpected disturbance could lead the recently formed island to the collapse; a traditional under-frequency load shedding scheme would act as backup protection system. Steady state frequencies and voltages are within normal operation limits except on bus 1 ($V_{MAX, 40s} = 1.048$ pu), which becomes a radial bus without load (Ferranti effect). Similarly, maximum and steady state frequency values in Island 2 are within normal operating limits, and the same observation applies to voltages. Time evolution of voltage magnitude and frequency on each bus, and lines loadability obtained by time-domain simulation, are shown in Figure 16.12, Figures 16.13 and 16.14, respectively.

16.5.3 Performance of Proposed ACIS

All simulations were performed using MATLAB® 2014a software running on Windows 7® (64-bit) platform, processor Intel i7® Q740@1.73 Ghz with 4 GB RAM memory. Total computation time is 0.12 s (7.2 cycles on 60 Hz). Power imbalance constraint limits and network splitting procedures are the most demanding, taking approximately 98% of total computation time. Overload assessment and control is fast and easy to implement,

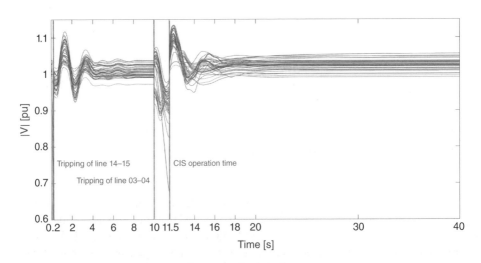

Figure 16.12 Proposed ACIS—Bus voltage magnitude [pu].

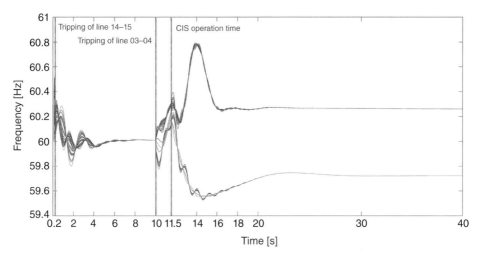

Figure 16.13 Proposed ACIS—Bus frequency [Hz].

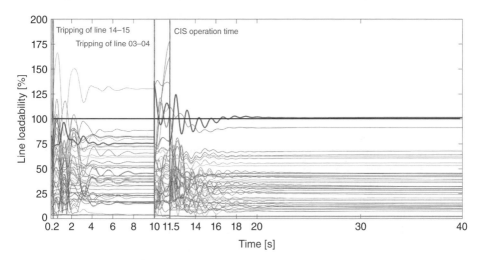

Figure 16.14 Proposed ACIS—Line loadability [%].

using only 2.4% of total computation time. A summary of computation times per phase of the methodology are shown in Table 16.3.

Reference [42] presents time delays associated with switching, protection, and control equipment installed in a real EPS to perform wide area control actions. Taking those values into consideration, total time needed to implement control actions dictated by the proposed CIS can be calculated as in (16.18):

$$
\begin{aligned}
t_{ACIS-total} &= t_{breakrers} + t_{adq} + t_{telecom} + t_{relay} + t_{ACIS-design} \\
&= 0.124 + 0.037 + 0.012 + 0.005 + 0.121 \\
&= 0.299 \quad [s] \\
&\cong 18 \quad [cycles \ 60 \, Hz]
\end{aligned}
\tag{16.18}
$$

Table 16.3 Computation times summary.

Phase	Computation time		
	[s]	[%]	[60Hz - cycles]
Power imbalance constraint limits determination	0.0417	34.4	2.5
Island 1	0.0205	16.8	1.2
Island 2	0.0190	15.8	1.1
Network splitting mechanism	0.0756	63.2	4.5
Overload assessment and control	0.0028	2.4	0.2
Total time	0.1205	100	7.2

Reference [3] studies tripping times of out of step relays when applied to an IEEE New England 10-machine 39-bus system, and through time-domain simulations and Monte Carlo methods, it determines that: minimum tripping time is 0.8342 s, mean tripping time is 1.2252 s with standard deviation of 0 3746 s. Total time needed to implement proposed CIS is less than minimum tripping time of out of step relays, and therefore the proposed approach is capable of avoiding system collapse by acting before uncontrolled system separation.

16.6 Conclusions and Recommendations

Through the development and integration of novel algorithms and procedures for graph partitioning and frequency behavior estimation, a new approach for power system islanding has been presented. The proposed scheme is able to avoid a system collapse by splitting the system into electrical islands with adequate generation-load balance. The proposed methodology is a contribution to CIS development, improving its ability to determine the necessary control actions to divide the power system and to ensure islands' survival.

The scheme changes its settings automatically according to received real-time data, thus adapting its behavior to system operational conditions. The proposed CIS incorporates a dynamic constraint into the network splitting procedure and allows controlling the frequency on each island based on user specified limits. To the author's knowledge, this is the first controlled islanding scheme that incorporates a dynamic mechanism to define the tolerance of the power imbalance constraint.

Results using the proposed approach show that the frequency on each island can be effectively controlled within user defined limits. This feature allows coordinating actions between traditional UFLS and proposed CIS. An important issue when designing real-time algorithms is to reduce calculation times as much as possible. It was shown that power imbalance constraint limits and network partitioning procedures consume most of the calculation time. Due to its characteristics, the graph partitioning procedure could be implemented using parallel computing, where each generator is assigned to one core. The main implementation issues are related to hardware deployment through the power network in order to open the circuit breakers of generators, loads, and transmission lines. The proposed load shedding/generation trip schemes consider load as a continuous variable, but in a real EPS it is a discrete variable.

Current developments in Wide Area Monitoring, Protection and Control (WAMPAC) systems, and high performance computing open a way for future implementation of the proposed CIS. Further work includes an enhanced procedure for overload assessment and control using PMU measurements to estimate shift factors, and taking into account the reactive components of power flows. These advancements would improve the performance and adaptability of the procedure.

References

1 V. Madani, D. Novosel, S. Horowitz, M. Adamiak, J. Amantegui, D. Karlsson, et al., "IEEE PSRC Report on Global Industry Experiences With System Integrity Protection Schemes (SIPS)," *Power Delivery, IEEE Transactions on*, vol. 25, pp. 2143–2155, October 2010.

2 J. Q. Tortos and V. Terzija, "Controlled islanding strategy considering power system restoration constraints," in *Power and Energy Society General Meeting, 2012 IEEE*, 2012, pp. 1–8.

3 J. C. Cepeda, "Evaluación de la Vulnerabilidad del Sistema Eléctrico de Potencia en Tiempo Real usando Tecnología de Medición Sincrofasorial," PhD, Instituto de Energía Eléctrica - IEE, UNSJ, San Juan, Argentina, 2014.

4 H. Song and M. Kezunovic, "A new analysis method for early detection and prevention of cascading events," *Electric Power Systems Research*, vol. 77, pp. 1132–1142, 2007.

5 C. G. Wang, B. H. Zhang, J. Shu, P. Li, and Z. Q. Bo, "A fast method for power system splitting boundary searching," in *Power & Energy Society General Meeting, 2009. PES '09. IEEE*, 2009, pp. 1–8.

6 M. A. Anuar, U. Dorji, and T. Hiyama, "Principal Areas for Islanding Operation Based on Distribution Factor Matrix," in *Intelligent System Applications to Power Systems, 2009. ISAP '09. 15th International Conference on*, 2009, pp. 1–6.

7 L. Yuanqi and L. Yutian, "Islanding cutset searching approach based on dispatching area," in *Power Engineering Conference, 2007. IPEC 2007. International*, 2007, pp. 950–954.

8 J. Ming, T. S. Sidhu, and S. Kai, "A New System Splitting Scheme Based on the Unified Stability Control Framework," *Power Systems, IEEE Transactions on*, vol. 22, pp. 433–441, 2007.

9 X. Wang, W. Shao, and V. Vittal, "Adaptive corrective control strategies for preventing power system blackouts," in *15th Power Systems Computation Conference*, Liège, Belgium, 2005.

10 H. You, V. Vittal, and W. Xiaoming, "Slow coherency-based islanding," *Power Systems, IEEE Transactions on*, vol. 19, pp. 483–491, 2004.

11 X. Guangyue, V. Vittal, A. Meklin, and J. E. Thalman, "Controlled Islanding Demonstrations on the WECC System," *Power Systems, IEEE Transactions on*, vol. 26, pp. 334–343, 2011.

12 S. M. Tabandeh and M. R. Aghamohammadi, "A new algorithm for detecting real-time matching for controlled islanding based on correlation characteristics of generator rotor angles," in *Universities Power Engineering Conference (UPEC), 2012 47th International*, 2012, pp. 1–6.

13 C. Juarez, A. R. Messina, R. Castellanos, and G. Espinosa-Perez, "Characterization of Multimachine System Behavior Using a Hierarchical Trajectory Cluster Analysis," *Power Systems, IEEE Transactions on*, vol. 26, pp. 972–981, 2011.

14 S. Kai, Z. Da-Zhong, and L. Qiang, "Splitting strategies for islanding operation of large-scale power systems using OBDD-based methods," *Power Systems, IEEE Transactions on*, vol. 18, pp. 912–923, 2003.

15 C. G. Wang, B. H. Zhang, Z. G. Hao, J. Shu, P. Li, and Z. Q. Bo, "A Novel Real-Time Searching Method for Power System Splitting Boundary," *Power Systems, IEEE Transactions on*, vol. 25, pp. 1902–1909, 2010.

16 P. A. Trodden, W. A. Bukhsh, A. Grothey, and K. I. M. McKinnon, "Optimization-Based Islanding of Power Networks Using Piecewise Linear AC Power Flow," *Power Systems, IEEE Transactions on*, vol. 29, pp. 1212–1220, 2014.

17 L. Juan, L. Chen-Ching, and K. P. Schneider, "Controlled Partitioning of a Power Network Considering Real and Reactive Power Balance," *Smart Grid, IEEE Transactions on*, vol. 1, pp. 261–269, 2010.

18 W. Xiaoming and V. Vittal, "System islanding using minimal cutsets with minimum net flow," in *Power Systems Conference and Exposition, 2004. IEEE PES*, vol. 1, pp. 379–384, 2004.

19 L. Ding, P. Wall, and V. Terzija, "Constrained spectral clustering based controlled islanding," *International Journal of Electrical Power & Energy Systems*, vol. 63, pp. 687–694, 2014.

20 J. Quiros-Tortos, R. Sanchez-Garcia, J. Brodzki, J. Bialek, and V. Terzija. (2014, Constrained spectral clustering-based methodology for intentional controlled islanding of large-scale power systems. *IET Generation, Transmission & Distribution*. Available: http://digital-library.theiet.org/content/journals/10.1049/iet-gtd.2014.0228

21 C. S. Chang, L. R. Lu, and F. S. Wen, "Power system network partitioning using tabu search," *Electric Power Systems Research*, vol. 49, pp. 55–61, 1999.

22 L. Liu, W. Liu, D. A. Cartes, and I.-Y. Chung, "Slow coherency and Angle Modulated Particle Swarm Optimization based islanding of large-scale power systems," *Advanced Engineering Informatics*, vol. 23, pp. 45–56, 2009.

23 J. C. Cepeda, J. L. Rueda, D. G. Colome, and D. E. Echeverria, "Real-time transient stability assessment based on centre-of-inertia estimation from phasor measurement unit records," *Generation, Transmission & Distribution, IET*, vol. 8, pp. 1363–1376, 2014.

24 N. Granda, Colome D., "Real-Time identification of coherent generators using PMU measurements," *Energia Technical Magazine*, vol. 8, pp. 76–84, 2012.

25 A. Denis Lee Hau, "A general-order system frequency response model incorporating load shedding: analytic modeling and applications," *Power Systems, IEEE Transactions on*, vol. 21, pp. 709–717, 2006.

26 X. Guangyue and V. Vittal, "Slow Coherency Based Cutset Determination Algorithm for Large Power Systems," *Power Systems, IEEE Transactions on*, vol. 25, pp. 877–884, 2010.

27 F. B. Zhan and N. C. E. , "Shortest Path Algorithms: An Evaluation Using Real Road Networks," *Transportation Science*, vol. 32, pp. 65–73, Feb. 1998.

28 P. M. Anderson and M. Mirheydar, "A low-order system frequency response model," *Power Systems, IEEE Transactions on*, vol. 5, pp. 720–729, 1990.

29 L. L. Fountain and J. L. Blackburn, "Application and Test of Frequency Relays for Load Shedding," *Power Apparatus and Systems, Part III. Transactions of the American Institute of Electrical Engineers*, vol. 73, pp. 1660–1668, 1954.

30 C. F. Dalziel and E. W. Steinback, "Underfrequency Protection of Power Systems for System Relief Load Shedding-System Splitting," *Power Apparatus and Systems, Part III. Transactions of the American Institute of Electrical Engineers*, vol. 78, pp. 1227–1237, 1959.

31 H. E. Lokay and V. Burtnyk, "Application of Underfrequency Relays for Automatic Load Shedding," *Power Apparatus and Systems, IEEE Transactions on*, vol. PAS-87, pp. 776–783, 1968.

32 C. Evrard and A. Bihain, "Powerful tools for various types of dynamic studies of power systems," in *Proceedings of 1998 International Conference on Power System Technology, POWERCON* 1998, pp. 1–6 vol. 1.

33 DIgSILENT GmbH, "DIgSILENT PowerFactory User Manual," ed. Gomaringen, Germany,: DIgSILENT GmbH, 2013.

34 Siemens Power Transmission & Distribution Inc, "PSS/E 30.2 Users Manual," ed. Schenectady, US: Siemens Power Transmission & Distribution Inc, 2005.

35 I. Egido, F. Fernandez-Bernal, P. Centeno, and L. Rouco, "Maximum Frequency Deviation Calculation in Small Isolated Power Systems," *Power Systems, IEEE Transactions on*, vol. 24, pp. 1731–1738, 2009.

36 IEEE-PES Task Force on Turbine-Governor Modeling, "Dynamic Models for Turbine-Governors in Power System Studies," IEEE, USA, PES-TR1, 2013.

37 L. Ljung. (2015). *System Identification Toolbox: User's Guide (Online version ed.)*. Available: http://www.mathworks.com/help/pdf_doc/ident/ident.pdf

38 ENTSO-E, "ENTSO-E Network Code on Load-Frequency Control and Reserves (LFCR)," 28 June 2013. [Online]: https://www.entsoe.eu/major-projects/network-code-development/load-frequency-control-reserves/Pages/default.aspx. Accessed on 31-08-2017.

39 P. A. Ruiz, A. Rudkevich, M. C. Caramanis, E. Goldis, E. Ntakou, and C. R. Philbrick, "Reduced MIP formulation for transmission topology control," in *Communication, Control, and Computing (Allerton), 2012 50th Annual Allerton Conference on*, 2012, pp. 1073–1079.

40 Y. C. Chen, A. D. Dominguez-Garcia, and P. W. Sauer, "Measurement-Based Estimation of Linear Sensitivity Distribution Factors and Applications," *Power Systems, IEEE Transactions on*, vol. 29, pp. 1372–1382, 2014.

41 Y. Fu and Z. Li, "A Direct Calculation of Shift Factors Under Network Islanding," *Power Systems, IEEE Transactions on*, vol. 30, pp. 1150–1551, 2014.

42 M. V. Flores, D. E. Echeverria, R. P. Barba, and G. Arguello, "Architecture of a systemic protection system for the interconnected national system of Ecuador " in *2015 IEEE Asia-Pacific Conference on Computer Aided System Engineering (APCASE)* Quito, Ecuador, 2015.

17

High-Speed Transmission Line Protection Based on Empirical Orthogonal Functions

Rommel P. Aguilar[1] and Fabián E. Pérez-Yauli[2]

[1] *Institute of Electrical Energy, Universidad Nacional de San Juan, Argentina*
[2] *Department of Electrical Energy, Escuela Politécnica Nacional, Quito, Ecuador*

17.1 Introduction

In order to provide an economical and efficient power supply, bulk power exchange between remote locations is essential. Thus, power transmission at extra-high voltage (EHV) levels is a common practice nowadays. However, as with any part of the power system, the elements that constitute the EHV transmission system are exposed to a number of factors that can cause critical faults. In fact, abnormalities introduced during fault events in EHV networks can not only produce catastrophic damage in the equipment but also lead the entire power system into unstable operation. Additionally, in spite of continuously growing energy demand, factors like de-regulated market, economics, rights-of-way, and environmental restrictions have limited the building of new transmission lines. As a result, existing transmission lines are forced to operate at high loadings, close to their operating limits. Consequently, it is imperative to develop protection devices as well as protection schemes with high-speed performance, which decrease the effect of faults on power system operation and also become an economic way to increase power transfer in a network without investment in new transmission lines.

Over the last three decades, a lot of research has been conducted in the development of ultra-high-speed protection algorithms. Particularly in the late 1970s and '80s, a number of researchers presented significant advances in this area. However, the available technology at that time was not adequate for the high computation requirements of such protection schemes, and therefore they could not take advantage of sophisticated techniques for transient analysis.

Despite the important technological advances in microprocessor based relays, the basic principle of transmission line protection has not changed for more than half a century [1]. This principle is based on extracting the power frequency components of current and voltage signals by means of so-called phasor estimation techniques [2]. These techniques use analog and digital filtering to attenuate the non-power frequency signals caused by fault generated transients and extract only information concerning the power frequency component. Thus, the success of phasor estimation techniques depends on how well the non-power frequency components are filtered out. However, in certain

Dynamic Vulnerability Assessment and Intelligent Control for Sustainable Power Systems, First Edition.
Edited by José Luis Rueda-Torres and Francisco González-Longatt.
© 2018 John Wiley & Sons Ltd. Published 2018 by John Wiley & Sons Ltd.

applications, particularly in EHV and ultra-high voltage (UHV) transmission lines, fault signals are prone to be highly contaminated with non-power frequency transients. Thus, the accuracy of phasor estimation is reduced [1, 3]. In fact, these algorithms suffer not only from accuracy problems but also from unwanted responses, noise, disturbance rejections, and reliability issues [3]. Additionally, the filtering process involves an inherent delay that limits the operation time to at least the length of the time window of the filter (normally one cycle of the fundamental frequency for phasor estimation techniques).

A solution to improving the operation time of transmission line protection is reducing the relay data window. However, for data windows shorter than one cycle, information on the steady-state power frequency component decreases whereas the transient content becomes more important. These shortcomings suggest that, for developing faster protection schemes, the analysis must be done over the entire frequency spectrum contained in electrical signals [4]. Therefore, identification of complex waveforms is required instead of the evaluation of periodic characteristics [1]. In this field, the term Transient Based Protection appears, which is based on analysis of the entire frequency spectrum contained in transient current and voltage signals appearing after a fault. This principle proposes that high frequency components contained in transient signals provide valuable information for achieving fast and accurate protection [4, 5].

The development of Electromagnetic Transients Programs (EMTP) and real-time digital simulators [6] established the fundamental basis for modeling and analysis of power system transients. In this field, more accurate models have been introduced over the years to represent the complexities of power systems. In fact, nowadays one can affirm with certainty that simulations using EMTP provide a faithful replica of real power systems. Therefore, these simulations allow gathering a (theoretically) infinite number of observations from a modeled power system. In this context, data mining and signal processing tools play an important role, since they allow relevant information to be obtained from simulation-generated data in order to achieve a meaningful understanding of the real phenomena.

It is well known that faults appear in a stochastic manner. Hence, for a power system exposed to a number of external factors, it is almost impossible to predict when, where, and how a fault will arise. However, despite the innumerable variables that influence the features of a fault, all the faults that can take place in a transmission line have the same mean where they appear and propagate, and then they share a common characteristic. Therefore, the patterns that govern fault generated transients help us to develop protection algorithms, based on the knowledge acquired from data mining and signal processing of a representative sample of fault generated transients.

In this context, the data mining technique known as Empirical Orthogonal Functions (a.k.a. Principal Components Analysis) emerges as a powerful tool for power system applications [7–11]. This technique is used to find a small set of orthogonal functions that gives a good interpretation concerning fault generated transients. Results show that fault signals decomposed in terms of these orthogonal basis functions exhibit well-defined patterns, which can be used for recognizing the main features of fault events such as inception angle, fault type, and fault location. Signals of less than a quarter of cycle can be used, hence high-speed protection algorithms can be designed utilizing these orthogonal basis functions.

The organization of this chapter is as follows: Section 17.2 introduces the theoretical background of Empirical Orthogonal Decomposition. Next, applications of Empirical

Orthogonal Functions for transmission line protection are provided in Section 17.3. A case study and evaluation of the protection scheme is presented in 17.4. Finally, conclusions are drawn in Section 17.5.

17.2 Empirical Orthogonal Functions

The term Empirical Orthogonal Functions (EOFs) was first mentioned by Edward Lorenz in 1956 as mechanism to find a simplified but optimal representation of climate data with time and space dependence [12]. Subsequently, this technique has been widely employed in oceanography and meteorology for data compression and pattern extraction [13, 14]. Contrary to other feature extraction techniques that employ exhaustive supervised training, EOFs are easy to implement and do not require adjustment of control parameters. They depend solely on the nature of the input data.

In scientific literature, the EOF technique is also referred to as Principal Components Analysis (PCA) [15–19]. Actually both techniques, EOFs and PCA, are the same. However, for signal processing, the term EOF is more appropriate than PCA, because the term EOF and the theory behind it were founded on the analysis of time series structures. EOFs allows decomposition of a signal in terms of orthogonal basis functions that give an optimal representation of the data. Therefore, it leads to a better understanding of the signal structure.

17.2.1 Formulation

Let the vector $x_i(t) = (x_{i,1}, x_{i,2}, \ldots, x_{i,p})'$ be a time series representing a p-length data window captured from the ith discrete signal of a set of n measured signals. Then this set of signals can be represented by the $n \times p$ data matrix X:

$$X = \{x_{i,j}\}_{i,j=1}^{n,p} = \begin{bmatrix} x_1(t)' \\ x_2(t)' \\ \vdots \\ x_n(t)' \end{bmatrix} = \begin{bmatrix} x_{1,1} & x_{1,2} & \cdots & x_{1,p} \\ x_{2,1} & x_{2,2} & \cdots & x_{2,p} \\ \vdots & \vdots & \ddots & \vdots \\ x_{n,1} & x_{n,2} & \cdots & x_{n,p} \end{bmatrix} \tag{17.1}$$

where n is the number of observed signals and p is the number of samples per signal.

For signal analysis it will be further assumed that each discrete signal (time series) is an observation, and each sample is a feature. The element $x_{i,j}$ is a generic element of X, which represents the value of the feature j in the observation i.

EOFs are based on the assumption that any observation $x_i(t)$ can be expanded in terms of orthogonal basis functions so that:

$$x_i(t) = q_{i1} f_1(t) + q_{i2} f_2(t) + \ldots + q_{ip} f_p(t) \tag{17.2}$$

where $f_1(t), f_2(t), \ldots, f_p(t)$ are the orthogonal basis functions, which are called the EOFs, and q_{ij} is the decomposition coefficient of $x_i(t)$ on $f_j(t)$. The number of EOFs is equal to the number of features (p), thus (17.2) expands $x_i(t)$ without loss of information [20]. However, p could be a large number; then for dimensionality reduction, it is desirable to select a small number of EOFs. Therefore, the series is truncated as follows:

$$x_i(t) = q_{i1} f_1(t) + q_{i2} f_2(t) + \ldots + q_{ir} f_r(t) + \varepsilon_i(t) \tag{17.3}$$

Here, $\boldsymbol{\varepsilon}_i(t) = (\varepsilon_{i1}, \varepsilon_{i2}, \ldots, \varepsilon_{ip})'$ is the error associated with the truncation, and r is the number of selected EOFs, $r < p$. In matrix form, (17.3) gives:

$$X = QF' + E \tag{17.4}$$

where $\boldsymbol{Q}_{n\times r}$ contains the coefficients q_{ij} for $i=1,2,\ldots n$, and $j=1,2,\ldots r$; $\boldsymbol{F}_{p\times r}$ is the matrix containing in its columns the r-EOFs $[\boldsymbol{f}_1(t), \boldsymbol{f}_2(t), \ldots, \boldsymbol{f}_r(t)]$; and $\boldsymbol{E}_{n\times p}$ contains in its rows the decomposition errors $[\boldsymbol{\varepsilon}_1(t), \boldsymbol{\varepsilon}_2(t), \ldots, \boldsymbol{\varepsilon}_n(t)]'$. The detailed expression for (17.4) is then:

$$X = \begin{bmatrix} \boldsymbol{x}_1(t)' \\ \boldsymbol{x}_2(t)' \\ \vdots \\ \boldsymbol{x}_n(t)' \end{bmatrix} = \begin{bmatrix} q_{1,1} & q_{1,2} & \cdots & q_{1,r} \\ q_{2,1} & q_{2,2} & \cdots & q_{2,r} \\ \vdots & \vdots & \ddots & \vdots \\ q_{n,1} & q_{n,2} & \cdots & q_{n,r} \end{bmatrix} \begin{bmatrix} \boldsymbol{f}_1(t)' \\ \boldsymbol{f}_2(t)' \\ \vdots \\ \boldsymbol{f}_r(t)' \end{bmatrix} + \begin{bmatrix} \boldsymbol{\varepsilon}_1(t)' \\ \boldsymbol{\varepsilon}_2(t)' \\ \vdots \\ \boldsymbol{\varepsilon}_n(t)' \end{bmatrix} \tag{17.5}$$

where the decomposition errors are formed by the truncated orthogonal basis functions $\boldsymbol{f}_{r+1}(t), \boldsymbol{f}_{r+2}(t), \ldots, \boldsymbol{f}_p(t)$:

$$E = \begin{bmatrix} \boldsymbol{\varepsilon}_1(t)' \\ \boldsymbol{\varepsilon}_2(t)' \\ \vdots \\ \boldsymbol{\varepsilon}_n(t)' \end{bmatrix} = \begin{bmatrix} q_{1,r+1} & q_{1,r+2} & \cdots & q_{1,p} \\ q_{2,r+1} & q_{2,r+2} & \cdots & q_{2,p} \\ \vdots & \vdots & \ddots & \vdots \\ q_{n,r+1} & q_{n,r+2} & \cdots & q_{n,p} \end{bmatrix} \begin{bmatrix} \boldsymbol{f}_{r+1}(t)' \\ \boldsymbol{f}_{r+2}(t)' \\ \vdots \\ \boldsymbol{f}_p(t)' \end{bmatrix} \tag{17.6}$$

From the above equation, it can be demonstrated that $EF = 0$ ($n\times r$ matrix of zeros), because of the fact that $\boldsymbol{f}_i(t)'\boldsymbol{f}_j(t) = 0$ for $i \neq j$ (definition of orthogonality). Additionally, without loss of generality it can be assumed that any orthogonal function $\boldsymbol{f}_j(t)$ has norm 1, which means that $F'F = I$ ($r\times r$ identity matrix). Therefore, the decomposition coefficients are obtained by:

$$Q = XF \tag{17.7}$$

That is:

$$q_{ij} = \boldsymbol{x}_i(t)'\boldsymbol{f}_j(t) \tag{17.8}$$

It is desirable to find a set of r-EOFs that retain as much as possible of the information of X. That is, E in (17.4) has to be minimized. The solution to the problem states that EOFs are obtained by:

$$F = eigenvectors(X'X) \tag{17.9}$$

Each eigenvector $\boldsymbol{f}_j(t)$ has an associated eigenvalue λ_j, which represents the amount of information that $\boldsymbol{f}_j(t)$ retains from X after the decomposition. Thus, those eigenvectors with higher associated eigenvalue acquire more information from input data. The term explained variability (EV) is used as a measure of the amount of information retained by each EOF.

$$EV_j = \frac{\lambda_j}{\sum \lambda_j} \times 100\% \tag{17.10}$$

The order of the EOFs is defined in accordance to their explained variability. That is, the first of the EOFs is the one having the greatest associated eigenvalue.

17.3   Applications of EOFs for Transmission Line Protection

Based on the characteristics and patterns of fault generated transients, this section introduces the proposed protection algorithms. Figure 17.1 presents the general protection scheme: the first stage consists of the formulation of the EOFs, which is the basis for pattern extraction of fault generated transients affecting the protected transmission line. This process is done offline and requires modeling of the protected transmission line and associated power system. Fault simulation is then performed to obtain a fault database, which is then used for attaining the EOF and the associated scores $X(q)$ that represent the fault database in the EOF domain.

The online algorithm begins with signal acquisition that consists of measuring of analog signals from the transducer and transformation to digital signals through an A/D converter. An algorithm is then implemented for event detection: once a potential fault is detected, the wavefront of this event is extracted from the signal; this wavefront must have the same sampling frequency and number of samples as the stated EOF. The detected event defined by the discrete signal $x(t)$ is then decomposed with the selected EOF, which gives a set of values that define the event in the EOF domain $x(q)$. This signal is then compared with precomputed faults that serve as reference for implementation of protection functions. Finally, the protection scheme has to decide which action should be executed.

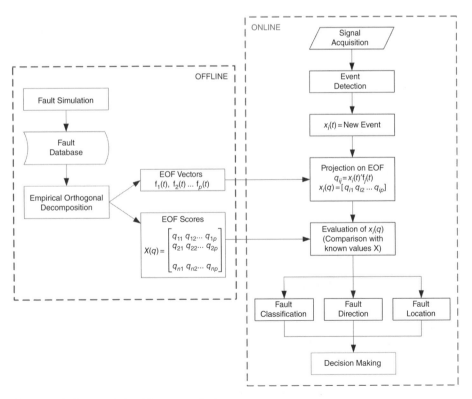

Figure 17.1 Proposed algorithm, general scheme.

17.3.1 Fault Direction

When the power system is not radial, faults can be fed by both line ends. Then defining the fault direction is indispensable for adequate fault location and guaranteeing selectivity in power system protection. For this protection function, decomposition using the two first EOFs is used; that is, fault current and voltage signals are represented by:

$$\Delta i(t) \approx q_{I1} f_1(t) + q_{I2} f_2(t) \tag{17.11}$$

$$\Delta v(t) \approx q_{V1} f_1(t) + q_{V2} f_2(t) \tag{17.12}$$

The directional function requires comparison of $\Delta i(t)$ and $\Delta v(t)$, then $f_1(t)$ and $f_2(t)$ obtained for forward fault currents are used. This is not a general rule—EOFs obtained from voltage signals can also be used and the results are similar.

The general scheme for directional evaluation is presented in Figure 17.2. After a fault is detected, the relay extracts the superimposed quantities of current and voltage signals sampled with the same data window and sampling frequency used for obtaining $f_1(t)$ and $f_2(t)$. Then, both discrete signals are decomposed with $f_1(t)$ and $f_2(t)$, which results in:

$$\Delta i(q) = \begin{bmatrix} q_{i1} \\ q_{i2} \end{bmatrix} = \begin{bmatrix} \Delta i(t)' f_1(t) \\ \Delta i(t)' f_2(t) \end{bmatrix} \tag{17.13}$$

$$\Delta v(q) = \begin{bmatrix} q_{v1} \\ q_{v2} \end{bmatrix} = \begin{bmatrix} \Delta v(t)' f_1(t) \\ \Delta v(t)' f_2(t) \end{bmatrix} \tag{17.14}$$

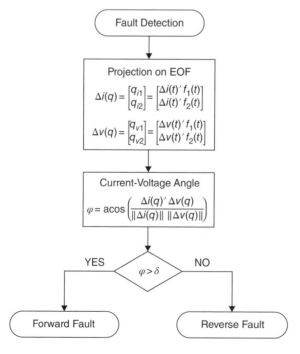

Figure 17.2 Directional protection, general scheme.

The angle between current and voltage vectors in the EOF space is given by:

$$\varphi = acos\left(\frac{\Delta i(q)'\Delta v(q)}{\|\Delta i(q)\|\,\|\Delta v(q)\|}\right) \tag{17.15}$$

Where $\|\Delta i(q)\|$ and $\|\Delta v(q)\|$ are the norm of $\Delta i(q)$ and $\Delta v(q)$, respectively, and $\Delta i(q)'\Delta v(q)$ is the dot product. Expressed in terms of the decomposition coefficients, (17.15) gives:

$$\varphi = acos\left(\frac{q_{i1}q_{v1} + q_{i2}q_{v2}}{\sqrt{(q_{i1}^2 + q_{i2}^2)}\sqrt{(q_{v1}^2 + q_{v2}^2)}}\right) \tag{17.16}$$

A recognized feature of forward faults is that they cause superimposed voltages and currents with different sign [3]. Thus, the rule to determine the fault direction is: φ close to 180° denotes fault in the forward direction and φ close to 0° denotes a reverse (backward) fault (see Figure 17.3). Hence, a forward fault has to fulfill:

$$\varphi > \delta \tag{17.17}$$

Where δ gives a confidence interval for the possible variations of φ, and must be set in accordance to precomputed training faults.

17.3.2 Fault Classification

In order to minimize the impact on power quality and continuity after a contingency, it is required to disconnect as few devices as possible from the power network. Single phase to ground faults have a higher probability of occurrence than multiphase faults. Thus, single pole tripping (SPT) schemes are implemented for improving selectivity in transmission system protection. SPT enhances the stability, power transfer capabilities, reliability, and availability of a transmission system during and after a single phase to ground fault [21, 22]. Furthermore, accurate SPT accompanied with automatic reclosing allows greater improvement of power system stability [23]. Therefore, in order to enable SPT and auto reclosing schemes, a method for classifying faults is essential.

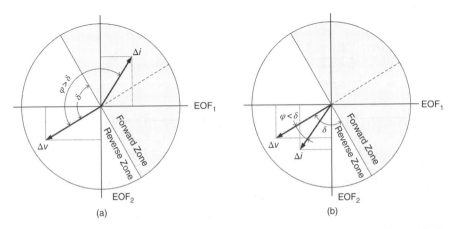

Figure 17.3 Representation of voltage and current signals in the EOF plane for: (a) forward fault and (b) backward fault.

17.3.2.1 Required EOF

For fault direction, data analysis of fault data is focused only on the first two EOFs, which normally contain more than 90% of data variability. However, information provided by $f_1(t)$ and $f_2(t)$ is not sufficient for fault classification, since faults of different types can have similar values in this space. Hence, in order to achieve an effective fault classification, those EOFs with smaller variances should be also considered, since they provide significant information of fault characteristics.

17.3.2.2 Fault Type Surfaces

A training database in the EOF space $X(q)$ contains the coefficients of training faults decomposed in terms of the r selected EOFs. $X(q)$ includes observations of the six possible fault types (ABC, AB, ABG, AC, ACG, and AG), which describes a perfectly defined r-dimension pattern, where fault location and fault inception angle patterns are relevant.

During the fault simulation stage, only a representative sample of possible faults is generated for training data. With the purpose of covering the entire fault type surface, formed in the r-dimensional space given by the selected EOF, the reference data is enlarged with estimated observations created by interpolation in accordance with the fault location and inception angle patterns.

17.3.2.3 Defining the Fault Type

Figure 17.4 presents the scheme for fault classification. After an event is detected, the relay extracts the superimposed current quantity sampled with the same data window and sampling frequency of the EOF. The discrete signal obtained is decomposed with the selected EOF, and the resultant vector is:

$$\Delta i(q) = \begin{bmatrix} q_1 & q_2 & q_3 & \cdots & q_r \end{bmatrix}' \tag{17.18}$$

The fault type is identified using k-nearest neighbor classification (kNN). Here, the evaluated signal $\Delta i(q)$ is assigned to the class of its nearest (most similar) sample in the database $Y(q)$. Different metrics can be used to find the k-nearest neighbor points, for instance: city block, Euclidean, or cosine distance. Cosine distance (17.19) is recommended since it evaluates the angle between the two vectors, and it is independent from vectors magnitude. This metric improves classification for ground contact faults, because fault resistance influences the current magnitude but not its angle.

$$d(\Delta i, y_0) = 1 - \frac{\Delta i \cdot y_0}{\|\Delta i\| \|y_0\|} = 1 - \frac{\sum_{i=1}^{r} (q_i \cdot q_{0i})}{\sqrt{\sum_{i=1}^{r} (q_i)^2} \sqrt{\sum_{i=1}^{r} (q_{0i})^2}} \tag{17.19}$$

where $y_0 = \begin{bmatrix} q_{01} & q_{02} & q_{03} & q_{04} & q_{05} \end{bmatrix}'$ is a reference fault current that belongs to the reference database $Y(q)$, and r is the number of selected EOFs. Additionally, when the evaluated event is a fault occurring in the protected transmission line, the minimum distance between $\Delta i(q)$ and $Y(q)$ is small; then, a maximum distance d_{ref} can be established as a first criterion to classify the event as an internal fault (inside the protected transmission line), non-fault event, or external fault.

In order to improve the security of this classification method, an additional criterion is included for choosing the more severe fault type in the case that the k-nearest neighbor classification faces a conflict—for instance, two surfaces with similar nearest neighbors to the evaluated fault.

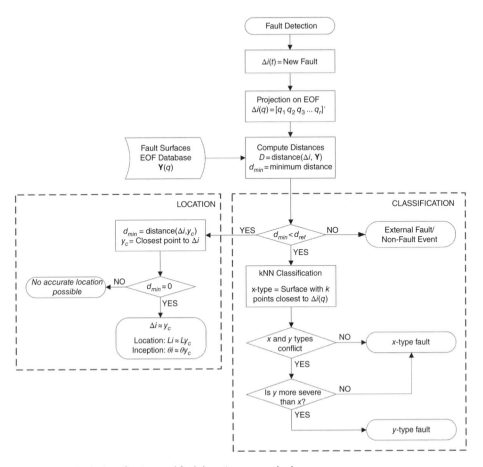

Figure 17.4 Fault classification and fault location, general scheme.

17.3.3 Fault Location

This function uses information from the classification function (see Figure 17.4). After the fault type is defined, the evaluated fault is assumed to have the same features as the nearest reference fault. This methodology obtains the reference fault that is most similar to the evaluated fault. This reference fault has inception angle and fault location previously defined and, if the distance metric is relatively small, the analyzed fault could directly be matched to the closer reference fault.

17.4 Study Case

17.4.1 Transmission Line Model and Simulation

Modeling the line to be protected and their neighbor lines is the first step to be carried out in order to get a fault database that describe the fault behavior of the transmission line, which will be then used for obtaining the EOF that describe the faults affecting the analyzed line and is required as reference for setting the protection functions.

Adequate models and software have to be used in order to get a precise emulation of the real transmission line. This part can be given as a limitation of the present approach, because of the difficulty of definition of an adequate model and software. Nevertheless, electromagnetic transient programs based on the EMTP method, proposed by Dommel in 1969 [24] and enhanced over the years with additional contributions [6], have demonstrated accuracy and reliability by a number of test cases [25]. Additionally, the representation of the frequency dependent parameters of a transmission line [26] denotes efficiency and accuracy for most simulation cases.

Because of the deterministic nature of the protection algorithm to be proposed, an important number of faults have to be generated for different conditions of location, fault inception angle (θ_0), fault resistance (R_f), and fault type, so that the algorithm will consider a representative sample of the possible faults that could appear in the studied transmission lines. Table 17.1 summarizes typical conditions employed for fault simulation.

Simulating a great number of faults is actually not difficult; a MATLAB application is developed in order to command the required simulations in the created EMTP model [27].

17.4.2 The Power System and Transmission Line

The Western System Coordinating Council (WSCC) 3-Machine 9-Bus test system [28] (see Figure.17.5) is evaluated in ATP/EMTP. The longest transmission line (line B6-B9, 180 km) is studied. Line B6-B9 and its adjacent lines B4-B6 (96 km) and B8-B9 (108 km) are modeled using the frequency dependent model of J. Martí [26], which provides an accurate and reliable time domain model for electromagnetic transients simulations. See model data in Appendix A. Fault signals are registered in relay R_6 located at bus B6.

17.4.3 Training Data

With the intention of establishing the accurate pattern exhibited by fault waveforms, a vast number of faults are computer-generated for different conditions of type, fault inception angle (θ_0), fault resistance (R_f), and location. The sampling technique aims to get enough information on the faults that could appear along the line; then the data obtained should serve as a good representation of the total population. The fault conditions used are detailed in Table 17.2 and Table 17.3.

17.4.4 Training Data Matrix

The training data matrix is composed by faults described in Table 17.2. From a data mining perspective, every discrete current or voltage fault signal $x_i(t)$ is an observation,

Table 17.1 Conditions for fault simulation.

Feature	Range	Detail
Fault Type	$1\phi g$, 2ϕ, $2\phi g$, 3ϕ	All fault types affecting the line
Inception Angle θ_0 (deg)	0, 30,..., 330	From 0° to 360°, with 30° step
Fault Location (km)	1, 10, 20, ... L	From 0 to line length, with 10 km step
Fault Resistance R_f (Ω)	0, 5, 40, 100	Or in accordance with typical values

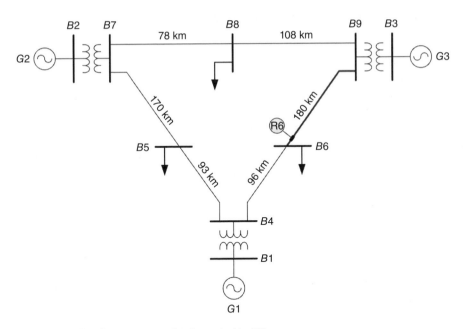

Figure 17.5 Nine-bus test system implemented in ATP.

Table 17.2 Forward faults (line B6-B9).

Fault Type	Inception Angle (deg)	R_f (Ω)	Fault Location from B6 (km)
ABC	0, 30,…, 330	–	1, 10, 20, …180
ABG	0, 30,…, 330	5	1, 10, 20, …180
AB	0, 30,…, 330	–	1, 10, 20, …180
ACG	0, 30,…, 330	5	1, 10, 20, …180
AC	0, 30,…, 330	–	1, 10, 20, …180
AG	0, 30,…, 330	40	1, 10, 20, …180

Table 17.3 Reverse faults (line B4-B6).

Fault Type	Inception Angle (deg)	R_f (Ω)	Fault Location from B6 (km)
ABC	0, 30,…, 330	–	1, 10, 20, …90
ABG	0, 30,…, 330	5	1, 10, 20, …90
AB	0, 30,…, 330	–	1, 10, 20, …90
ACG	0, 30,…, 330	5	1, 10, 20, …90
AC	0, 30,…, 330	–	1, 10, 20, …90
AG	0, 30,…, 330	40	1, 10, 20, …90

and the samples of each signal are their features. This set of signals is represented by the $n{\times}p$ data matrix X shown in (17.1).

17.4.4.1 Data Window

Defining the time window length is a crucial aspect in the design of a protection algorithm. The longer the time window, the more information it provides, whereas a shorter time window yields high-speed decision making. In the present analysis, time windows in the order of 600 µs are used, which is long enough to observe the evident patterns of fault generated transients and avoids reflected waves from the remote bus affecting the wave shape of signals measured by relay R_6 (see Figure. 17.6). A methodology such as that proposed in [29] can be used to detect the initial wavefront.

17.4.4.2 Sampling Frequency

In a transmission line, frequencies in the order of some hundreds of kHz are expected in fault generated transients. By virtue of Nyquist's theorem, the sampling frequency should be twice that frequency if the entire frequency spectrum is taken into account, which means an additional cost in the sampling process during A/D conversion. However, the sampling frequency should be distinguished depending on the protection function to be implemented, for instance a directional function is associated with the two firsts EOFs only, which are related to low frequency patterns and therefore they are scarcely affected by high frequency components [14, 16]. Hence, not too high a sampling frequency is required for pattern based directional protection. On the other hand, fault classification and location can require the identification of high frequency patterns, thus a high sampling frequency is necessary in this function. In the following sections, various sampling frequencies are studied in order to show the effect of this variable on the signal patterns.

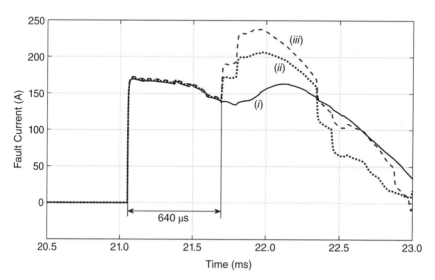

Figure 17.6 Effect of reflected waves from remote node (Node B4), for the following configurations: one power transformer and (i) two transmission lines, (ii) three transmission lines, (iii) four transmission lines. For twice the traveling time of the neighbor line (640 µs), there is no significant change in the signal waveform.

17.4.5 Signal Conditioning

17.4.5.1 Superimposed Component

The pre-fault component is removed from fault signals, by means of (17.20), in order to obtain the superimposed component introduced by the fault:

$$\Delta x(t) = x(t) - x_S(t) \tag{17.20}$$

where:

$\Delta x(t)$ is the generated deviations from the steady-state signals;
$x(t)$ is the discrete signal obtained from the model;
$x_S(t)$ is the pre-fault signal.

A common practice to derive the incremental magnitudes is using the signals measured exactly one cycle before the actual measurement. This can be achieved, for instance, by using a delta filter [30]. Another practice is using high-pass filters to suppress the steady-state components so that only the high frequency components will be present in the relaying signals [3].

The data matrix of fault signals (voltages or currents) is formed with the superimposed components $\Delta x(t)$:

$$\Delta X = \left[\Delta x_1(t) \; \Delta x_2(t) \; \dots \; \Delta x_n(t)\right]' \tag{17.21}$$

17.4.5.2 Centering the Variables

An essential condition for applying EOFs is removing the mean of each variable from the data set in order to center the data on the origin. This process does not change the EOFs, because they are obtained from the variance–covariance matrix, which is computed from the centered data. However, for a new signal that is going to be compared with predefined faults, it must be centered using the mean vector of the training data matrix.

17.4.5.3 Scaling

In some cases, such as for implementing the pattern based directional function, data can be scaled to set data in a common scale independent from measurement units. In this case, the coefficients of the training matrix are scaled between -1 and 1 by dividing by the maximum absolute coefficient; this last operation does not affect the data shape.

For protection functions such as fault classification and location, data scaling is not recommended since the fault magnitude contributes to adequate fault classification.

17.4.6 Energy Patterns

An alternative for categorizing a transient event is by evaluating the energy associated to voltage and current waveforms. In signal processing, the energy E_i of a p-length discrete signal $x_i(t) = (x_{i1}, x_{i2}, \dots, x_{ip})'$ is computed by:

$$E_i = \sum_{j=1}^{p} (x_{ij})^2 \Delta t \tag{17.22}$$

Where Δt is the time interval between each sample of the discrete signal. The mean power P_i and the RMS value in the interval $T = p \times \Delta t$ are given by:

$$P_i = \frac{E_i}{p\Delta t} = \frac{1}{p} \sum_{j=1}^{p} (x_{ij})^2 \tag{17.23}$$

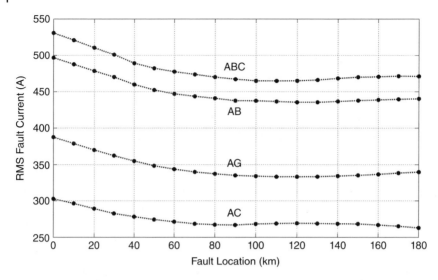

Figure 17.7 RMS fault current, fault type, and fault location pattern, $\theta_0 = 90°$, data window = 600 μs.

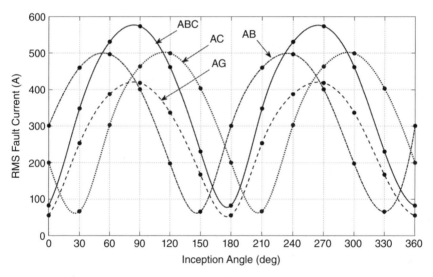

Figure 17.8 RMS fault current, fault type, and fault inception angle pattern, location = 1 km, data window = 600 μs.

$$rms_i = \sqrt{P_i} = \sqrt{\frac{1}{p} \sum_{j=1}^{p} (x_{ij})^2} \qquad (17.24)$$

The RMS value as a measure of energy was evaluated for the group of fault current signals generated with the conditions detailed in Table 17.2. Results highlight that faults of the same type follow a defined pattern governed by a fault inception angle pattern and a fault location pattern (see Figures 17.7 and 17.8).

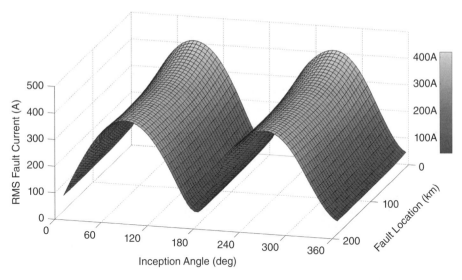

Figure 17.9 RMS fault current, fault inception angle and location pattern for AG faults, $R_f = 40\,\Omega$.

Combination of both characteristics (location and inception angle) for single phase to ground faults (AG) with $R_f = 40\,\Omega$ is presented in Figure 17.9. Here is can be seen that faults of the same type, in this case single phase to ground faults, have a defined place in the demarked surface.

The data window plays an important role in the patterns defined by the RMS values of fault currents. Longer data windows have more information and hence allow major differentiation between fault signals. For example, see in Figure 17.10 how the location of a fault can be distinguished using longer data windows. Nonetheless, longer signals (in terms of time) contain transient components altered at remote nodes, and therefore

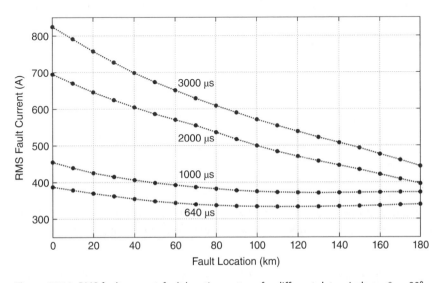

Figure 17.10 RMS fault current, fault location pattern for different data windows, $\theta_0 = 90°$.

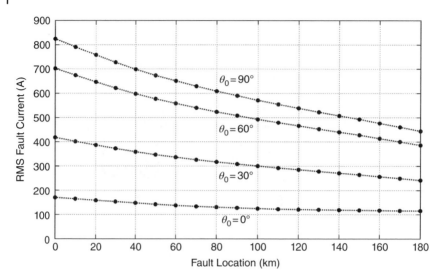

Figure 17.11 RMS fault current, fault location pattern for different θ_0, data window $= 3$ ms.

they are affected by the configuration of those nodes (see Figure 17.11). Although this aspect is not a major limitation, it has to be considered in the design of the protection algorithm.

In general, the energy associated with faulted signals is higher than the energy of non-fault signals. However, some faults can generate signals with relatively low associated energy: such is the case of high impedance faults (HIF) or faults that could occur with small inception voltage. On the other hand, some non-fault transients can produce signals with high energy: for instance, transmission line or capacitor bank switching. A mechanism to distinguish faults arising on the protected device and external faults is also necessary since both can produce high energy signals. Here, EOF decomposition plays an important role because EOFs can be used to decompose energy patterns into an n-dimension space that retains the dominant patterns of data processed.

17.4.7 EOF Analysis

17.4.7.1 Computing the EOFs

The training data matrix is composed of the faults detailed in Table 17.2. That is, the data matrix contains 6 fault types, 12 fault inception angles, and 19 fault locations. That gives a total of 1368 (6×12×19) faults. Each current signal is a 600 μs discrete-time series. The number of samples in each signal will depend on the sampling frequency. Different sampling frequencies (f_s) are studied in order to show the incidence of this variable in the EOF. The left singular vectors and their related eigenvalues of the data matrix are computed. Explained variability EV of the first eight EOF is summarized in Table 17.4.

From the results presented in Table 17.4, we can draw out some important features of the EOFs:

a) The two first EOFs (EOF$_1$ and EOF$_2$) represent the dominant pattern since they attain more than 90% of EV in all the cases. This is in fact one of the advantages of EOF

Table 17.4 Explained variability of the first eight EOFs for different sampling frequencies, data window 600 µs.

f_s	Δt	EV(%)[a]							
		EOF$_1$	EOF$_2$	EOF$_3$	EOF$_4$	EOF$_5$	EOF$_6$	EOF$_7$	EOF$_8$
1 MHz	1 µs	79.08	12.90	3.96	2.28	0.53	0.45	0.22	0.14
200 kHz	5 µs	79.70	12.50	4.04	2.17	0.62	0.38	0.19	0.11
100 kHz	10 µs	82.80	11.25	3.63	1.58	0.31	0.15	0.09	0.05
50 kHz	20 µs	88.86	8.84	1.71	0.26	0.12	0.06	0.04	0.02
20 kHz	50 µs	94.37	4.83	0.58	0.10	0.05	0.03	0.01	0.01

a) Variability percentage from original data retained by each EOF.

decomposition: that most of the variation presented in the original signals is retained in the first EOF.

b) The *EV* of EOF$_1$ is greater for lower sampling frequencies. This means that the main EOF (EOF$_1$) is associated with a low frequency pattern.

From the last remark, we can state a relationship between the EOF order and the frequency. From experiences with different transmission lines it has been observed that the first EOF associated with larger eigenvalues is related to low frequencies patterns. That is, the higher the order of the EOF, the higher its frequency (see Figure 17.12). This statement is also mentioned in [14] where Hannachi et al. maintain that "high/low power are

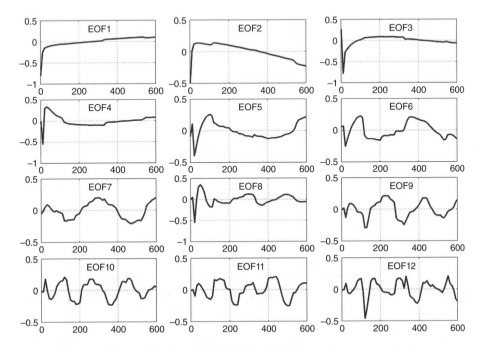

Figure 17.12 Coefficients of the first twelve EOFs, $f_s = 100$ kHz .

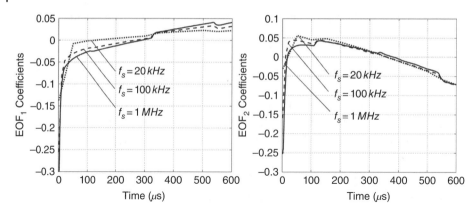

Figure 17.13 Effect of sampling frequency in the two first EOFs.

associated respectively with low/high frequency variability. Hence low frequency and large scale patterns tend to capture most of the variance observed in the system." The aforementioned property makes the first EOFs (principal components) immune to noise as is also stated in [16].

Figure 17.13 shows the waveform of EOF_1 and EOF_2 for different sampling frequencies. As can be seen, they are barely affected by the sampling frequency. This occurs because these EOFs represent low frequency patterns and therefore they are not altered by the high frequencies of fault generated transients.

17.4.7.2 Fault Patterns Using EOF

Consider $f_s = 50\,kHz$. For this sampling frequency, it can be observed in Table 17.4 that the first three EOFs have more than 99% of the overall explained variability. That is, the entire group of analyzed faults can be decomposed in terms of only three EOFs with less than 1% loss of information:

$$x_i(t) \approx q_{i1}\boldsymbol{f}_1(t) + q_{i2}\boldsymbol{f}_2(t) + q_{i3}\boldsymbol{f}_3(t) \tag{17.25}$$

Expression (17.8) shows that coefficients q_{ij} are computed by correlating the ith signal with the associated jth EOF. The coefficients q_{ij} also presents fault location and fault inception angle patterns similar to energy patterns (see Figures 17.17 and 17.18), presented in section 14.5. However, the orthogonal decomposition allows representing these patterns in an optimal multidimensional space. For instance, in Figure 17.14, faults generated along the line with the same fault inception angle $\theta_0 = 30°$ are projected into a 2-dimensional space conformed by the first two EOF. In this projection, different fault types are well-differentiated.

Results show that even though the fact that faults may appear under (theoretically) infinite stochastic conditions (i.e., fault location, inception angle, fault resistance, and number of involved phases), the waveforms of faulted signals follow a well-defined deterministic behavior. Therefore, it is not necessary to consider the infinite possible fault conditions in order to attain generalized EOFs. In fact, just a small representative sample of different fault conditions is required to achieve a good representation of the total population.

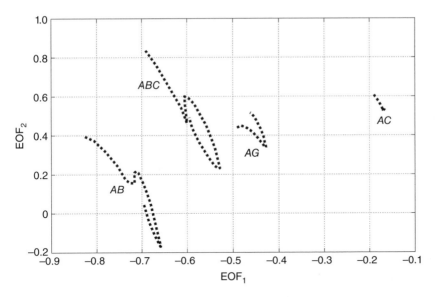

Figure 17.14 Different fault types in the two first EOF, fault location from 0 to 180 km, $\theta_0 = 30°$.

17.4.8 Evaluation of the Protection Scheme

17.4.8.1 Fault Direction

Phase A to ground (AG) fault currents projected in the EOF plane are presented in Figure 17.15. Here, twelve clusters, corresponding to the twelve θ_0 used in the training data, are defined. Each point in Figure 17.15a represents a fault with specific θ_0 and fault location. The corresponding forward fault voltages are presented in Figure 17.15b: it can be observed that forward voltage signals in the EOF plane appear around 180° away from the respective current signal.

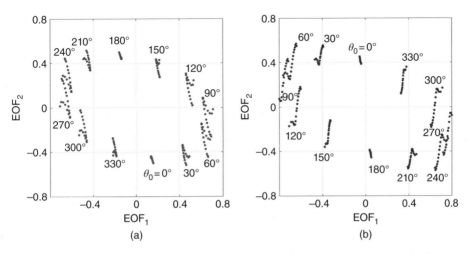

Figure 17.15 Forward single phase to ground faults (AG), $R_f = 40\,\Omega$, representation on the first two EOFs, (a) current signals, (b) voltage signals.

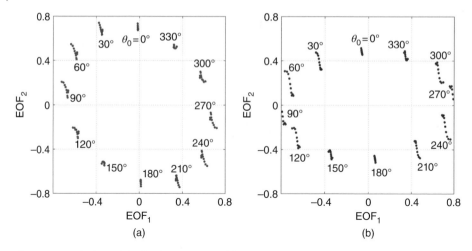

Figure 17.16 Backward single phase to ground faults (AG), $R_f = 40\,\Omega$, representation on the first two EOFs, (a) current signals, (b) voltage signals.

On the other hand, backward faults cause voltages and currents with similar angle in the EOF plane denoted by $f_1(t)$ and $f_2(t)$. Figure 17.16 presents the current and voltage signals generated by backward single phase to ground faults (AG).

17.4.9 Fault Classification

In order to obtain reference surfaces for fault classification, a cubic spline data interpolation is employed. Fault location patterns are interpolated from 0 to 180 km with a 2 km step; Angle patterns are interpolated from 0 to 355° with 5° step. Examples of fault location and inception angle interpolation are presented in Figures 17.17 and 17.18. After interpolation, original data grows from 1368 observations to a database with 39 312

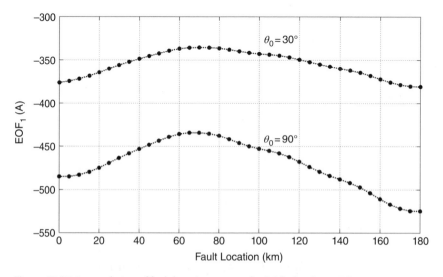

Figure 17.17 Interpolation of fault location pattern for AG faults, $R_f = 40\,\Omega$.

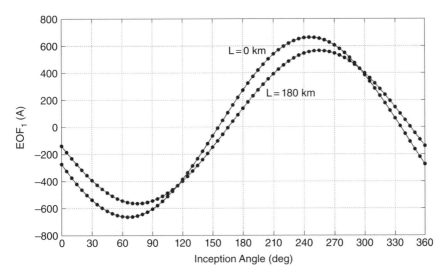

Figure 17.18 Interpolation of inception angle pattern for ABG faults, $R_f = 5\,\Omega$.

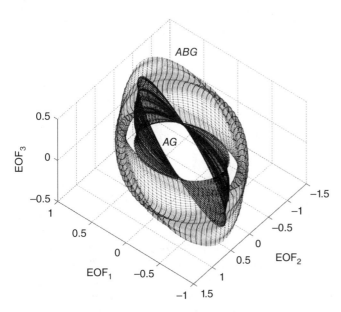

Figure 17.19 Three-dimensional representation of ABG and AG current faults using the three first EOFs. Fault location from 0 to 180 km, fault inception from 0° to 355°.

observations per fault type, which are stored in matrix $Y(q)$. Figure 17.19 shows the three-dimensional surface of ABG and AG faults for inception angles from 0° to 355°, plotted with interpolated data.

17.4.9.1 Classification

The faults detailed in Table 17.5 were evaluated with the proposed classification scheme. At first, the training stage considered different classification parameters in order to find

Table 17.5 Test faults, line B6-B9.

Fault Type	Inception Angle (deg)	$R_f (\Omega)$	Fault Location from B6 (km)
ABC	0, 30,…, 330	–	5, 15, 25, …175
ABG	0, 30,…, 330	5	5, 15, 25, …175
		15	1, 5, 10, …180
AB	0, 30,…, 330	–	5, 15, 25, …175
ACG	0, 30,…, 330	5	5, 15, 25, …175
		15	1, 5, 10, …180
AC	0, 30,…, 330	–	5, 15, 25, …175
AG	0, 30,…, 330	40	5, 15, 25, …175
		0, 20, 100, 300	1, 5, 10, …180

Table 17.6 Classification results, confusion matrix.

	ABC	ABG	AB	ACG	AC	AG	Accuracy
ABC	213	0	0	4	0	0	98.16%
ABG	0	632	8	0	0	4	98.14%
AB	2	28	208	2	0	3	85.60%
ACG	1	0	0	638	6	16	96.52%
AC	0	0	0	16	210	26	83.33%
AG	0	0	0	0	0	1943	100.00%
Sensitivity	98.61%	95.76%	96.30%	96.67%	97.22%	97.54%	97.07%

the best kNN classifier. For the present case study, the optimal classification parameters are as follows. Sampling frequency: 200 kHz; selected EOF: from 1 to 12; distance metric: cosine; number of nearest neighbors: 1. Results are presented in Table 17.6: it can be seen that accuracy for AG faults is 100%, which means that no multiphase fault was classified as a single phase to ground fault AG, which is important for security. Nevertheless, accuracy for multiphase faults is reduced; most confusion appears between faults that involve the same phases, that is, AB with ABG or AC with ACG. Almost all the AG faults that where classified as two phase faults correspond to faults with fault resistance $R_f = 300\,\Omega$. Recall that the reference database only considered AG for $R_f = 40\,\Omega$. Hence, classification accuracy could be improved by incorporating one or two more AG surfaces for different fault resistances.

17.4.10 Fault Location

Results of fault location are summarized in Table 17.7. For the 3960 test faults of Table 17.5, 2630 (66.41%) were located with 100% accuracy (0 km error), and only 131 faults (3.31%) present inaccuracy greater than 5 km. Most inaccuracies greater than 5 km are found in single phase to ground faults with $R_f = 300\,\Omega$ and $\theta_0 = 0°$. Some examples of fault location are shown in Table 17.8: it can be seen that greater inaccuracies

Table 17.7 Results of fault location.

Error (km)	Number of Faults	%
0	2630	66.41
1	751	18.96
2	202	5.10
3	123	3.11
4	95	2.40
5	28	0.71
>5	131	3.31

Table 17.8 Some results of fault classification.

Evaluated Fault				Nearest Reference Fault				Location Error (km)
Fault Type	θ_0	R_f (Ω)	L (km)	Fault Type	θ_0	R_f(Ω)	L (km)	
ABC	90°	–	15	ABC	90°	–	13	2
ABG	90°	5	25	ABG	85°	5	25	0
ABG	60°	15	65	ABG	55°	5	66	1
AB	30°	–	95	ABG	50°	5	93	2
ACG	0°	15	165	ACG	355°	5	166	1
AC	60°	–	115	AC	60°	–	115	0
AG	60°	20	135	AG	55°	40	133	2
AG	120°	0	85	AG	120°	40	88	3
AG	150°	100	170	AG	155°	40	171	1
AG	60°	300	95	AG	55°	40	96	1
AG	90°	300	135	AG	125°	40	156	21
AG	30°	300	125	ACG	50°	5	142	17
AG	150°	300	115	AG	145°	40	102	13

appear for AG faults $R_f = 300\,\Omega$. As aforementioned, accuracy of fault location can be improved by incorporating more AG surfaces for different fault resistances.

17.5 Conclusions

Empirical orthogonal functions appear to be a promising alternative for signal processing applied to fault diagnosis. Results show that in spite of the fact that faults may appear under infinite stochastic conditions such as fault location, inception angle, fault resistance, pre-fault current, and number of involved phases, the waveforms of faulted signals follow a well-defined deterministic pattern. The proposed methodology can obtain precise information on a fault by evaluating a very short signal (600 µs for the presented case study), which is really valuable for implementation in high-speed protection schemes.

The proposed protection scheme uses kNN for fault classification and location, which is a very simple classification method. In fact, we highlighted that classification performance can be improved. Readers are encouraged to evaluate more sophisticated classification methods such as support vector machines and other kernel based classifiers.

Appendix 17.A

Study Cases: WECC 9-bus, ATPDraw Models and Parameters

Figure 17A.1 shows the WECC 9-bus power system. Lines B4-B5, B5-B7, and B7-B8 are implemented with the 3-phase distribution parameter transposed line Clarke model; transmission lines B4-B6, B6-B9, and B8-B9 are modeled in a more detailed fashion. These lines have three transposition points and eight line segments for simulating faults along the line—see Figure 17A.2. Every line segment uses the JMarti model with the data

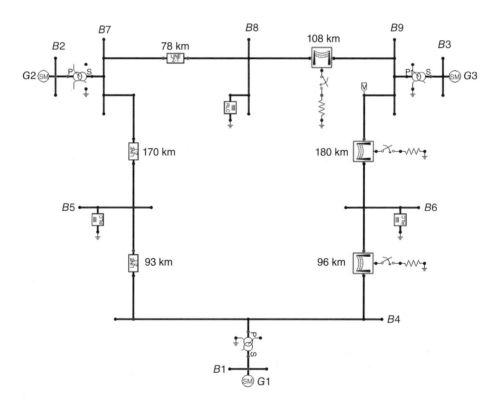

Figure 17A.1 WECC 9-bus test system, ATPDraw Model.

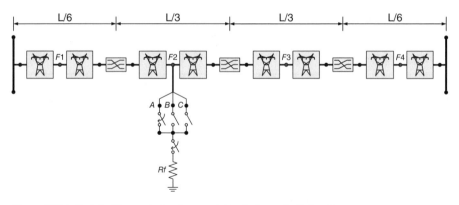

Figure 17A.2 Detailed transmission line model including a fault circuit.

Table 17A.1 ATPdraw line/cable model, lines: B4-B6, B6-B9, and B8-B9.

Model:	JMarti	Ground resistivity (Ωm):	100
Type:	Overhead line	Initial freq. (Hz):	0.1
Number of phases:	3	Decades:	7
Transposed:	Inactive	Points/dec:	100
Auto bundling:	Inactive	Freq. matrix (Hz):	3000
Skin Effect:	Active	Freq. SS (Hz):	60
Segmented ground	Inactive	Use default fitting	Active
Real transf. matrix	Inactive		

Table 17A.2 ATPdraw line/cable data, line B4-B6.

#	Ph.No.	Rin (cm)	Rout (cm)	DC Res. (Ω/km)	Horiz (m)	Vtower (m)	Vmin (m)
1	1	0.4134	1.2408	0.0918	−6.5	18	7.1
2	2	0.4134	1.2408	0.0918	0	18	7.1
3	3	0.4134	1.2408	0.0918	6.5	18	7.1
4	0	0.4626	0.771	0.3002	−4	22.35	14.5
5	0	0.4626	0.771	0.3002	4	22.35	14.5

presented in Table 17A.1. Particular data for each line are presented in Tables 17A.2, 17A.3 and 17A.4.

The resultant impedances and admittances for lines B4-B6, B6-B9, and B8-B9 are presented in Table 17A.5.

Table 17A.3 ATPdraw line/cable data, line B6-B9.

#	Ph.No.	Rin (cm)	Rout (cm)	DC Res. (Ω/km)	Horiz (m)	Vtower (m)	Vmin (m)
1	1	0.492	1.148	0.1141	−6.5	18	8.6
2	2	0.492	1.148	0.1141	0	18	8.6
3	3	0.492	1.148	0.1141	6.5	18	8.6
4	0	0.4626	0.771	0.3002	−4	22.35	14.5
5	0	0.4626	0.771	0.3002	4	22.35	14.5

Table 17A.4 ATPdraw line/cable data, line B8-B9.

#	Ph.No.	Rin (cm)	Rout (cm)	DC Res. (Ω/km)	Horiz (m)	Vtower (m)	Vmin (m)
1	1	0.527	1.581	0.0561	−6.5	18	7.1
2	2	0.527	1.581	0.0561	0	18	7.1
3	3	0.527	1.581	0.0561	6.5	18	7.1
4	0	0.4626	0.771	0.3002	−4	22.35	14.5
5	0	0.4626	0.771	0.3002	4	22.35	14.5

Table 17A.5 Line impedances and admittances.

Line	Sequence	Impedance (ohm/km)	Admittance (mho/km)	Surge Impedance (ohm)	Surge Velocity (km/s)
B4 – B6	0	0.2643+j0.9821	j2.4178×10⁻⁶	648.5–j7.5	2.425×10⁵
	1	0.0934+j0.5033	j3.2944×10⁻⁶	394.2–j5.3	2.915×10⁵
B6 – B9	0	0.2891+j0.9685	j2.3901×10⁻⁶	650.2–j8.3	2.451×10⁵
	1	0.1157+j0.5069	j3.2539×10⁻⁶	399.7–j6.4	2.917×10⁵
B8 – B9	0	0.2281+j0.9699	j2.4917×10⁻⁶	632.4–j6.6	2.408×10⁵
	1	0.0581+j0.4850	j3.4260×10⁻⁶	377.6–j3.4	2.919×10⁵

References

1 Q. H. Wu, Z. Lu, and T. Ji, *Protective Relaying of Power Systems Using Mathematical Morphology*, 1st ed. Liverpool: Springer, 2009.

2 M. S. Sachdev et al., "*Understanding Microprocessor-Based Technology Applied to Relaying*," Power System Relaying Committee, Atlanta, 2009.

3 A. T. Johns and S. K. Salman, *Digital Protection for Power Systems*, 1st ed. London: Peter Peregrinus, 1997.

4 Z. Q. Bo, F. Jiang, Z. Chen, X. Z. Dong, G. Weller, and M. A. Redfern, "Transient Based Protection for Power Transmission Systems," in *IEEE PES Winter Meeting*, Singapore, 2000, vol. 3, pp. 1832–1837.

5 Z. Q. Bo, G. Weller, F. T. Dai, and Q. X. Yang, "Transient Based Protection for Transmission Lines," in *Proceedings of International Conference on Power System Technology (POWERCON)*, 1998, vol. 2, pp. 1067–1071.

6 N. Watson and J. Arrillaga, *Power Systems Electromagnetic Transients Simulation*, 1st ed. London: Institution of Engineering and Technology, 2002.

7 A. R. Messina and V. Vittal, "Extraction of Dynamic Patterns From Wide-Area Measurements Using Empirical Orthogonal Functions," *IEEE Trans. Power Syst.*, vol. 22, no. 2, pp. 682–692, May 2007.

8 P. Esquivel and A. R. Messina, "Complex Empirical Orthogonal Function analysis of wide-area system dynamics," in *2008 IEEE Power and Energy Society General Meeting - Conversion and Delivery of Electrical Energy in the 21st Century*, 2008, pp. 1–7.

9 R. P. Aguilar, F. E. Perez, E. A. Orduna, and N. J. Castrillon, "Empirical orthogonal decomposition applied to transmission line protection," in *2011 IEEE PES Conference on Innovative Smart Grid Technologies (ISGT Latin America)*, 2011, pp. 1–7.

10 R. Aguilar, F. Perez, E. Orduna, and C. Rehtanz, "The Directional Feature of Current Transients, Application in High-Speed Transmission-Line Protection," *IEEE Trans. Power Deliv.*, vol. 28, no. 2, pp. 1175–1182, 2013.

11 J. C. Cepeda and D. G. Colomé, "Benefits of empirical orthogonal functions in pattern recognition applied to vulnerability assessment," in *Transmission Distribution Conference and Exposition - Latin America (PES T D-LA), 2014 IEEE PES*, 2014, pp. 1–6.

12 E. N. Lorenz, "Empirical Orthogonal Functions and Statistical Weather Prediction," Dept. of Meteorology, MIT, Massachusetts, Scientific Report 1, 1956.

13 I. T. Jolliffe, *Principal Component Analysis*, 2nd ed. New York: Springer-Verlag, 2002.

14 A. Hannachi, I. T. Jolliffe, and D. B. Stephenson, "Empirical Orthogonal Functions and Related Techniques in Atmospheric Science: a Review," *Int. J. Climatol.*, vol. 27, no. 9, pp. 1119–1152, Jul. 2007.

15 E. Vazquez, I. I. Mijares, O. L. Chacon, and A. Conde, "Transformer Differential Protection Using Principal Component Analysis," *IEEE Trans. Power Deliv.*, vol. 23, no. 1, pp. 67–72, Jan. 2008.

16 P. Jafarian and M. Sanaye-Pasand, "A Traveling-Wave-Based Protection Technique Using Wavelet/PCA Analysis," *IEEE Trans. Power Deliv.*, vol. 25, no. 2, pp. 588–599, Apr. 2010.

17 R. Aguilar, F. Pérez, and E. Orduña, "High-Speed Transmission Line Protection Using Principal Component Analysis, a Deterministic Algorithm," *IET Gener. Transm. Distrib.*, vol. 5, no. 7, pp. 712–719, Jul. 2011.

18 J. A. Morales and E. Orduna, "Patterns Extraction for Lightning Transmission Lines Protection Based on Principal Component Analysis," *IEEE Lat. Am. Trans.*, vol. 11, no. 1, pp. 518–524, Feb. 2013.

19 Y. Wang, J. Zhou, Z. Li, Z. Dong, and Y. Xu, "Discriminant-Analysis-Based Single-Phase Earth Fault Protection Using Improved PCA in Distribution Systems," *IEEE Trans. Power Deliv.*, vol. 30, no. 4, pp. 1974–1982, Aug. 2015.

20 D. Randall, "*Empirical Orthogonal Functions*," Department of Atmospheric Science, Colorado State University, Colorado, USA, Jan. 2003.

21 F. Calero and D. Hou, "Practical considerations for single-pole-trip line-protection schemes," in *58th Annual Conference for Protective Relay Engineers, 2005.*, 2005, pp. 69–85.

22 D. W. P. Thomas, M. S. Jones, and C. Christopoulos, "Phase Selection Based on Superimposed Components," *Transm. Distrib. IEE Proc. - Gener.*, vol. 143, no. 3, pp. 295–299, May 1996.

23 W. M. Al-hassawi, N. H. Abbasi, and M. M. Mansour, "A neural-network-based approach for fault classification and faulted phase selection," in *Canadian Conference on Electrical and Computer Engineering, 1996*, 1996, vol. 1, pp. 384–387 vol. 1.

24 H. W. Dommel, "Digital Computer Solution of Electromagnetic Transients in Single-and Multiphase Networks," *IEEE Trans. Power Appar. Syst.*, vol. PAS-88, no. 4, pp. 388–399, Apr. 1969.

25 A. Ametani, N. Nagaoka, Y. Baba, and T. Ohno, *Power System Transients: Theory and Applications*. CRC Press, 2013.

26 J. R. Marti, "Accuarte Modelling of Frequency-Dependent Transmission Lines in Electromagnetic Transient Simulations," *IEEE Trans Power App Syst*, vol. PAS-101, no. 1, pp. 147–157, Jan. 1982.

27 G. D. Guidi and F. E. Perez, "MATLAB program for systematic simulation over a transmission line in alternative transients," presented at the International Conference on Power Systems Transients, Vancouver, Canada, 2013.

28 P. M. Anderson and A. A. Fouad, *Power System Control and Stability*, 2nd ed. John Wiley & Sons, 2002.

29 R. Aguilar, F. Perez, and E. Orduna, "Ultra-sensitive disturbances detection on transmission lines using principal component analysis," in *24th IEEE Canadian Conference on Electrical and Computer Engineering (CCECE)*, Ontario, Canada, 2011, pp. 93–98.

30 G. Benmouyal and J. Roberts, "Superimposed quantities: Their true nature and application in relays," in *30th Annual Western Protective Relay Conference*, Spokane, 2003.

18

Implementation of a Real Phasor Based Vulnerability Assessment and Control Scheme: The Ecuadorian WAMPAC System

Pablo X. Verdugo[1], Jaime C. Cepeda[2], Aharon B. De La Torre[1], and Diego E. Echeverría[1]

[1] *Research and Development Engineer, Technical Development Department, Operador Nacional de Electricidad, CENACE, Quito, Ecuador*
[2] *Head of Research and Development, Technical Development Department, Operador Nacional de Electricidad, CENACE, Quito, Ecuador*

18.1 Introduction

With each passing year, the operation of electric power systems experiences several technical challenges associated with the new paradigms of management and planning such as the inclusion of deregulated markets, the interconnection with neighboring regional systems, the diversification of energy sources, the inclusion of environmental restrictions, and even the massive penetration of nonconventional generation plants. Under these conditions, security and reliability can be seriously compromised since unexpected disturbances may cause violations to the security limits established for the electric power system leading to the outage of important system elements and even partial or total blackouts [1].

Real time supervision of the static and dynamic security of the power system plays a fundamental role inside the applications used in control centers. The main purpose of real time monitoring is to present an early warning to the system operators so they can execute adequate control actions and thus mitigate potentially harmful stress conditions in the system. In this context, besides SCADA/EMS functionalities, it is necessary to include complementary technological solutions to evaluate and improve the security of the power system in real time; synchrophasor measurement systems (PMU/WAMS) are part of these technologies [2]. In addition, other Smart Grid applications have also been designed in order to perform self-healing and adaptive reconfiguration actions based on system-wide analysis with the objective of reducing the risk of power system blackouts.

Phasor Measurement Units (PMUs) are devices that allow the estimation of current and voltage synchrophasors in different areas of an electric power system. Their high precision, response speed, and time synchronization make PMUs viable equipment for monitoring the steady and dynamic states of an electric system and also for applications of protection and control within a Wide Area Monitoring, Control, and Protection (WAMPAC) system. The Ecuadorian National Interconnected System (SNI from the

Dynamic Vulnerability Assessment and Intelligent Control for Sustainable Power Systems, First Edition.
Edited by José Luis Rueda-Torres and Francisco González-Longatt.
© 2018 John Wiley & Sons Ltd. Published 2018 by John Wiley & Sons Ltd.

acronym in Spanish) is rather small, and up until 2003 it had no permanent connections with neighboring power systems. Nowadays, the Ecuadorian SNI maintains a synchronous connection with the Colombian system, which is a little less than three times larger in terms of load demand. Several issues have been presented in this interconnected system's operation, such as: oscillatory instability, large congestions in transmission power lines, and partial collapses due to single contingencies, to name but a few. It is in this connection that the need to implement a reliable monitoring system arose. Therefore, since 2010, the Ecuadorian ISO, CENACE, has undertaken a project to implement a WAMPAC infrastructure in Ecuador that facilitates the real time monitoring, supervision, and control of the Ecuadorian system through the use of synchrophasor measurements. This initiative was based on the current availability of synchronized phasor measurement technology for monitoring, in real time, highly stressed operating conditions that might eventually cause problems of steady-state angle or voltage stability, as well as oscillatory stability risks. Within the past few years, CENACE has concluded the first stage of the project, which consists in the installation of 30 PMUs around the SNI along with a complex optical fiber communication infrastructure and a sophisticated phasor data concentrator (PDC), administrated by the WAProtector™ software developed by the Slovenian Company ELPROS. WAProtector™ manages the PMU data via the intranet communication network, and it connects to the PMUs using the IEEE C37.118 communication protocol [3]. This platform constitutes CENACE's Wide Area Monitoring System (WAMS). Likewise, a smart infrastructure based on synchrophasor technology for real time control of N-2 critical contingencies has been also implemented in Ecuador. This redundant control scheme is made up of 58 relays with built-in PMU functionality. Together, these two technological infrastructures constitute the current Ecuadorian WAMPAC System.

18.2 PMU Location in the Ecuadorian SNI

One of the criteria used for determining the location of PMUs is system observability, whilst aiming to maximize redundancy by using the fewest PMUs without compromising the desired observability of the system. Several authors have developed mathematical models to determine optimal PMU placement in an electric system using optimization methods [4]. Another criterion for PMU placement relies on placing PMUs at the most important or most significant substation buses, although this is considered subjective since it depends on the information that is required for a specific analysis. As a starting point, PMU placement in the SNI was defined based on monitoring requirements, taking into consideration the most relevant buses in the Ecuadorian power system and the operating criteria. Once the initial WAMS was implemented and the first results were gathered, it was possible to either identify new areas for PMU placement or relocate the ones previously installed with the aim of obtaining full system observability not only related to steady-state conditions but also to dynamic system behavior. Therefore, the necessity of monitoring different areas that are considered to have a relatively high operative relevance, in order to perform a precise and reliable evaluation of the system performance, has been defined in a procedure for deciding further PMU locations. In general terms, the main objectives for installing PMUs in the SNI, which have been defined in CENACE's procedure, are:

- Supervising the real time dynamic operation of the SNI to allow the operators to take preventive actions when facing instability risks (early warning).
- Having more precise information and tools to perform stability analysis and determine the presence of poorly damped oscillation modes.
- Performing post-mortem analyses in order to evaluate the system behavior and then improve the procedures to restore the energy supply after fault events.
- Tuning of Power System Stabilizers (PSS) and validation of control systems models.
- Monitoring the most congested transmission corridors via steady-state angle stability and voltage stability assessment tools.

Once each PMU is located, the measured data is delivered to the control center where WAProtector enables PMU management and further power system analysis. Inside the WAProtector server, data analysis and real time assessment of system security, including the evaluation of steady-state and oscillatory stability phenomena, are carried out. The main applications available in WAProtector are: steady-state angle stability (phase angle difference), steady-state voltage stability of transmission corridors, oscillatory stability, islanding detection, harmonic distortion analysis, historic data, and system event analysis, among others. WAProtector's applications enable early-warning alerts when pre-specified thresholds are exceeded; however, these limits must be configured by the user. Therefore, it is necessary to specify an adequate reference framework to allow the monitoring of specific power system thresholds as regards each one of the available supervisory applications. In this connection, three methodologies were proposed by CENACE for determining adequate thresholds regarding: (i) phase angle difference, (ii) voltage profile power transfer of transmission corridors, and (iii) oscillatory issues. These limits are the referential framework for assessing stability in real time, and constitute indicators that give the operators real time early-warning signals of the possible risk of system stress conditions.

18.3 Steady-State Angle Stability

Angle stability refers to the ability of the synchronous machines of an interconnected power system to remain in synchronism after the system has been subjected to a disturbance. To maintain synchronization, it is necessary to keep or restore the balance between the mechanical and electromagnetic torques of each synchronous machine [5].

The angular difference between two buses of a power system is a direct measure of the transfer capacity between these nodes. Figure 18.1 illustrates two areas (A and B) of a power system interconnected through a set of electric branches.

Figure 18.1 Power transfer between two system buses.

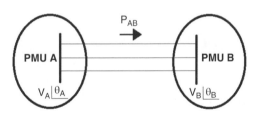

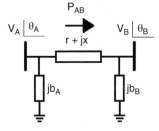

Figure 18.2 "π" equivalent of system branches.

Assuming the "π" equivalent model for the set of branches between the two areas, presented in Figure 18.2, the power transfer between A and B is given by (18.1):

$$P_{AB} = \frac{V_A^2 r - r V_A V_B \cos(\theta_A - \theta_B) + x V_A V_B \sin(\theta_A - \theta_B)}{r^2 + x^2} \tag{18.1}$$

where V_A and V_B are the voltage phasor magnitudes at buses A and B, respectively; θ_A and θ_B represent voltage phasor angles at buses A and B, respectively; x is the series reactance between buses A and B; and r corresponds to the series resistance.

Considering that $x \gg r$ in a high-voltage power system, (18.1) can be reduced to (18.2), whose graphic representation is depicted in Figure 18.3.

$$P_{AB} = \frac{V_A V_B \sin(\theta_A - \theta_B)}{x} \tag{18.2}$$

Mathematically, maximum power transfer occurs when $\sin(\theta_A - \theta_B) = 1$, that is when $\theta_A - \theta_B = 90°$. Nevertheless, due to the complexity of the power system that can cause congestion in the transmission network, it is usually not possible to reach this theoretical limit. Thus, the maximum difference between nodal phase angles depends on the actual system topology, and it has to be computed for each electric power system. Considering that $\theta_A - \theta_B$ usually presents small values in electric power systems that satisfy the

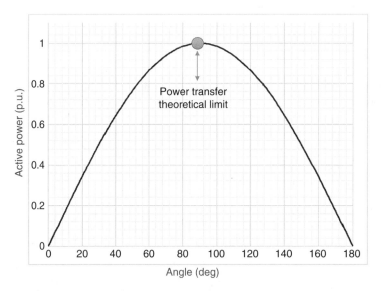

Figure 18.3 Power–angle curve.

steady-state angle stability constraint, the power flow transferred by the branch between buses A and B, as shown in (18.3), is proportional to the angle difference between these two buses:

$$P_{AB} \approx \frac{V_A V_B(\theta_A - \theta_B)}{x} \Rightarrow P_{AB} \propto (\theta_A - \theta_B) \tag{18.3}$$

In this connection, the actual power transfer limit between buses A and B depends on the maximum phase angle difference between these nodes, and vice versa. Therefore, to determine the phase angle difference maximum limit (defined as steady-state angle stability limit), it is necessary to reach the power transfer limits through the transmission corridor. Thus, the monitoring of phase angle differences between the system buses gives the operator an early warning of potential system congestion. WAProtector has an application that allows monitoring the angle difference of phasor voltages at buses supervised by PMUs through advanced visualization applications. Figure 18.4 shows an example of the graphic display via dynamic contour plots of the angle difference in the SNI in real time. It is possible to observe how the contour of the Ecuadorian map turns into a light-gray like tone in correspondence with an increase of the congestion of the network and thus the growth of the angular difference. Additionally, the software allows setting the limit values of angle separation between buses, with the aim of providing

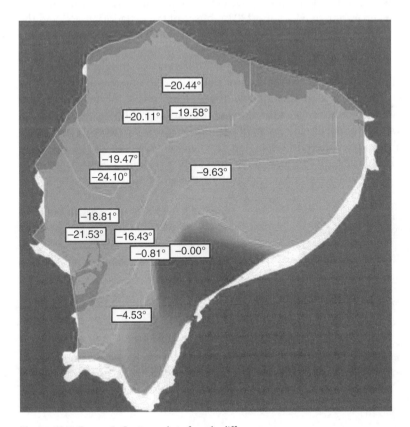

Figure 18.4 Dynamic Contour plot of angle differences.

the operator an adequate warning. These values correspond to the limits of steady-state angle stability, which should be properly determined.

As a starting point, it is necessary to define a reference bus (in this case, the Molino 230 kV substation, where one of the largest generation plants in Ecuador, Paute, is located), in order to give the operator an adequate reference as to provide an early warning of the risk of congestion, under the premise of maintaining the steady-state system security. The proposal consists of performing static simulations to assess the most critical power transfer conditions in the SNI, considering the potential occurrence of N-1 contingencies and the gradual increase of nodal loads on particular system areas. The proposed mechanism of analysis is to simulate, by means of power system simulator software (DIgSILENT PowerFactory in this case), the behavior of the power transfers under different operating conditions (high and low hydrological scenarios with low, medium, and high demand conditions), including contingency analysis and the gradual increase of load. The approach presented in Figure 18.5 is proposed to this end. The methodology begins with the definition of scenarios. Subsequently, case studies that include the operational conditions to be analyzed are prepared. Then, a DPL (DIgSILENT Programming Language) script is executed that controls an iterative N-1 contingency analysis of several transmission elements associated with gradual increases of load in specific areas (50 previously defined critical contingencies are simulated). This analysis allows reaching high-stress system conditions, and thus determining the maximum angular separation between the 230 kV buses of the Ecuadorian SNI and the Molino substation. Interested readers can find further explanation of this methodology in [2].

The mean (μ) of the established maximum angular differences is the alert limit. Moreover, the alarm limit corresponds to the sum of the mean and one standard deviation (σ) of the sample. This is based on the criterion that at least 50% of the analyzed values should be included within the range of the standard deviation around the mean ($\mu + \sigma$), known as Chebyshev's inequality [6]. As an illustration, Figure 18.6 presents the resulting histogram of the angle differences between Pascuales and Molino 230 kV buses for high hydrological scenarios. Additionally, Table 18.1 summarizes the results of alert and alarm limits for all SNI buses with PMUs for high hydrological scenarios.

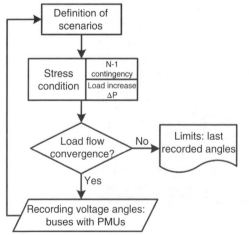

Figure 18.5 Methodology for determining the steady-state angle stability limits.

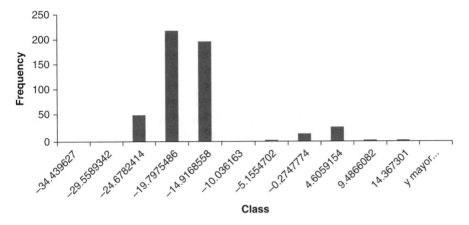

Figure 18.6 Histogram of angle differences between Pascuales and Molino buses for high hydrological scenarios.

Table 18.1 Alert and alarm limits for SNI buses as regards Molino bus—high hydrological scenarios.

230 kV Bus	Alert Limit	Alarm Limit
Milagro	−15.02°	−20.74°
Pascuales	−18.86°	−25.65°
Pomasqui	−24.27°	−35.62°
Quevedo	−21.61°	−29.68°
Santa Rosa	−23.42°	−33.98°
Totoras	−11.86°	−18.20°
Zhoray	−0.98°	−2.16°

18.4 Steady-State Voltage Stability

Voltage stability refers to the ability of a power system to maintain steady voltages in all buses in the system after being subjected to a disturbance from a given initial operating condition [5]. Computation of Power–Voltage (P-V) curves and the corresponding available transfer capability are the tools most commonly used to analyze the power system steady-state voltage stability [7]. The emerging synchronized phasor measurement technology has allowed the development of novel methodologies to monitor the power system voltage stability in real time. One of the most promising techniques is the so-called *Thevenin equivalent* method, which allows computing the proximity of the actual operational state to the voltage collapse via the determination of the P-V curve in real time. This tool is mainly being used for monitoring the voltage stability of transmission corridors since it allows determination of the power system relative strength in regards to the load buses [7]. This type of application is offered by WAProtector.

Nevertheless, one of the main challenges is to determine the adequate reference early-warning indicators that allow configuring the real time monitoring application.

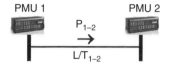

PMU 1   PMU 2   **Figure 18.7** Transmission corridor monitored by PMUs.

In this connection, a methodology for determining the voltage profile power transfer limits of the monitored transmission corridors is proposed. These transfer limits are the referential framework for assessing voltage stability in real time, and constitute the early-warning indicators for the system operators. The Thevenin equivalent method allows estimating, in real time, the P-V curve of transmission corridors whose sending bus (B1) and receiving bus (B2) have PMUs installed. Figure 18.7 illustrates a transmission corridor monitored by PMUs.

The receiving end is considered as a load bus ($\overline{Z}_{app}$), which can be calculated with the voltage and current phasor measurements obtained from PMU 2. This end is connected to a Thevenin equivalent consisting of an equivalent voltage generator ($\overline{V}_{th}$) in series with an equivalent impedance ($\overline{Z}_{th}$). Based on this approach, the transmission corridor of Figure 18.7 can be represented by the Thevenin equivalent depicted in Figure 18.8.

Each parameter of the Thevenin equivalent can be easily determined by using the phasor measurements of PMU 1 and PMU 2, as shown by (18.4) and (18.5) [7]:

$$\overline{Z}_{th} = \frac{\overline{V}_1 \overline{V}_1 - \overline{V}_2 \overline{V}_2}{\overline{V}_1 \overline{I}_1 - \overline{V}_2 \overline{I}_2} \tag{18.4}$$

$$\overline{V}_{th} = \overline{V}_1 \left(\frac{\overline{V}_2 \overline{I}_1 - \overline{V}_1 \overline{I}_2}{\overline{V}_1 \overline{I}_1 - \overline{V}_2 \overline{I}_2} \right) \tag{18.5}$$

where $\overline{V}_1$ and $\overline{I}_1$ are the voltage and current phasors measured at PMU 1 (sending end), and $\overline{V}_2$ and $\hat{I}_2$ are the voltage and current phasors measured at PMU 2 (receiving end).

By using the Thevenin equivalent parameters and the computed receiving end load power (i.e., the transfer power), it is possible to determine V_2 (voltage magnitude of receiving end) as a function of the transfer power, as shown by (18.6) [7]:

$$V_2 = \sqrt{\frac{V_{th}^2}{2} - (Q_l X_{th} + P_l R_{th}) \pm \sqrt{\frac{V_{th}^4}{4} - V_{th}^2 (Q_l X_{th} + P_l R_{th}) - (P_l X_{th} - Q_l R_{th})^2}} \tag{18.6}$$

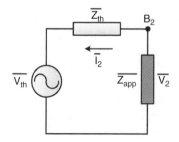

Figure 18.8 Thevenin equivalent of a transmission corridor.

where P_l and Q_l are the load's active and reactive power, R_{th} and X_{th} are the real and imaginary parts of the Thevenin equivalent impedance (i.e., equivalent resistance and reactance), and V_{th} is the magnitude of the Thevenin equivalent voltage.

The expression (18.6) constitutes the mathematical representation of the P-V curve of the transmission corridor determined via the Thevenin equivalent method. This curve can be easily determined using real time measurements of voltage and current phasors of PMUs installed at each end of the corridor. WAProtector has an application that computes, in real time, the P-V curve of transmission corridors using the Thevenin equivalent method. Similarly to the steady-state angle stability application, this voltage stability function requires the adequate setting of predefined alert and alarm thresholds, which will give the operators an early-warning signal. These security limits correspond to the so-called voltage profile stability transfer limits, which are determined by applying the existing relationship between P-V curves and the voltage profile stability band, as stated in [8].

Figure 18.9 illustrates a transmission corridor P-V curve obtained via the Thevenin equivalent method in combination with the voltage profile stability band (shaded area). This voltage profile stability band is formed by the upper and lower limits of voltage values for normal voltage operation. The voltage profile stability transfer limit is then calculated considering the first intersection of the P-V curve with the stability band [8], that is, point A in Figure 18.9. For SNI, two bands have been defined based on the concept of normal (−3%) and emergency (−6%) operation [7]. Thus, the alert and alarm threshold values are the result of the P-V curve intersection with the normal and emergency bands, respectively (i.e., normal and emergency voltage profile stability transfer limits).

In order to iteratively consider the most critical power transfer conditions in the SNI, taking into account the potential occurrence of N-1 contingencies as well as high and low

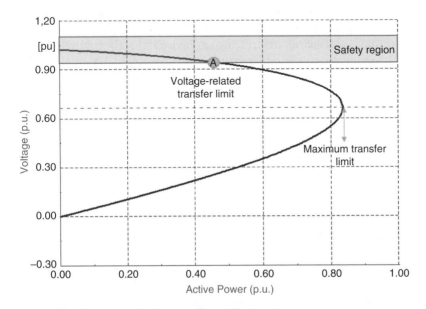

Figure 18.9 P-V curve and voltage profile stability band.

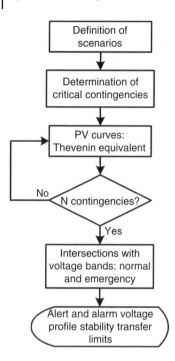

Figure 18.10 Methodology for determining the voltage profile stability transfer limit of transmission corridors.

hydrological scenarios, the methodology depicted in Figure 18.10 has also been included into DIgSILENT PowerFactory via DPL programing scripts. All the formulation of the Thevenin equivalent method has been included in the DPL script in order to compute P-V curves for multiple scenarios. Interested readers can find further explanation of this methodology in [7].

As an illustration, the P-V curves of the double circuit Totoras–Santa Rosa 230 kV transmission line, per circuit and per unit based on 300 MW, considering maximum load and a high hydrological scenario, are presented in Figure 18.11. The continuous P-V curve corresponds to a normal condition scenario whereas the dotted P-V curve results from the worst voltage-related contingency. The intersections of the worst contingency P-V curve (dotted) with the lower limits of the bands (i.e. −3% and −6%) correspond to the alert and alarm voltage profile stability transfer limits, respectively.

With the objective of getting the most representative values of alert and alarm transfer limits, the means of the values obtained for each hydrological scenario are computed. Table 18.2 presents a summary of the alert and alarm limits (mean values) for the Totoras–Santa Rosa 230 kV transmission line considering both high and low hydrological scenarios.

18.5 Oscillatory Stability

The rotor angle stability problem involves the study of the electromechanical oscillations inherent in power systems. Oscillatory instability occurs when there is lack of damping torque [5]. Oscillation problems may be either local or global in nature [9]. Local problems (local mode oscillations) are associated with oscillations among the rotors of

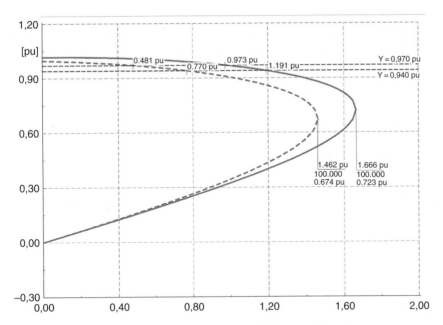

Figure 18.11 P-V curves of the Totoras–Santa Rosa 230 kV transmission line.

Table 18.2 Alert and alarm voltage limits of Totoras–Santa
Rosa 230 kV transmission line per circuit.

	High hydrology	Low hydrology
Alert limit (MW)	254.4	149.4
Alarm limit (MW)	340.8	200.1

a few generators close to each other. These oscillations have frequencies in the range of
0.7 to 2.0 Hz. Global problems (inter-area mode oscillations) are caused by interactions
among large groups of generators. These oscillations have frequencies in the range of 0.1
to 0.7 Hz. There is another type of oscillatory problem caused by controllers of different
system components, named control modes [9]. These types of oscillations present a large
associated range of frequencies. From these types of modes, the Ecuadorian National
Interconnected System presents the so-called very low frequency control modes (in the
range of 0.01 to 0.1 Hz), which usually appear in systems with high penetration of hydroelectric plants and are associated with improper tuning of hydroelectric unit governors.

It is well-known that the damping ratio associated with system dominant modes can
vary significantly depending on different factors, such as, for instance, grid strength,
generator operating points, loading conditions, and the number of power transfers.
Thus, continuous evaluation of the system's damping performance should be performed
in order to capture the operating conditions that could entail major damping concerns.
In this connection, WAProtector possesses a modal identification algorithm that
allows the identification of critical oscillatory modes from the PMUs power flow
measurements. This tool allows estimation of the frequency, damping, magnitude, and

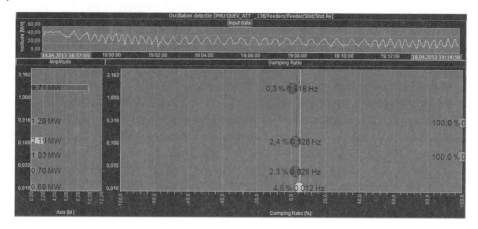

Figure 18.12 Oscillatory event recorded by WAProtector.

phase of the critical modes that are observable in a given power flow signal record. The real time data analysis functionality of WAProtector is also complemented by the option of obtaining hourly statistical information of the most predominant oscillatory modes detected by each PMU in order to perform post-operative analysis. Figure 18.12 shows WAProtector's oscillatory application for the active power signal recorded in the PMU installed at Quevedo substation after the outage of the Quevedo–San Gregorio 230 kV transmission line. After this event, power oscillations of large amplitude were recorded, while the transmission line was out of service.

By using the modal identification algorithm, the active power signal immersed modes are determined during the event period, showing the appearance of a local mode at a frequency of 1.918 Hz, damping of 0.3% and a large amplitude of about 9.71 MW. Therefore, this event caused a sustained oscillatory phenomenon that was alerted to the operators via the WAMS application. WAProtector also allows calculating the damping hourly average of dominant oscillatory modes, according to frequency ranges. For this purpose, the modal identification algorithm permanently updates the required windows and calculations, which are recorded in the appropriate databases. It is worth mentioning that the modal identification algorithm of WAProtector constantly regulates itself in order to determine the modes presented in the power signal in the range of 0 to 10 Hz. Through the methodology depicted in the flowchart presented in Figure 18.13, initially introduced by the authors in [10], a statistical analysis of the information obtained from the recorded data of the PMUs located in the power system is performed in order to characterize the oscillatory behavior of the SNI and to determine the most typical amplitudes of oscillations (used to define the alert and alarm amplitude limits).

From the complete information, only modes that present a damping lower than the predefined alert and alarm thresholds are analyzed. These thresholds have been agreed as 5% (alert limit) and 3% (alarm limit) according to the SNI small signal stability periodical study [10]. The proposal corresponds to a statistical analysis carried out via histograms and computation of means and standard deviations of the modes. As an example, Figure 18.14 presents a histogram of the amplitudes of the modes with a damping smaller than 5%, registered within one month.

Figure 18.13 Methodology for determining amplitude oscillation limits.

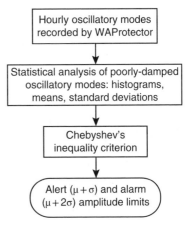

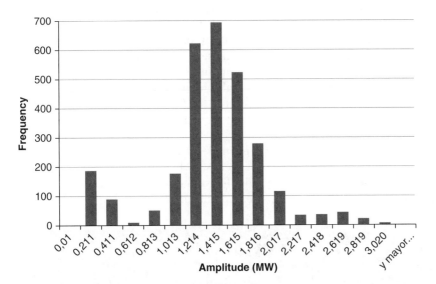

Figure 18.14 Histogram of inter-area mode amplitude.

This statistical information of oscillatory modes is then employed to determine the limits of oscillation amplitudes that will be used as an early-warning reference for real time monitoring as the alert and alarm settled thresholds. With the statistical information, the alert and alarm limits of amplitudes are determined based on the mean (μ), standard deviation (σ), and the above mentioned Chebyshev's inequality criterion [6], which specifies that at least 50% of the analyzed values should be in the range of one standard deviation around the mean ($\mu \pm \sigma$), while 75% of the analyzed values should be within the range covered by two standard deviations ($\mu \pm 2\sigma$). In this context, the values presented in Table 18.3 have been determined as the alert and alarm limits for the magnitude of the SNI registered oscillations, for the low frequency control modes, inter-area modes, and local modes.

Table 18.3 Alert and alarm amplitude limits for SNI oscillatory modes.

	Control modes	Inter-area modes	Local modes
Alert limit (MW)	2.5	1.8	0.4
Alarm limit (MW)	3.5	2.4	0.6

18.5.1 Power System Stabilizer Tuning

As was previously mentioned, the Ecuadorian SNI maintains a permanent 60 Hz synchronous connection with the Colombian power system through four 230 kV transmission lines. It is to be considered that due to voltage control tactics, some of these transmission lines are disconnected in low demand scenarios, increasing the impedance in the connection between the two countries. Several offline analyses and constant monitoring of the oscillatory stability of the Ecuador–Colombia interconnected power system, via the Ecuadorian WAMS, have shown that there is a permanent presence of a poorly damped inter-area mode whose frequency varies within the range of 0.4 to 0.5 Hz. To be more accurate, this mode is mostly seen around a frequency of 0.45 Hz. Figure 18.15 shows the damping and the amplitude of this inter-area mode from WAProtector's Oscillatory Stability tool, as registered by the PMU connected closest to the Colombian system, during a severe event in the SNI. For didactic purposes, the Ecuadorian 230 kV transmission corridor and its connection to Colombia is presented in Figure 18.16.

It is known that the damping of inter-area modes declines with an increase in the congestion of transmission lines, and consequently with an increase in the angle difference between areas. In this regard, in order to avoid unstable oscillatory conditions in the Ecuadorian system, in some scenarios it has been necessary to limit the power

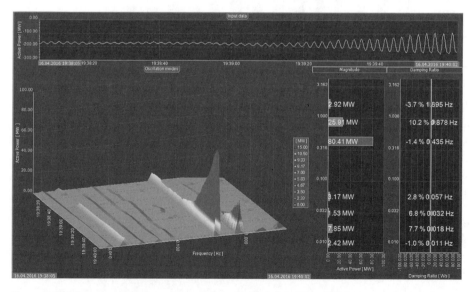

Figure 18.15 Presence of the inter-area mode in the Ecuador–Colombia interconnected system.

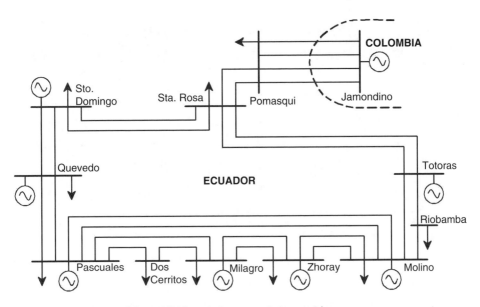

Figure 18.16 Diagram of the 230 kV Ecuadorian transmission corridors.

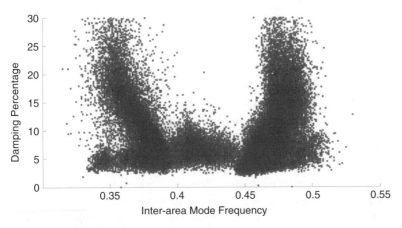

Figure 18.17 Scatter plot of the frequency and damping percentage of the inter-area modes observed in 2015.

transfer between Colombia and Ecuador, particularly from Ecuador to Colombia, and to avoid disconnecting the interconnection transmission lines. It is easy to see then that improving the damping of the 0.45 Hz inter-area mode will benefit the operation of the Ecuadorian SNI both technically and economically. In order to understand the criticality of the problem, a scatter plot of the frequency and damping of all inter-area modes that appeared during 2015 is presented in Figure 18.17; this has been extracted from the PMUs' available data via the application of the modal identification algorithm of WAProtector. As was previously mentioned for the Ecuadorian system, a mode is considered poorly damped when its damping percentage is smaller than 5%. It is possible to observe that there are two different inter-area modes in the Ecuador–Colombia

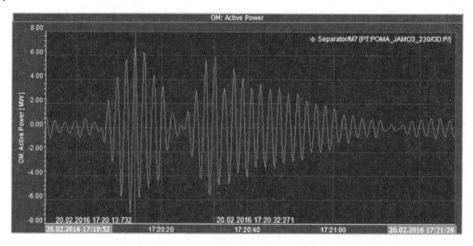

Figure 18.18 Oscillations in the Ecuadorian power system caused by a loss of generation event.

interconnected power system. One of them corresponds to the 0.45 Hz inter-area mode, which involves the participation of generation units in Colombia and in Ecuador, and the other one whose frequency oscillates between 0.35 and 0.4 Hz involves the action of generators connected in the Ecuadorian system alone.

As stated, in order to improve the oscillatory stability in the Ecuador–Colombia interconnected power system, which will ultimately allow increasing the power transfer between these countries, a PSS tuning is done with focus on the 0.45 Hz inter-area mode. According to [12], the phase of oscillations detected in different parts of a system show the differences in the damping contribution from different generators. Thus, generators with leading phase provide less damping to the oscillation mode, while generators with lagging phase provide more damping. In this context, through a phase analysis of signals registered in different PMUs connected in the Ecuadorian power system decomposed in different frequency ranges using the WAProtector modal identification algorithm, it is possible to determine where the 0.45 Hz is more observable, and which generators present a larger participation in this mode—in other words, it is possible to establish the oscillation source location. To illustrate this procedure, Figure 18.18 presents an oscillation detected in the Pomasqui substation. In this particular case, the 0.45 Hz inter-area mode is excited due to a response of the Paute generation plant to a loss of generation event in the Ecuadorian system. The phase analysis of this oscillation using the modal identification algorithm is presented in Figure 18.19. Through the phase analysis, it is possible to determine that, as the inter-area mode was excited, the signals observed in Pomasqui (the closest substation to Colombia), Santa Rosa, and Molino present the highest phase lead; and the signal in Quevedo, a substation not included in the path of the Ecuador–Colombia inter-area mode, presents a phase lag, therefore there are no oscillation sources located in this area. It is then possible to conclude that generators in Colombia and Paute hydroelectric power plant are the biggest contributors to the 0.45 Hz inter-area mode. Paute is one of the largest generation plants in Ecuador and consists of two plants: plant AB connected to the Molino 138 kV substation and plant C connected to the Molino 230 kV bus.

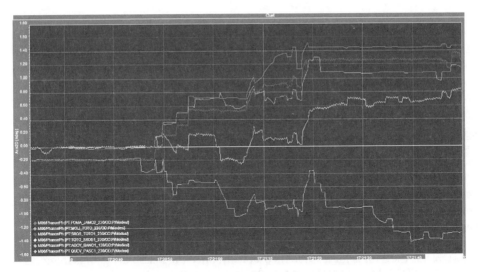

Figure 18.19 Phase analysis of the angle of signals obtained from PMUs.

Once it had been determined that the Paute generation plant presents the largest contribution to the 0.45 Hz inter-area mode, from the Ecuadorian side, and considering some technological limitations regarding the structure of the PSSs installed in phase C and the fact that these devices were tuned to improve the damping of a local mode inherent to the plant, a tuning process was carried out for the existing PSSs in phase AB. It is worth mentioning that within this plant, only unit U2 has a digital excitation system. In general terms, Figure 18.20 presents the methodology for carrying out the PSS tuning of the Paute plant's AB generators based on synchrophasor measurements, which also includes the previously presented step.

Figure 18.20 Procedure for tuning Paute plant AB stabilizers.

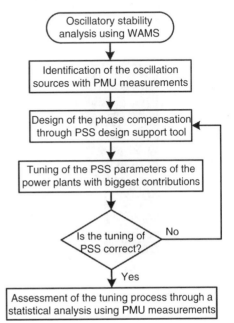

For theoretically tuning the power stabilizers in the Paute AB generation plant, the Generator Exciter Power System (GEP) approach was used along with the Single Machine Infinite Bus methodology [13]. The PSS installed in Paute's unit U2 resembles a PSS2B standard, a dual input PSS that is capable of using both generator speed and power, however the speed input was disabled since Ecuador is prone to large frequency deviations due to the predominantly hydro generation and relatively low inertia of the power system [14]. The phase compensation parameters were determined to provide a phase shift in PSS+AVR+Gen as close to −90° as possible, for the main mode frequency, and the washout filter parameters were defined to reduce the low frequency modes and to allow full observability of the 0.4–0.5 Hz inter-area mode.

It is to be noted that the washout transfer function also introduces a phase lead that needs to be accounted for in the design of the phase compensation [13]. Once the parameters were computed, a complete set of field tests using real time information from PMUs connected to each generation unit in Paute was carried out. The gain of the PSS was set based on online tests in order to avoid instability of control modes that would become less stable as the gain is increased. Theoretically, once the instability gain is found, it is reduced to 1/3 of this value [15]. It should be mentioned that during the tuning exercise it was not possible to find control mode instability; therefore the gain was limited not by control stability issues, but rather by the allowable MVAr swings in the generation unit [14].

Once the PSS tuning of the Paute AB generation plant was concluded, it was necessary to carry out a monitoring stage in order to identify the effectiveness of the process. For this purpose, the same WAProtector modal identification algorithm was used in order to determine the damping of the inter-area mode before and after the tuning, considering a one month test period. It is possible to determine that the tuning of the PSSs of the Paute AB generation plant had a positive effect on damping the 0.45 Hz inter-area mode, as can be seen in the bar plot presented in Figure 18.21. From this data, it has been determined

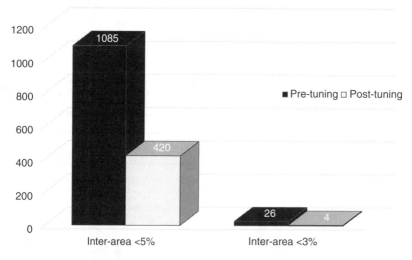

Figure 18.21 Bar plot comparing the appearances of the 0.45 Hz inter-area mode pre-tuning and post-tuning.

that the number of appearances of this mode with a damping percentage below 5% has decreased by about 61%.

Interested readers can find further explanation of this methodology in [16].

18.6   Ecuadorian Special Protection Scheme (SPS)

Special Protection Schemes (SPS) are designed in order to detect abnormal conditions and carry out corrective actions that mitigate possible consequences and allow an acceptable system performance. However, the conditions that lead the system to black-outs are not easy to identify mainly because the process of system collapse depends on multiple interactions. In this connection, some Smart Grid applications have been designed in order to perform timely self-healing and adaptive reconfiguration actions based on system-wide analysis, with the objective of reducing the risk of power system blackouts. Real time dynamic vulnerability assessment (DVA) has to be done first in order to decide and coordinate the appropriate corrective control actions, depending on the event evolution. Emerging technologies such as PMUs and WAMS offer a new benchmark for performing post-contingency DVA that could be used to trigger SPSs in order to implement enhanced corrective control actions [17].

In the Ecuadorian case, ensuring efficiency in the daily operation of the SNI has become a paramount concern due to several factors that have increased the potential sources for system disturbances and occasionally lowered the operational security and flexibility. Among these factors are weakly meshed topology and unsymmetrical distribution of generation and load, especially for distant portions of the eastern and southern areas that are radially connected to the main 230 kV transmission ring [18, 19]. Component failures and protections operation due to bottlenecks in key transmission corridors have sometimes led to adverse consequences such as partial or widespread disruptions. Occasionally, overloading of certain transmission facilities occur, especially due to high power transfers from bulk generation located in the central and southern areas to the main consumption centers in the northern (i.e., Quito) and southwestern (i.e., Guayaquil) areas; this can entail congestion, unexpected generation outages, load shedding, and/or poor dynamic performance. In this connection, the Special Protection Scheme (SPS) portrayed in this section was designed to prevent out-of-step conditions between Ecuador and Colombia or the separation of both grids, which can lead to a grid collapse in Ecuador. The study results, based on different operational scenarios that consider double contingencies of the 230 kV main trunk transmission lines, and subsequent statistical and regression analyses, have led to the establishment of tables of mitigation actions that must present a quick response (less than 200 ms) in real time to predefined contingencies. These tables represent functions that allow determination of the place where the contingency occurs and the loads that are in the affected and nearby circuits, in order to execute adequate countermeasures. The mitigation actions comprise the automatic shedding of computed quantities of generation and load at pre-specified locations or regions. The Ecuadorian SPS is defined as an automatic system, composed of a set of elements of protection, control, and communication networks, which acts upon the occurrence of predefined events (the untimely disconnection of critical SNI branches), executing a post-contingency corrective action composed of calculated load shedding and disconnection of generation, in order to avoid or mitigate problems of instability. The SPS is a wide area controller of stability of a power system that is capable of responding in a very short time (in the order of milliseconds) [20]. The Ecuadorian SPS is based on synchrophasor technology that makes up a redundant control scheme made up by 58 local relays with built-in PMU functionality installed in the main SNI substations. Inside the so-called Mitigation Matrix there is a set of equations, for each of the strategies listed in Table 18.4 and represented in Figure 18.22, that allows calculating the right amount of load and generation that must be disconnected in order to avoid system instability, on occurrence of one of the eight defined N-2 contingencies, taking into account the pre-fault power flow in the transmission corridors. The reader can find further information about the SPS implemented in Ecuador in [21].

Table 18.4 Implemented strategies in the SPS.

Strategy	N-2 Transmission lines contingencies
1	Santa Rosa–Totoras
2	Santo Domingo–Santa Rosa
3	Santo Domingo–Quevedo and Quevedo–Baba
4	Quevedo–Pascuales
5	Totoras–Molino and Molino–Riobamba
6	Milagro–Zhoray
7	Molino–Pascuales
8	Santo Domingo–Quevedo and Santo Domingo–Baba

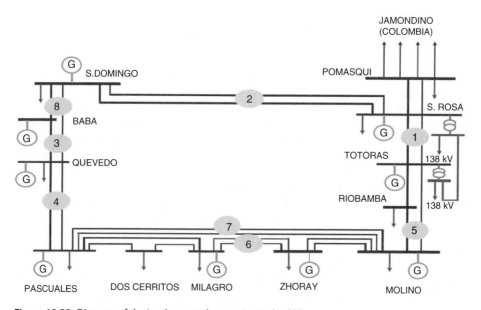

Figure 18.22 Diagram of the implemented strategies in the SPS.

18.6.1 SPS Operation Analysis

On May 6, 2015, the Molino–Pascuales 230 kV transmission line got disconnected (strategy 7) with a pre-fault power flow of 414 MW. According to the analysis performed, the strategy programmed in the SPS for this contingency should be activated with a power flow of 350 MW. In accordance with the expressions defined for this strategy, as depicted in (18.7) and (18.8), the execution of the mitigation actions produced the disconnection of 114 MW of generation and 123 MW of load, causing around 89.2 MWh of energy not supplied. It is worth mentioning that it only took 141.7 ms, after the detection of the switching of the transmission line, for the SPS to execute all mitigation actions.

$$DP = k_{7\_1}(P - Pset_{7\_2}) + Pset_{7\_3} \tag{18.7}$$

$$DPLoad = k_{7\_2}DP_{actual} \tag{18.8}$$

where DP and $DPLoad$ are the active power values that need to be disconnected from generators and loads, respectively, P is the measured pre-fault power flow through the transmission line, DP_{actual} is the real disconnected amount from generation, and k_{7_1}, k_{7_2}, $Pset_{7\_2}$ and $Pset_{7\_3}$ are the previously defined constants via the above mentioned regression analysis. These computations are done inside a master smart PDC that manages the local relays.

In order to evaluate the effectiveness of the SPS, a post-mortem analysis was performed via simulation to estimate the consequences in the SNI operation in the face of the occurrence of this contingency without the SPS. It was possible to determine that after the contingency of the Molino–Pascuales 230 kV transmission line, there would have been a large reactive power requirement in the Guayaquil area, mainly caused by the loss of the reactive compensation supplied by Paute generation plant and by the reactive power consumption of nearby transmission lines due to the sudden increase of the active power. This reactive power deficit would have certainly caused the switching of two major generation plants in the Guayaquil area due to over-excitation and thus a further reactive power deficit and subsequent uncontrollable voltage decay. In this context, it is possible to establish that without the SPS, the outage of the Molino–Pascuales transmission line would have caused a voltage collapse in several SNI substations. From the available data it was determined that this voltage collapse would have caused the outage of about 893.4 MW of load. In raw numbers, and assuming the feasibility of restoring the power supply to the affected areas in approximately 1 or 1.5 hours, the energy not supplied would have reached an estimated value between 893 MWh and 1340 MWh. In Ecuador, the cost of energy not supplied is 1533.00 USD/MWh (defined by the Regulator), and so, as shown in Table 18.5, the actuation of the SPS was able to

Table 18.5 Energy Not Supplied cost with and without the actuation of SPS.

	Energy not supplied (MWh)	Cost from energy not supplied (USD MM)
Without SPS	[893.4, 1340.1]	[1.37, 2.05]
With SPS	89.2	0.14
Economic savings		[1.2, 1.9]

prevent the loss of an economic amount between USD 1.2 million and USD 1.9 million, in one event alone.

18.7 Concluding Remarks

Three methodologies for determining adequate thresholds regarding: (i) phase angle difference, (ii) voltage profile power transfer of transmission corridors, and (iii) oscillatory issues have been presented in this chapter. These limits are calculated periodically and are later set into the WAProtector application as the referential framework for assessing stability in real time. Additionally, a practical procedure for determining the source locations of poorly damped low frequency power system oscillations using WAMS records was presented. For this purpose, the WAProtector's modal identification algorithm was used in order to determine the Ecuadorian system oscillatory behavior. Afterwards, the tuning of PSSs and further assessment was also carried out based on the results obtained from the WAMS application. It is feasible to determine that the PSS tuning process performed in the Ecuadorian SNI, whose main focus was to improve the damping of the Ecuador–Colombia inter-area mode, had successful results, considering the decrease in the appearances of poorly damped 0.45 Hz inter-area modes. A similar procedure will be followed for tuning additional PSSs in the Ecuadorian system with the aid of a real time digital simulator recently implemented in CENACE.

The Ecuadorian Special Protection Scheme (SPS) was designed to prevent partial or total collapses in the SNI due to the occurrence of eight defined double contingencies in the 230 kV trunk transmission corridors. Since its commissioning, it has had three satisfactory actuations and it has been established that during the outage of the Molino–Pascuales 230 kV transmission line, the SPS was able to prevent the loss of an economic amount between USD 1.2 million and USD 1.9 million, taking into consideration the cost of energy not supplied (CENS) in Ecuador. Due to the growth of the SNI, the SPS will be redesigned, using also the recently implemented real time digital simulator.

References

1 J. Cepeda, J. Rueda, G. Colomé and I. Erlich, "Data-Mining-Based Approach for Predicting the Power System Post-contingency Dynamic Vulnerability Status," *International Transactions on Electrical Energy Systems*, vol. 25, issue 10, pp. 2515–2546, October 2015.

2 J. Cepeda and P. Verdugo, "Determinación de los Límites de Estabilidad Estática de Ángulo del Sistema Nacional Interconectado," *Revista Técnica* **"energía"**, No. 10, January 2014.

3 IEEE Power Engineering Society, "IEEE Standard for Synchrophasors for Power Systems," IEEE Std. C37.118.12011, December 2011.

4 A. De La Torre and J. Cepeda, "Implementación de un sistema de monitoreo de área extendida WAMS en el Sistema Nacional Interconectado del Ecuador SNI," *Ingenius*, vol. 10, pp. 34–43, 2013.

5 P. Kundur, J. Paserba, V. Ajjarapu, et al, "Definition and Classification of Power System Stability," IEEE/CIGRE Joint Task Force on Stability: Terms and Definitions. IEEE Transactions on Power Systems, vol. 19, August 2004.

6 J. Han and M. Kamber, *Data Mining: Concepts and Techniques*. Second edition, Elsevier, Morgan Kaufmann Publishers, 2006.

7 J. Cepeda, P. Verdugo and G. Argüello, "Monitoreo de la Estabilidad de Voltaje de Corredores de Transmisión en Tiempo Real a partir de Mediciones Sincrofasoriales," Revista Politécnica, vol. 33.

8 Y. Nguegan, Real-time identification and monitoring of the voltage stability margin in electric power transmission systems using synchronized measurements. Universität Kassel, Doctoral Thesis, June 2009.

9 P. Kundur, *Power System Stability and Control*. McGraw-Hill, Inc., Copyright 1994.

10 J. Cepeda, and A. De La Torre, "Monitoreo de las oscilaciones de baja frecuencia del Sistema Nacional Interconectado a partir de los registros en tiempo real," Revista Técnica *"energía"*, No. 10, January 2014.

11 L. Grigsby, *Power system stability and control*, Boca Raton: Taylor & Francis Group, 2007.

12 N. Al-Ashwal, D. Wilson and M. Parashar, *"Identifying Sources of Oscillations Using Wide Area Measurements"*, CIGRE US National Committee, 2014.

13 D. Wilson, N. Al-Ashwal and K. Hay, *"Power System Stabiliser Tuning: Part 1 System Analysis & Plant Selection"*, ALSTOM – PSYMETRIX Consulting Project, 2015.

14 D. Wilson, N. Al-Ashwal and B. Berry, *"Power System Stabiliser Tuning: Part 3 Final Report"*, ALSTOM – PSYMETRIX Consulting Project, 2015.

15 J. Agee, S. Patterson, R. Beaulieu, M. Coultes, R. Grondin, I. Kamwa, et al., *"IEEE tutorial course power system stabilization via excitation control"*, IEEE Power and Energy Society, June 2007.

16 P. Verdugo, J. Cepeda, A. De La Torre, D. Echeverría, and H. Flores, *"Oscillation Source Location and Power System Stabilizer Tuning Using Synchrophasor Measurements"*, IEEE Transmission and Distribution Latin America (T&D-LA) 2016, Morelia, México, September 2016.

17 J. Cepeda, "Real Time Vulnerability Assessment of Electric Power Systems Using Synchronized Phasor Measurement Technology", National University of San Juan, Doctoral Thesis, December 2013.

18 Corporación Eléctrica del Ecuador-TRANSELECTRIC, "Plan de expansión de transmisión período 2013–2022", Quito, March 2012.

19 Agencia de Regulación y Control del Ecuador-ARCONEL, "Plan Maestro de Electrificación 2013–2022", Quito, Available: http://www.conelec.gob.ec.

20 S. Wang, and G. Rodriguez, *"Smart RAS (Remedial Action Scheme)"*, Innovative Smart Grid Technologies (ISGT), IEEE, Gaithersburg, MD, 2010.

21 M. Flores, D. Echeverría, R. Barba and G. Argüello, "Architecture of a Systemic Protection System for the Interconnected National System of Ecuador", *2015 Asia-Pacific Conference on Computer Aided System Engineering (APCASE)*, IEEE, Quito, July 2015.

Index

Dynamic Vulnerability Assessment and Intelligent Control for Sustainable Power Systems, First Edition.
Edited by José Luis Rueda-Torres and Francisco González-Longatt.
© 2018 John Wiley & Sons Ltd. Published 2018 by John Wiley & Sons Ltd.